Phase and Phase-Difference Modulation in Digital Communications

The Artech House Signal Processing Library

Alexander Poularikas, *Series Editor*

Hilbert Transforms in Signal Processing, Stefan L. Hahn

Phase and Phase-Difference Modulation in Digital Communications, Yuri Okunev

Principles of Signals and Systems: Deterministic Signals, Bernard Picinbono

Statistical Signal Characterization, Herbert L. Hirsch

Statistical Signal Characterization Algorithms and Analysis Programs, Herbert L. Hirsch

Surface Acoustic Waves for Signal Processing, Michel Feldman and Jeannine Henaff

For a complete listing of the *Artech House Telecommunications Library*, turn to the back of this book.

Phase and Phase-Difference Modulation in Digital Communications

Yuri Okunev

Artech House
Boston • London

Library of Congress Cataloging-in-Publication Data
Okunev, IU. B. (IUrii Borisovich)
  Phase and phase-difference modulation in digital communications / Yuri Okunev
    p.   cm. — (Artech House signal processing library)
  Includes bibliographical references (p.   ) and index.
  ISBN 0-89006-937-9 (alk. paper)
    1. Digital communications.   2. Phase modulation.   I. Title.   II. Series.
TK5103.7.047   1997
621.382'16—DC21   97-21324
                                                                   CIP

British Library Cataloguing in Publication Data
Okunev, Yuri
  Phase and phase-difference modulation in digital communications
  1. Phase modulation   2. Digital communications
  I. Title.
  621.3'815364

  ISBN 0-89006-937-9

Cover design by Deborah Dutton and Joseph Sherman Design

International Standard Book Number: 0-89006-937-9
Library of Congress Catalog Card Number: 97-21324

10 9 8 7 6 5 4 3 2 1

Contents

Preface

You are holding my first book written in English. It was made possible with the aid of many people to whom I am sincerely grateful.

First of all, I would like to thank the American scientists whose energetic support and inexhaustible benevolence followed me during this difficult period in my professional career: Dr. B. Gold, Dr. Yu. Goldstein, Dr. E. Finkelstein, Prof. S. Fleisher, Dr. A. McRae, Prof. E. Plotkin, Dr. W. Pritchard, Prof. J. Proakis, Dr. A. Rizkin, Prof. L. Roytman, Dr. F. Tunik, Prof. A. Yabrov, and Prof. A. Zayezdny.

The content of this book is based primarily on the results obtained during my work at the Digital Communications Research Laboratory of the St. Petersburg State University of Telecommunications. This laboratory was distinguished not only by its significant scientific results but also by the high standard of relationships among its coworkers. I express a deep gratitude to my friends and colleagues at the laboratory for many years of fruitful work together.

I am pleased to give due credit to Dr. M. Lesman, Dr. B. Kagan, T. Chitova, and S. Skvortsova for their assistance in preparing the manuscript.

Finally, I would like to thank the reviewer of this book for valuable and constructive comments, as well as the staff at Artech House for their tremendous and laborious work, which successfully united high professionalism with a tolerant and delicate attitude towards the author for whom English is a second language.

Introduction

PHASE MODULATION IN DIGITAL COMMUNICATIONS: HISTORICAL BACKGROUND

Serious development of digital communications systems and phase modems began almost simultaneously at the beginning of the 1960s. Certainly, there was an element of randomness in this fact, caused by the unpredictable results of human creativity. However, in general the intertwining of digital communications and phase (phase-difference) modulation was fortunate. Digital communications, primarily digital radio communications, in their state of the art would have been impossible without the invention of new phase-modulated (PM) signals and the synthesis of the optimum algorithms for this type of signal processing. At the same time phase modem theory and engineering have been developing rapidly and have accompanied every new stage of digital transmission technology.

Considerable success has been achieved in this area of telecommunications. The potential noise immunity of the perfect receiver is practically realized in contemporary phase demodulators, and we have to develop special methods and extremely precise devices to measure the energy losses of a real demodulator when compared to a perfect one. The theoretical telecommunication channel capacity is no longer considered remote, unattainable, and merely the subject of intellectual exercises for scientists; in fact, engineers are in the process of accomplishing it. Achievement of the information transmission rate close to the desired channel capacity has become a real task for technical projects in the field of digital communications.

All of this indicates that digital communications is on the threshold of a new stage of its development. In solving the problems that occur at this stage of creating the digital communications of the future, the phase modulation and phase-difference modulation (PDM), as they were before, will continue to be of great importance. It is not difficult to predict the main features of phase modems of developing digital systems: multiposition (multiphase and multiamplitude) signals, multidimensional signal-code structures, the optimum adaptive algorithms of coherent and incoherent processing, approximations of the optimum algorithms of reception as a whole, digital implementation on the basis of special digital signal processors, and universal microprocessors.

The history of digital communications and phase modulation is rather instructive. The first wire communications systems were characterized by many features of modern digital communications, such as a source of the discrete message (a telegraph key) and discrete signals in a communication channel, which were dots, dashes, and pauses. The first radio communications systems—wireless telegraph— were also primitive systems of digital transmission.

After telephony had appeared and been developed, the analog transmission systems pressed digital systems first in wire and then in radio communications. The position of analog communications was strengthened due to the appearance of radio broadcasting and television.

So we move on to the 1950s, when new sources of the discrete messages appeared; for example, electronic digital computers and various digital control systems. These new sources resulted in data transmission systems, which together with new multichannel telegraphy systems had begun to compete seriously with analog communications systems. However, a radical turning point that benefited digital communications took place when (to the clear advantages of digital transmission of the analog signals) the technical and economic opportunity arose to realize the transmission of analog signals in a digital form.

The advantages of using digital methods for analog signal transmission are well known and substantiated in detail in the technical literature. These advantages increased according to the development of the microelectronics technology. From the beginning of the 1960s, alongside the data transmission systems there developed digital telephone systems, including wire channel telephony, based on pulse-code modulation, digital radiotelephony in the HF and UHF bands, and satellite multi-channel digital telephony. Then the digital transmission methods were applied in radio and television broadcasting. Multiprogram stereophonic digital broadcasting and digital television enlarge the sphere of digital transmission applications.

To develop the new communications systems mentioned, a considerable increase in the speed (bit rate) of information transmission over communications channels with limited frequency and energy resources was necessary. In some cases the bit rate increase had to be accompanied by a noise immunity increase. The digital transmission systems based on phase-difference modulation corresponded in the greatest degree to these new requirements, which appeared in the beginning of the 1960s.

Phase-difference modulation was invented in Russia by Prof. N. T. Petrovich in 1954 [1].[1] Three years were required for the author to achieve recognition for his invention. Therefore, the first widely available publications on PDM were made simultaneously in Russia [2] and in the United States [3] in 1957.

The discovery of phase-difference modulation was made possible through vast amounts of research in the field of phase telegraphy, synchronous detecting, and noise immunity theory. Even in the 1930s many original methods that aimed at realizing reception of signals with phase manipulation were suggested. However the absolute phase of a signal eluded researchers—they were able to determine it only with an accuracy up to discrete values, corresponding to the transmitted discrete signal gradations. Despite this failure, the search for a technical solution to the problem proceeded. This search was stimulated by the serious advantages that could accrue through the use of phase modulation compared with the frequency and amplitude modulations; these advantages were proven in the 1940s by the potential noise immunity theory [4].

The dramatic struggle of scientists and engineers to solve the mysterious "phase problem" was described in detail not only in special technical monographs, but also in the popular literature [5]. Therefore there is no reason to discuss this problem in detail. It is only reasonably interesting to note that this struggle, having concluded with a resolute victory in 1954, resulted in the approval of phase modulation through its denial. It appeared that phase modulation could be realized only by refusing it, that is, by refusing an absolute phase as an information carrier and refusing the obligatory strict coherence of a reference oscillation in a demodulator. Moreover, it has appeared that the signal with phase-difference modulation can be received with the high noise immunity by means of incoherent processing, which does not need the coherent reference oscillation at all.

Thus, as happens in science, the "phase problem" was solved in quite another way—by PDM. However the researchers were rewarded by the grace and extreme simplicity of this solution. Moreover, PDM had allowed researchers to take the first step on the path to developing new, differential methods of information transmission. In 1964 within the framework of this approach, second-order phase-difference modulation was suggested [6], providing the possibility to transmit information by phase-modulated signals with an unknown carrier frequency. The invention of high-order PDM stimulated development of the general theory of phase-difference modulation [7].

The basic direction of subsequent PDM theory development included algorithms and performance analysis of incoherent processing of phase-modulated signals, first of all the optimum incoherent and autocorrelated algorithms. Algorithms of the optimum incoherent reception of PDM signals were synthesized on

---

1. Petrovich called this new transmission method *relative phase telegraphy* (RPT). Later this method was called the relative phase modulation (RPM) or phase-difference modulation (PDM), which is equivalent to DPSK.

the basis of the highly developed general theory of incoherent processing, but the autocorrelated reception theory was actually created just within the studies on PDM. Another important direction of theoretical research in the field of phase-modulated signal reception consisted of synthesizing the optimum and suboptimum quasicoherent demodulators and in studying their noise immunity performances. In later years theoretical research in the field focused on synthesis of effective multiposition phase-modulated signals and signal-code structures on their base, and also on development of the feasible and simultaneously close to optimum algorithms of multidimensional signal reception as a whole. Theoretical studies in the field of PDM always had the practical application tendency. At the same time, practice followed theoretical achievements, and almost parallel progress in practical and theoretical developments was typical for the PDM technology.

As far as it is possible to judge under the available publications, the first industrial PDM system appeared in 1957 in the United States [8], that is, simultaneously with the first theoretical publications on this problem. It was "Kineplex"—a 20-channel modem with the four-phase PDM that transmitted 3000 bps within the 300- to 3400-Hz telephone channel. The modem had to be used for data transmission over HF radio channels and wire channels. It is interesting to note that the first PDM modem and the following multitone PDM modems realized optimum incoherent reception—PDM technology began with incoherent processing!

In the 1960s and 1970s the technology of the multitone modems with multiposition PDM, which was intended for use in data transmission, multichannel telegraphy, and digital vocoder telephony transmission over multipath radio channels, developed rapidly. Multitone PDM modems MS [9] and a number of "Kineplex" modifications were developed. The appearance of the multitone modems with PDM has resulted in radical changes in HF radio communications. Before them the data transmission rate in HF radio channels did not exceed several hundred bits per second. The PDM modems gave users the opportunity to transmit information at rates of 1200, 2400, and 4800 bps, which allowed engineers to implement new operation modes in automated control and communications systems.

Another improvement of digital radio communications was connected with the practical use of the second-order PDM in the middle of the 1970s. It resulted in effective data transmission by phase-modulated signals in Doppler channels at the communications of rapidly moving objects and in other radio channels with uncertain signal frequency.

In the 1970s and 1980s the technology of the multichannel modems with orthogonal subchannel signals and multilevel PM and PDM was developed for data transmission via voice frequency channels of wire communications lines. In these applications both incoherent and coherent methods of reception are used and are realized by means of digital signal processing techniques. Due to information transmission by a great number of subcarriers, the elementary symbol duration in multichannel modems exceeds the duration of transient processes, appearing in a communications channel. Therefore these modems are insensitive to signal linear distortions and allow the information to be transmitted at a high bit rate over switched wire channels without adaptive correction of their amplitude frequency and phase frequency characteristics.

Together with multichannel modems, single-channel modems with multilevel PM, PDM, or amplitude-phase (phase-difference) modulation (QAM) have been widely developed. These modems are intended for transmission of information in a digital form over wire and radio channels. The problem of providing a high bit rate in channels with significant linear distortions is solved in these modems (1) by adaptive correction of the system frequency characteristics, (2) by forming the specially shaped signals, (3) by interference cancellation, and (4) by the algorithms of reception as a whole. In single-channel modems, developed for transmitting digital information via the standard telephone wire channels, transmission rates were achieved step by step: 1200, 2400, 4800, 9600, 14,400, 19,200, 28,800, and 33,600 bps. Modems have also been developed that have higher transmission rates for wide-bandwidth wire channels. In the high-speed single-channel modems, as a rule, multiposition amplitude-phase modulation (QAM) and various modifications of the coherent method of signal processing are used. It is interesting to note that in the first single-channel PDM modem, which appeared in 1961 [10], the autocorrelated method of reception was used.

Since 1959 PDM has been used in space communications and in satellite transmission systems with repeaters installed in the satellites. At present the digital satellite systems use almost exclusively the PM and PDM modulation techniques. The information transmission rates of modems with PM and PDM, used in satellite communications systems, reach tens and hundreds of millions of bits per second. Extremely high transmission rates are required for multichannel telephone satellite communications systems with time division multiple access (TDMA), as well as for digital TV systems. High digital transmission rates are also realized in microwave radio relay communications systems on the basis of using multiposition signals with amplitude-phase modulation.

At present the phase transmission methods are widely used in rapidly developing wireless telecommunications systems, including cellular networks, mobile satellite systems, and personal communications systems (PCS). Plans have been made to use the PM and PDM signals in optic and fiber optic communications channels. Modems with PDM have also been developed for domestic digital tuners, operating in multiprogram stereophonic digital broadcasting systems. Thus, phase modems are being applied in practically all radio-frequency bands and in almost all channels and systems of wire and radio communications.

Putting PM and PDM to practical use was not always a smooth process. Alongside the objective difficulties, which always accompany a new development, researchers also came across subjective difficulties. The first adherents of PM and PDM remember well that to their colleagues, who had greatly contributed in the frequency modulation technology, the signal phase seemed to be too "fragile" for information transmission, especially in radio channels, where it is subjected by random fluctuations. These fears have appeared to be unreasonable. Long-term experience with development and operation of the PM and PDM digital communications systems has demonstrated that there are practically no situations or conditions in which it would be impossible to carry out the qualitative information transmission, varying phase-modulated signal parameters and applying this or that method of reception from a powerful arsenal of optimum processing algorithms.

CONTENTS AND ORGANIZATION OF THE BOOK

This book contains discussions on the following topics:

- General principles of digital transmission of information by phase-modulated signals;
- Theory of phase modulation and phase-difference modulation, and algorithms of PM and PDM signals processing;
- Calculation of phase modem noise immunity;
- Applications of PM and PDM modems in digital communication systems.

The developed theory of PDM is based on the mathematical means of high-order finite difference calculus and stochastic analysis and synthesis of the optimum signal processing algorithms. We synthesize and study in detail the optimum algorithms of coherent, incoherent, and autocorrelated processing of PM and PDM signals. The results are illustrated by numerous schemes and block diagrams. Then we derive formulas for calculating the error probability of PM and PDM signal processing algorithms. Comparative analysis of the bit error rate for all phase modems is performed. The results are illustrated by means of plots and tables that allow engineers to estimate the performance of digital communication systems. The book contains much information on schemes of optimum and suboptimum demodulators of multilevel PM, PDM, and QAM signals and their application consideration.

This book is intended for researchers and engineers in the area of telecommunications, radio and wireless engineering, data transmission, and optimum signal processing. We assume that the readers have basic knowledge in the field of digital communications.

The book consists of six chapters. Chapters 1 and 2 prepare the reader for an exhaustive comprehension of the main part of the book. Chapters 3, 4, and 5 are the core of the book. Chapter 6 allows the reader to estimate the studied techniques. Every chapter includes specific references. A bibliography at the end of the book indicates related fundamental literature. A list of abbreviations and symbols facilitates understanding of the text.

Chapter 1 is devoted to principles of digital transmission by phase-modulated signals. It is addressed first of all to those readers who are dealing with PM and PDM systems for the first time. Section 1.1 of this chapter can be used as a brief introduction to the study of digital communications and modem functionality. The next sections, Sections 1.2, 1.3, and 1.4, contain simple and understandable descriptions of the phase and phase-difference modulation technology, methods of forming and receiving PM and PDM signals, and the concept of high-order difference modulation and its invariance properties. These topics are developed in detail in the following chapters, but this chapter introduces some important notions, terminology, and basic classification of PDM signal processing algorithms. This classification is of particular importance to this book's organization. Section 1.5 is an introduction into the fast growing area of multilevel two-dimensional signals that are based on PM and PDM. Section 1.6 is an especially interesting section dealing with a comparatively new and very important direction in digital communications, namely, signal-code structures. The corresponding complex signals are as a rule the combinations of simple multiphase signals. In this section we consider two signal-code structures, based on PM or PDM signals, and the corresponding algorithms of reception as a whole. By means of these examples, we explain why and how a combination of the phase-modulated signal-code structure and processing as a whole enable us to improve noise immunity in comparison with the perfect element-by-element PM signal processing.

Chapter 2 deals with special problems of differential modulation. The first three sections of this chapter contain a theory of high-order difference modulation: characteristics of discrete difference transforms, the notion of high-order difference modulation, algorithms of the high-order difference processing, and characteristics of the high-order PDM. The theory is based on mathematical means of finite difference calculus. In Section 2.3 we discuss characteristics of the high-order PDM and demonstrate how to use this signal in channels with instable carrier frequency, or in channels with the Doppler effect at communication with mobile or nonstationary objects. The optimum keying code and mapping algorithms for multiphase PM and PDM signals are considered in Section 2.4. We show in this section that the well-known Gray code is not the unique optimum keying code. On the basis of a combinatorial analysis, we have found a typical structure and characteristics of other optimum codes. This typical structure is used for synthesis of general algorithms of multiphase signal decoding in Section 2.5. General algorithms of PDM signal decoding represent transmitted binary symbols as a function of trigonometric transformation of the received phase difference. We have derived such algorithms for arbitrary multiphase signals mapped by the Gray code. These general algorithms are the basis for finding the specific optimum algorithms of PM and PDM signal processing in the following three chapters.

Chapters 3, 4, and 5 contain a detailed description of three basic methods of PM and PDM signal processing that correspond to the hierarchy of optimum algorithms in the statistical theory of communications: coherent, optimum incoherent, and autocorrelated processing. These chapters have similar structures: we discuss the essence and application of a method, the synthesis of basic algorithms for various PM and PDM signals, the block diagrams of the corresponding modems and their modifications, and the implementation aspects and application peculiarities of real modems. The chapters contain optimum and quasioptimum algorithms for PM and PDM signals processing, the basic schemes of the corresponding modems, their practically useful modifications, and application recommendations. These three chapters are addressed first of all to designers and elaborators of modems; they will facilitate the ability to make the best decision on the basis of all system characteristics and requirements.

Chapter 3 addresses the problems of coherent and quasicoherent processing of PM, PDM, and QAM signals. After an introductory section we synthesize the optimum algorithms of PM and PDM signal processing that provide the minimum error rate. These algorithms are realized by means of quasicoherent adaptive demodulators, and the main problem of the coherent processing is how to implement the adaptation process to reach the maximum possible noise immunity. To solve this problem, we introduce a notion of reduced projections of a received signal onto orthogonal reference signals with an arbitrary initial phase controlled by demodulator decision feedback. On the basis of this approach, development succeeded for a number of quasicoherent demodulators for multilevel PM, PDM and QAM signals that are close to the optimum ones. Received schemes are the practically effective base for digital implementation of the coherent signal processing, and we hope engineers will use them. The chapter concludes with Section 3.6, which includes a detailed analysis of carrier recovery techniques, which are widely used in high-frequency modems. This section contains the most exhaustive classification of methods and units of selecting the coherent reference oscillations.

Chapter 4 is on optimum incoherent processing of PDM signals. As we have already mentioned, PDM allows us to use incoherent processing for phase signal detection, and in a number of cases the incoherent PDM demodulators do not concede or even exceed the coherent PDM demodulators by the noise immunity. The main portion of this chapter is a discussion of the synthesis of the optimum incoherent processing algorithm for multilevel PDM signals. On the basis of this general algorithm, we have developed effective schemes of optimum incoherent demodulators for two-, four-, and eight-phase signals. These results are recommended for use in practice because some of the developed demodulators actually have the same error probability as the ideal coherent receiver. Practical possibilities of this approach are extended due to a number of demodulator modifications based on the use of active and passive filtering techniques. These modifications are considered in Section 4.3. In Section 4.4 we develop important algorithms and schemes for optimum incoherent demodulators with second-order PDM (PDM-2) signals, and the unexpected fact is that the noise immunity of these demodulators is higher than the noise immunity of the optimum incoherent demodulators of first-order PDM (PDM-1) signals and is almost indistinguishable from the noise immunity of the perfect coherent demodulators. This result is additionally developed in Section 4.5. In this section we introduce a concept and algorithms of optimum multisymbol incoherent detection of PDM signal and indicate how to achieve the minimum possible error rate without carrier recovery. On the basis of this section's results, we have come to the fundamental conclusion that we should increase the noise immunity by means of increasing the interval of optimum incoherent processing of signals without redundancy.

Chapter 5 is on autocorrelated processing of PDM signals. Autocorrelation processing takes a special place in the hierarchy of the optimum algorithms. This hierarchy is opened by a coherent processing algorithm, which is optimum when the transmitted signal variants are completely known. Then comes optimum incoherent algorithm; it is optimum when signal variants have a random and uniformly distributed initial phase. If we go further, reducing the a priori information of the receiving signal, we can synthesize other reception methods, optimum under the appropriate conditions. This hierarchy of the optimum methods is finished by processing algorithms of the unknown shape signals, and the algorithms of autocorrelated processing, studied in this chapter, refer to them. This statement and peculiarities of autocorrelation processing are considered in Section 5.1. Then in Sections 5.2 and 5.3 we derive autocorrelation algorithms and block diagrams of multiphase PDM-1 and PDM-2 demodulators. A reader can find here many options and practical recommendations on how to implement a system in the best way. Two important sections, Sections 5.4 and 5.5, deal with the new technology of carrier frequency invariant demodulators based on autocorrelation processing of PDM-2 signals. In these sections we propose autocorrelation PDM demodulators that are absolutely or relatively insensitive to carrier frequency variations. This result will be interesting for researchers and engineers in the field of data transmission over the Doppler channels.

Chapter 6, on noise immunity of PM and PDM systems, is devoted to the calculation of error probabilities in digital systems using various phase signals and algorithms of their reception. It contains exact analytical expressions and approximate estimations of the error probability as functions of the signal-to-noise ratio for binary and multiposition PM and PDM signals when using coherent, quasicoherent, optimum incoherent, and autocorrelation signal processing. In addition, we give a number of diagrams and tables that allow the reader to find quickly a desired error rate estimate and compare this result with the results corresponding to other methods of transmission and reception.

In this chapter we explain to readers the most effective and conventional approaches and mathematical methods of the error probability calculation. From this point of view, Sections 6.1 and 6.2 may be considered a brief introduction to noise immunity theory. These sections' fundamental results are developed in Sections 6.5 and 6.6 for the cases of multilevel PM and PDM signals. Sections 6.3 and 6.7 deal with specific problems of the noise immunity of high-order PDM. Section 6.3 contains general formulas of error probability and error propagation for binary differential systems of the arbitrary order. Section 6.7 represents error rate estimates for carrier frequency invariant demodulators of the PDM-2 signals. Here we demonstrate how to reach the invariance to carrier frequency variations at the minimum energy loss.

Section 6.4 shows that the optimum multisymbol incoherent processing of PDM signals as a whole allows us to reach the potential minimum error probability without any estimation of an unknown initial carrier phase, that is, without carrier recovery. This section and Section 6.8 also discuss the problem of error probability convergence for multisymbol incoherent processing and quasicoherent processing of PM and PDM signals. Chapter 6 should be useful both to readers who want to get concrete reference information and to readers who intend to work in the field of noise immunity theory and its applications.

References

[1] Petrovich, N. T., "Telegraph Wire and Radio Communications Method Using Phase-Keyed Signals," USSR Patent 105692, 1954.
[2] Petrovich, N. T., "New Methods of Phase Telegraphy." *Radiotechnika*, No. 10, 1957, pp. 47–54.
[3] Doelz, M. L., E. T. Heald, and D. L. Martin, "Binary Data Transmission Techniques for Linear Systems," *Proc. IRE*, Vol. 45, No. 5, 1957, pp. 656–661.
[4] Kotelnikov, V. A., *Potential Noise Immunity Theory*, Moscow: Gosenergoizdat, 1956.
[5] Petrovich, N. T., *Who Are You?* Moscow: Molodaya Gvardiya, 1970.
[6] Okunev, Yu. B., "Discrete Message Transmission Method with Relative Phase Keying," USSR Patent 177450, 1964.
[7] Zayezdny, A. M., Yu. B. Okunev, and L. M. Rakhovich, *Phase-Difference Modulation*, Moscow: Svyaz, 1967.
[8] Mosier, R. R., and R. G. Claubaugh, "KINEPLEX, A Band-Width Efficient Binary Transmission System," *Commun. Electron.*, No. 34, 1958, pp. 723–727.
[9] Ginzburg, V. V., et al. *Digital Information Transmission Equipment MS-5*, Moscow: Svyaz Publ., 1970.
[10] Golomb, S., et al. *Digital Communications with Space Applications*, Englewood Cliffs, NJ: Prentice Hall, 1964.

CHAPTER 1

Principles of Digital Transmission by Phase-Modulated Signals

1.1 DIGITAL TRANSMISSION SYSTEM AND MODEM FUNCTIONS

Digital and analog methods of information transmission through communications channels are two concepts used in the development of communications systems and radio engineering information systems. During the past few decades, these techniques successfully competed with each other, either being ahead when new message sources and new kinds of information engineering appeared or lagging behind because of unpredictable changes in circuit technology. At present, the scale is definitely leaning toward the use of digital transmission methods.

In digital message transmission, a useful signal in a communications channel is discrete, that is, a finite set of values is accepted. This property can be considered to be definitive of digital transmission methods in contrast to analog ones, at which the useful signal in a communications channel is a continuous one and can accept any value within given limits.

In a digital system the possible variants of the signal (or the values of information signal parameter) can be numbered, comparing each of these variants with a digit (from here the name of the method) with a binary, quaternary, decimal one, etc., depending on a number of possible signal variants in a communications channel.

The discreteness of the information signal parameters predetermines the main properties and advantages of digital transmission in comparison with analog transmission, the main point of which is a higher noise resistance. The effect of noise resistance increasing at digital transmission is easy to explain. At analog transmission any interference results in more or less distortion of a transmitted signal as any received signal is admissible. Unlike with digital transmission, any interference can distort a transmitted signal variant if only the interference transforms it into another permitted signal variant, that is, if the received combination of a transmitted signal and an interference appears to be more like any other admissible transmitted variants.

Figure 1.1 illustrates the transmitted signals S and the interference ξ submitted as two-dimensional vectors. This representation corresponds to the narrowband signal and interference with the fixed carrier frequency. Every vector on the plane represents a signal or an interference with the corresponding amplitude and phase. Let the signal S_1 be transmitted and the interference be ξ_1; hence the received signal is $x_1 = S_1 + \xi_1$.

In the analog system the receiver determines this signal to be a transmitted one without any alternatives, and the actual distortion of the transmitted signal is characterized by the length of the interference vector of ξ: $|\xi_1| = |x_1 - S_1|$. In the digital system the receiver cannot determine the transmitted signal x_1 as having been transmitted because there is no such signal among the admissible signals. If, for example, only two variants of a signal S_1 and S_2 are admissible (two-position system), the receiver will choose the nearest of admissible signals S_1. As a result there is not any distortion of the transmitted signal.

Note that the correct transmission will take place until the sum of the transmitted signal S_1 and the interference appear to be above the dashed line. Also, only if the received signal appears under the dashed line will signal S_2 be determined instead of transmitted signal S_1. In all other cases, when the sum of the transmitted signal and interference does not fall outside the limits of the top half-plane, the digital system will give the correct result while the analog system is characterized by significant signal distortions.

The main feature of digital transmission systems, that is, signal discreteness in a channel, results in an essential change in the role and functions of many of the elements in traditional message transmission systems. Let's consider the most general scheme of a digital transmission system (Fig. 1.2), which reflects typical signal conversion in a digital transmission trunk [1–6].

The system includes sources of digital and/or analog messages (SDM and SAM) and receivers of digital and/or analog messages (RDM and RAM).

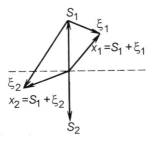

Figure 1.1 The vector diagram of transmitted and received signals.

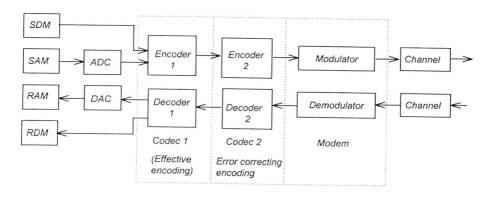

Figure 1.2 The general scheme of a digital transmission system.

When transmitting from the digital signal source (telegraph, data source, computer, digital telemeter sensor, etc.), its output symbols come directly to the input of encoder 1, which carries out effective encoding of the digital symbols. Sometimes the operative memory is installed between the SDM and encoder 1 to match the incoming and outcoming symbols rate. This step is necessary when, for instance, the information is coming from an SDM by separate packets or when the data of several low-speed sources are transmitted by an integrated digital information stream.

When the signals are transmitted from the analog source (telephony, broadcasting, TV, telemetry, etc.), the output signal is sent to an analog-to-digital converter (ADC), which converts the analog signal into a digital code, in particular, a binary code. The signal is then converted in encoder 1. The analog-to-digital conversion of a continuous signal into a discrete one is executed by quantization on time and level.

The operation of quantization on time means that the initial continuous signal is replaced by its samples in discrete intervals. If the signal is in a band from zero to Δf Hz (or can be linearly transformed to such a low-frequency equivalent), then according to the sampling theorem it can be precisely restored by its samples, taken in an interval $\Delta t = 1/2\Delta f$. In an ADC the quantization interval is usually chosen as such or a little shorter; the quantization frequency, determined as a number of signal samples transmitted per second, is equal to

$$f_q = 1/\Delta t = 2\Delta f$$

The quantization operation on level means that the transmission of every signal sample is interchanged with a transmission of a discrete level number that is most closely related to the true value of the given sample. The higher the required accuracy of analog signal transmission and the wider the range of its change (dynamic range of a signal), the higher the number of discrete quantization levels that should be provided in the ADC. If a number of quantization levels is equal to K, then $\log_2 K$ binary symbols will be required for the transmission of each sample. Thus, the necessary and sufficient transmission rate, in bits per second, of an analog signal with spectrum width Δf and an accuracy appropriate to K quantization levels in the digital form is

$$I = 2\Delta f \log_2 K \tag{1.1}$$

For example, at digital speech transmission with a band of $\Delta f = 4$ kHz quantization levels $K = 256$ are used to achieve quite high quality. It gives the required transmission rate of $I = 64$ Kbps.

The signals of high-quality stereophonic broadcasting are characterized by a spectrum width up to 20 kHz and a dynamic range up to 100 dB. Hence, not less than 40,000 samples per second with not less than 60,000 quantization levels (16-digit binary encoding) are required for signal conversion of a single monoprogram into digital form. As a result the actual rate of digital stream at the stereophonic program transmission is 1280 Kbps. For the transmission of a video signal at the $\Delta f = 6$ MHz band at $K = 1024$ the digital stream rate should be higher: $I = 120$ Mbps.

Note that broadband communications channels are required for digital data transmission at such high rates. Starting from the point that up-to-date modems provide a specific transmission rate of 2 bps/Hz of a channel frequency band, the channel band F necessary for analog signal transmission with the band Δf at K quantization levels could be evaluated as follows:

$$F = \Delta f \log_2 K \tag{1.2}$$

An important conclusion follows from this simple equation: The band demanded for digital transmission of an analog signal with K quantization levels is $\log_2 K$ times more than the spectrum width of this signal. This circumstance is an essential shortcoming of digital methods of transmission. To overcome it, it should be useful to convert a continuous signal into a digital code using various means of data compression.

The terms *data compression* and *effective encoding* can be treated as synonyms. The terms mean that the redundancy is eliminated from the original signal, and it is then encoded as economically as it is possible, that is, with a minimum number of binary symbols. Redundancy elimination is executed by encoder 1 (see Fig. 1.2). This encoder is called a source encoder, an effective encoder, or an effective encoding system encoder. The effective encoding is applied when transmitting analog as well as discrete messages. The theory of effective encoding of signals from various sources is a branch of modern information theory [7]. The results of the theory descend from the first Shannon theorem [8], which proved that the minimum average number of binary symbols necessary for data transmission from some source is equal to its entropy.

The effective encoding of signals from discrete sources, approximating a specified Shannon limit, is achieved by symbol combination in large blocks and by the application of special nonuniform codes, for example, Huffman codes and Shannon–Fano codes.

For effective signal encoding of analog sources, their natural redundancy, for instance, correlation of adjacent elements of the analog messages, and the human properties of sight and hearing are used. For this purpose the following methods are used: interpolation and extrapolation of speech and broadcasting signals, including the spline functions; selection of the most significant for the perception spectral components from these signals; a reduced description of signals in the spaces of various basic functions; and so on. For analog message transmission, the effective encoders are frequently combined with an ADC and become a single unit for analog-to-digital conversion and initial signal compression.

The application of the effective encoding methods essentially reduces the required channel transmission rate. For example, the methods of synthetic (vocoder) telephony allow for a speech signal with satisfactory legibility by means of a digital flow with the rate of 1200 bps instead of 64 Kbps at linear pulse code modulation (PCM) [9].

Methods of effective compression of hi-fidelity stereophonic broadcasting signals have been developed that allow the transmission of a single stereophonic program in the 20-Hz to 20-kHz band with a dynamic range up to 100 dB by 256-Kbps digital flow. Methods for digital TV signal compression that lower the rate to 24 Mbps and less have also been developed. The theory and equipment of effective encoding are rapidly developing, especially for analog message sources in connection with the increasing usage of digital methods of transmission.

Let's return to Figure 1.2. At the output of encoder 1 at the transmission of the discrete as well as analog messages, the signal is introduced as a sequence of discrete symbols incoming to encoder 2 at the rate determined by the rate characteristics of the source and encoder 1. As a rule, this is a uniform sequence of binary symbols in the form of bipolar pulses.

Encoder 2 is called a channel encoder or an encoder for error correction encoding. This encoder converts the incoming sequence of discrete symbols into another sequence providing the opportunity to detect and correct the errors when receiving. Such transformation is called *noise immune encoding,* and its theory goes back to the second Shannon theorem: The message from any source with information efficiency not exceeding the communication channel capacity can be encoded and then decoded with an arbitrary small error probability [8]. This theorem is fundamental for the development of the constructive theory of error correcting encoding. In the framework of this theory, feasible algorithms for error correction encoding and decoding have been developed [10–12].

Note that the operations of effective and error correction encoding are in some ways contradictory: The first encoder eliminates the natural redundancy of signals, and the second one introduces an artificial redundancy that adds the additional verifying symbols to information symbols. These symbols enable errors to be detected and corrected at the receiving end of a communications system. The specified contradiction of the operations is caused by the fact that the natural redundancy of a signal source is inconvenient to use for increasing the interference resistance in automatic systems. Besides, the redundancy introduced at error correction encoding is usually insignificant when compared with the natural redundancy of signals.

The existence of two stages of encoding—effective and error correcting—is not mandatory. Sometimes (e.g., at data transmission) the signals from the discrete message source come directly to the input of encoder 2 because they do not have natural redundancy. In other cases, the system does not have encoder 2, and the discrete signal coming from encoder 1 is direct modulator input; the required noise immunity is achieved without special error correction encoding. In a number of cases a signal from a discrete message source or a discrete signal from ADC output goes directly to modulator input without both stages of encoding.

In any case a uniform stream of discrete symbols becomes modulator input (see Fig. 1.2). As a rule the symbols are binary two-pole pulses of a direct current. The modulator matches a discrete encoder output and a continuous channel input.

In classical radio engineering, the modulation is understood to be a process of HF oscillation parameter control, for example, the control of amplitude, frequency, or phase of the harmonic carrier.

In the modern theory of digital communications the modulation is treated more widely—as a process of converting the elements of a discrete message into elementary signals for transmission by a continuous communications channel. In some cases this conversion is the changing of the HF oscillation parameters. According to this treatment when the discrete symbol appears at the modulator input, one of the admissible variants of signals appears at its output. The modulator output is directly connected to the continuous communications channel input, which is characterized by frequency, power and statistical parameters. The latter ones reflect the properties of channel interference.

The output modulator signals should correspond to the channel parameters for providing the most favorable (in the sense of message transmission rate and reliability) conditions of their transmission by a channel. Therefore such modulator parameters as frequency carrier, modulation method, and element duration are chosen in accordance with the frequency and power characteristics of the communications channel as well as taking into account the possibilities of the equipment at the receiving site of the communications system.

The modulator as a functional unit of a system is completely determined by the table of conformity between a set of the input discrete symbols and a discrete set of output signals. The modulator can be also given by a modulation method—

amplitude, frequency, phase, or phase difference—as well as by the table of conformity between discrete symbols at the input and discrete values of a modulated parameter at the output. Besides the signal modulation method, the modulator is characterized by an operating or transmission rate that demonstrates how many elementary signals (discrete symbols) or bits per second the modulator can transmit to a communications channel.

The channel is also part of a digital system (see Fig. 1.2). Besides the signal propagation media the channel includes various means of formation, conversion, radiation, and amplification of a signal. For example, satellite communication channels include a transmitting ground station with a powerful transmitter and transmitting antenna, satellite communication payload, a complex transceiver system, and a receiving ground station with a receiving antenna, sensitive receiver, and other components. In a satellite communications channel the signal is numerously transformed by filtering, up and down frequency transfer, amplification, etc. In wire communications systems the channel is formed by physical lines (communications cables), multiplexers, switches, and so on. In this case the signal is also subjected to some conversion including spectrum transfer, filtering, and other linear, nonlinear, and parameter conversions.

From a communications channel output, the signal goes to a demodulator where the received signal elements are converted into discrete symbols. The received signal is represented as a mixture of a useful signal, distorted because of linear and nonlinear conversions in the channel, and interferences. In such conditions exact restoration of the transmitted symbols is impossible. The demodulator has to process the signal with minimum error probability.

The demodulator receiving the signal in the given interference condition with minimum error probability is called the *optimum demodulator* (for the given conditions). The design of modern demodulators is based on the statistical theory of optimum signal receiving. In this theory, the optimum algorithms of signal processing for different interference conditions are synthesized [1–7].

When the received signal is correct, the demodulator forms a series of discrete symbols that are the same as at the modulator input. The part of the digital transmission system from the modulator input to demodulator output is called the *discrete communications channel* in contrast to the part of the system from the modulator output to demodulator input, which is called a *continuous channel*. The statistical characteristics of a discrete channel are defined by the characteristics of a continuous channel. They permit us to develop a compact statistical model of a discrete channel. On the basis of this model, we can choose the optimum method of error correction encoding.

From the output of the discrete channel, the symbols come to error correcting decoder 2. Using the redundancy incorporated into the transmitting sequence, this unit detects or/and corrects the errors that have appeared in the discrete channel. In the first case either the faulty combination is excluded from consideration or with the help of a feedback communications channel a transmission of a code combination in which the error has been detected can be repeated. In the second case the corrected information symbols of the code combination are sent to source decoder 1. The decoder executes the functions inverse to the ones of coder 1, that is, it restores the original message in digital form. The digital-to-analog converter (DAC) converts the received digital signal to an initial analog message.

The encoder and decoder are usually developed as a unit called a *codec*. Besides, in two-way (duplex) communications systems an encoder of a direct channel and a decoder of a reverse channel are structurally made as a combined unit. The same is true of the modulator–demodulator pair, which is called a *modem.*

As follows from previous discussion, the modem converts discrete symbols into signal elements for transmission by a continuous communications channel. It also converts the received signals into discrete symbols. The last operation is referred to as *element-by-element reception.* The demodulator should divide a received signal into elementary signals (chips). This is fulfilled by a clock synchronization unit that is part of the demodulator. The demodulator identifies each elementary signal with one of a number of discrete symbols of a message source or a coder.

The role of modems in communications systems continues to become more significant. When the data transmission rate was quite low and the requirements for precision rather moderate, modulators and demodulators were considered to be minor components of a transmitter or receiver. However, in accordance with the increasing requirements of speed and noise immunity as well as in connection with the occurrence of new efficient methods of modulation and algorithms for optimum signal processing, the significance of modems has increased and the actual volume of their equipment has been enlarged. Such major communications characteristics as data transmission rate and noise immunity have, to an extent, been defined by how successfully modem parameters have been chosen and how well the modem has been implemented.

At a certain stage of radio engineering development it became clear that from the technological and system point of view the improvement of a modem is more expedient than increasing the transmitter power or perfecting the communications channel characteristics. The expenditure for modem modification often results in a greater advantage than the expenditure for other parts of a communications system improvement.

As a result, up-to-date modems are quite complex units which in a number of cases are designed as separate units and racks. There is the tendency to further increase the role and complexity of modems and codecs in the process of their integration into one unit: codem which realizes encoding, modulating, demodulating, and decoding without division into separate operations. In this case the demodulator and decoder receive code combinations (signal-code structures) as a whole and represent themselves (algorithmically and structurally) as a single unit. In system engineering, being able to optimize as a whole gives greater gains that optimizing by parts. This general principle is also applicable for the optimization of encoding, modulation, demodulation, and decoding: The optimization of a codem as a whole can give greater advantages than the optimization of a modem and codec made up separately.

In this book, special attention is paid to phase modems, and foremost to phase-modulated signal demodulators because they are one of the basic means for achievement of the high rates of speed and noise immunity. The phase modems provide new opportunities for increasing these parameters in combination with signal-code structures and signal reception as a whole.

As can be understood from the following sections, we are dealing with a wide range of algorithms and phase modem schemes operating in various conditions, which enables us to make the best decision when designing a particular system for the digital transmission of information from the point of view of obtaining both the given external characteristics of a system in the whole and its realization on the basis of up-to-date digital and analog integrated circuits.

1.2 PHASE AND PHASE-DIFFERENCE MODULATION

Modems are distinguished by a kind of a signal carrier and a modulation method. The carrier and the modulation method completely determine a modem as a unit that converts discrete symbols into signals transmitted over a communication channel. Various signals can be used as carriers: simple harmonic oscillations, complex signals presented as the harmonic oscillations sum, spread-spectrum signals, composite signals, radio and video pulses of different forms, and pulse sequences.

Modulation methods are determined by what carrier signal parameter is changed under the effect of an information (modulating) signal. Usually the name of the type of modulation corresponds to the name of the modulated parameter of a carrier: amplitude, frequency, pulse-duration, pulse-phase, structural (the modulation of a spread spectrum signal form), and so on. Single harmonic oscillation or the sum of harmonic oscillations of different frequencies are most frequently used as carrier signals in radio and wire communications channels. A harmonic oscillation is completely defined by three parameters: amplitude, frequency, and initial phase, which is why the main methods of harmonic carrier modulation are distinguished as follows:—amplitude (AM), frequency (FM), and phase (PM) modulation.

The important parameter of a modem and modulation method is the number of signal variants at the modem output or a number of modulated parameter levels. If we comment on "m-level" or "m-ary" phase modulation, it means that every signal element at the modulator output has one of m admissible initial phases. If all m signal variants are equiprobable, the modulator efficiency, that is, the amount of information produced by the modulator per unit of time, is directly proportional to the binary logarithm of m:

$$N = \log_2 m$$

(1.3)

This value is called *modulation multiplicity* because it shows how many binary digits are in every signal element or how many times more information capacity of the system is in comparison with the two-position (single valued) system at the same elementary signal duration. Most frequently a number of phase levels are chosen to be equal to the degree of 2, then multiplicity N is an integer. For example, N-multiple phase modulation means that each elementary signal at a modulator output contains N bits of information, and the signal phase at the channel input has $m = 2^N$ admissible levels. If the duration of an elementary signal is equal to T seconds, the element generating rate (modulation rate) is equal to $1/T$ elements per second. This rate is measured in bauds, so 1 baud corresponds to the transmission of one elementary signal per second. Respectively, when signals are equiprobable, the data rate at the modulator output (bps) is

$$I = \log_2 m/T = N/T \tag{1.4}$$

As is known from information theory, the rate of information transmission over a discrete channel formed by a modem and a continuous channel depends on the statistical characteristics of the channel, in particular on error probability. At small probabilities of an error, this rate approximates to (1.4).

Figure 1.3 shows the signals for the elementary methods of two-position (single-valued) modulation of a harmonic carrier. Figure 1.3(a) shows that at the modulator input there is a binary signal in the form of bipolar pulses representing binary 0 and 1. Figure 1.3(b–d) shows the signals at an ideal modulator output when using two-position amplitude, frequency, and phase modulation.

The represented signals are comparatively broadband. The index of "the broadband" or "the narrowband" is the ratio f_0/F (carrier frequency to spectrum width of the modulated signal). Here, $f_0 = 2/T$ and $F = 1/T$, so the index of the narrowband $\delta = f_0/F = 2$, that is, is equal to a number of carrier periods in one elementary signal. For very narrowband radiosignals, δ is of several decades and more. Narrowband signals have a number of advantages concerning their realization.

The modulated signals shown in Figure 1.3 illustrate the main peculiarity of digital transmission systems—the principle of signal discreteness in a communications channel. For every one of the indicated modulation methods only two signal variants are admitted in a channel.

At the amplitude modulation these variants are recorded within the interval $0 < t \leq T$ as follows:

$$S_1(t) = a \sin(\omega_0 t + \varphi_0)$$
$$S_2(t) = 0 \tag{1.5}$$

For signals with AM shown in Figure 1.3, $\omega_0 = 4\pi/T$, and $\varphi_0 = 0$. In contrast to the FM and PM signals, the variants of the AM signals are not of equal power.

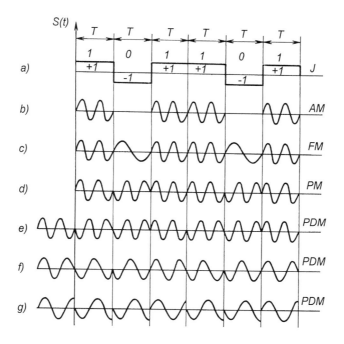

Figure 1.3 Elementary methods of two-position (single-valued) modulation of harmonic carrier.

At frequency modulation the signal variants in the interval $0 < t \leq T$ are the following:

$$S_1(t) = a \sin(\omega_1 t + \varphi_1)$$
$$S_2(t) = a \sin(\omega_2 t + \varphi_2) \tag{1.6}$$

The parameters of the FM signals shown in Figure 1.3 are equal to $\omega_1 = 2\pi/T$, and $\omega_2 = 4\pi/T$. These signals are concerned with the class of intense orthogonal signals because the integral from their product on an interval $(0,T)$ is equal to 0 at any phases φ_1 and φ_2.

FM signals are not always orthogonal. Strict orthogonality of the signals $S_1(t)$ and $S_2(t)$ is reached only when the circular frequencies ω_1 and ω_2 are harmonics of the frequency $\omega_0 = 2\pi/T$, that is, when

$$\omega_1 = \kappa_1 2\pi/T$$
$$\omega_2 = \kappa_2 2\pi/T \tag{1.7}$$

where κ_1 and κ_2 are integers. In systems with narrowband FM signals, the orthogonality condition (1.7) is sometimes replaced by a less strict condition of the approximate orthogonality (κ- integer):

$$|\omega_2 - \omega_1| = \kappa 2\pi/T \qquad (1.8)$$

In the systems with broadband FM signals the condition of approximate orthogonality can be achieved by increasing the difference between ω_1 and ω_2 frequencies:

$$|\omega_2 - \omega_1| \gg 2\pi/T \qquad (1.9)$$

The interest in the intense orthogonal signals has been challenged by some circumstances. First, these signals can be completely selected by quite simple units—correlators at any arbitrary initial phase of reference oscillations. Secondly, at incoherent element-by-element reception, the intense orthogonal two-position signals provide the highest noise immunity [1].

Thus, the discrete two-position FM signals have different modifications distinguished by frequencies ω_1 and ω_2; at certain ratios between ω_1 and ω_2 the FM signals become intensely orthogonal. The performance of systems with FM also depends on frequencies ω_1 and ω_2. At optimum incoherent element-by-element reception, the minimum error probability is achieved at ω_1 and ω_2 satisfying the conditions of strict orthogonality (1.7). But at coherent reception the minimum error probability is achieved [1], when

$$|\omega_2 - \omega_1| \approx 1.43\pi/T$$

The FM signals shown in Figure 1.3(c) are referred to as a class of FM signals with a continuous (without break) phase in which the initial phase of the next signal element is equal to the phase by which the previous element has been finished in spite of its frequency. The FM signals with a continuous phase require special modulators providing "the continuity" of the phase. However they can provide a little stronger noise immunity than the signals with arbitrary (independent) initial phases of elements because of using the information about the connections between the phases of elements at the reception of the signal as a whole [13].

Let's return to Figure 1.3 and record the formulas for signal variants with two-position phase modulation in the interval of one elementary signal $0 < t \leq T$:

$$\begin{aligned} S_1(t) &= a \sin(\omega t + \varphi_0) \\ S_2(t) &= a \sin(\omega t + \varphi_0 + \pi) = -a \sin(\omega t + \varphi_0) \end{aligned} \qquad (1.10)$$

Figure 1.3(d) shows an example of the signals of (1.10) when $\omega_0 = 4\pi/T$, $\varphi_0 = 0$. In a general case, as can be seen from (1.10), the variants of a PM binary signal have an arbitrary initial phase and differ from each other by the 180-degree phase shift, that is, by a sign. Therefore the signals with two-position PM are called *antipodal signals* in contrast to the orthogonal signals with two-position FM.

Interest in PM digital transmission of messages was aroused because (according to the fundamental result of the theory of the potential noise immunity) of all the two-position signals the antipodal signals, that is, the binary PM signals, are potentially most noise resistant. The quantitative correlation is the following: It is possible to achieve the given error probability at PM when the signal power is twice less than at FM and four times less than at AM. However there are some difficulties in its realization.

Note that the real AM and FM discrete signal demodulators are almost exclusively incoherent. Any information of an initial phase is not used in the operation algorithms of these demodulators. They are invariant to the initial phase. Such a realization of AM and FM signal demodulators is quite natural because at these methods of modulation the initial phase is not the information parameter. In contrast to that at the PM the initial phase is the information parameter and, hence, the incoherent methods of a reception are not possible.

Thus, a demodulator of signals with absolute PM can be coherent only. This means that the samples of signal variants coinciding quite precisely with transmitted signals by the frequency and initial phase should be generated and stored in this demodulator.

Let's consider the simplest modem with two-position PM (Fig. 1.4). Here the modulator consists of a driving oscillator (O) and multiplier. The harmonic carrier $f_0(t)$ with a fixed frequency generated by the oscillator is multiplied by the information signal $J(t)$ as bipolar pulses of dc. Figure 1.3(a) shows the general view of such a signal. As a result radiopulses, differing from each other by a phase of 180 degrees, are generated at the modulator output [see Fig. 1.3(d)]. Demodulation of the received PM signal is fulfilled by a phase detector consisting of a multiplier and integrator or low-pass filter (LPF); in the multiplier the received signal $S(t)$ is multiplied by the reference oscillation $f'_0(t)$, and the low-pass filter selects the constant component of the product $S(t) f'_0(t)$. The reference oscillation frequency is equal to the received signal frequency. If the reference oscillation $f'_0(t)$ coincides with the phase of the received signal element $S(t)$, the signal at the LPF output has positive polarity, but if $f'_0(t)$ and $S(t)$ are counterphased the LPF output signal is negative.

Therefore, for the coherent demodulation of a signal with PM the reference oscillation of a phase detector must coincide with the phase with such a PM signal variant at the demodulator input, which corresponds to a binary modulating signal of positive polarity. Otherwise, if the reference oscillation coincides with a signal

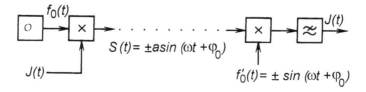

Figure 1.4 The simplest phase modem.

corresponding to a modulating signal of negative polarity, the output binary signals of the demodulator will be opposite to those actually transmitted.

The simple but important condition of operation of a PM signal coherent demodulator can be executed if it is possible to do the following: (1) generate a reference oscillation that coincides with a phase with any of two input PM signals, and (2) determine to which of the transmitted binary symbols this reference oscillation corresponds.

The first task can be solved rather successfully by using special methods and units for the selection of coherent reference oscillations. (For more information, see Chapter 3.) Without any detail here, it should be noted that these methods permit us to generate a reference oscillation which with rather high accuracy coincides by frequency and phase with one of the PM signal variants at the demodulator input.

The second task, in contrast to the first one, can not be solved in principle if we do not introduce some additional features into the transmitting signal for the one-to-one conformity between transmitted binary symbols and signal variants at the demodulator input. In fact, if the transmitting PM signal variants are equiprobable and do not contain any redundancy, elementary signals coming to the demodulator input have no information about which binary symbol corresponds to each of the two signal variants.

Such a paradoxical peculiarity of the phase-modulated signal, called *initial phase uncertainty* or *decision ambiguity,* can be explained, on the one hand, by an arbitrary and unknown phase shift that is added to the transmitted phase in a communications channel, and, on the other hand, by the fact that the signal phase always results in the interval 2π and the signals differed by a phase of 2π are actually indistinguishable.

The indicated decision ambiguity is quite typical just for PM. At AM a received signal also differs from a transmitted one. However, if at the modulator output a binary symbol corresponded to a signal with greater amplitude, the same symbol will correspond to the signal variant with a greater amplitude at the demodulator input. Here there is no ambiguity. The same is true of FM: If one of two frequencies is greater than another at the modulator output, it will be greater after all conversions, and it is impossible to confuse signal variants.

The inability to state a one-to-one conformity between transmitting symbols and signal variants at the demodulator input at absolute phase modulation has become the main reason for transition to the relative phase or phase-difference modulation.

The theoretical substantiation of difference methods of modulation that enable the influence of unpredictable changes in signal parameters to be reduced or removed is discussed in the following chapter. Here we merely note that the problem of phase uncertainty is decided very simply by the PDM by means of calculation of adjacent chip phase difference. The initial phases of elementary signals at the demodulator input are arbitrary and differ from the transmitted signals by the unknown value. That is why, when using PM, decision ambiguity concerning transmitted symbol is possible. However, the phase differences of adjacent chips are always constant and equal to 0 or 180 degrees as well as at the modulator output, which excludes the decision ambiguity. Therefore, if the information of transmitted binary symbols is contained not in the absolute phase of chips but in the phases difference between the chips, it is possible to avoid the decision ambiguity peculiar to a PM signal (having saved practically all advantages of this signal). In short, this is the PDM concept.

Figure 1.3(e) shows the signal with PDM corresponding to the binary symbol transmission process shown in Figure 1.3(a). As can be seen, this signal consists of the same modulated by the phase of 180-degree chips as the PM signal shown in Figure 1.3(d). The difference is that at PDM the phase of these chips depends not only on the transmitting symbol as at PM but also on the initial phase of the previous chip. The principle of initial phase generation is simple: If a binary symbol 0 is transmitted by this chip, its phase is the same as the previous chip phase (the phase difference is equal to zero). If a symbol 1 is transmitted by this chip, its phase is changed by 180 degrees in comparison with the phase of the previous chip (phase difference is equal to 180 degrees).

If an integer of carrier periods with frequency ω fills into the chip of T duration the phase difference of 180 degrees between two chips is equally matched to the phase hop of 180 degrees on the boundary between them. The PDM signal shown in Figure 1.3(e) corresponds to this situation. The situation can be different at other relations between ω and T. For example, if an odd number of carrier semiperiods fills into a chip, the phase hop of 180 degrees on the boundary of two chips occurs at a zero phase difference of these chips, and, in contrast, the unchanged phase on the boundary of chips corresponds to the phase difference of 180 degrees. The PDM signal shown in Figure 1.3(f) corresponds to this last case. Figure 1.3(g) shows the third example of a signal with two-position PDM. Here, when an odd number of a quarter of a carrier periods fills into a chip, the required phase differences of 0 and 180 degrees are established at the ±90-degree phase hops on a boundary of chips. Thus, the generation of a two-position PDM signal is not always accompanied by 180-degree phase hops at the boundaries of chips and is not always reduced to the generation of such hops. It is important for the phase difference between adjacent chips to be 0 or 180 degrees depending on a transmitted binary symbol; that is, the signal variants need to be antipodal (counterphase).

At PDM the transmitted binary symbol is determined by two chips, and all methods of PDM signal reception are based on comparison of the initial phases of two adjacent chips. Therefore, one additional chip is needed to transmit the first binary symbol in the system with PDM. This extra chip is transmitted just before the beginning of the data transmission and is considered to be a starting point [see Figure 1.3(e–g)]. The occurrence of one starting chip at PDM (or, what is the same, the vanishing of the first of transmitting symbols) is treated as though it is payment for the elimination of the decision ambiguity at the phase-modulated signal receiving side.

The PDM signals, in contrast to the signals with AM, FM, and PM, are recorded in the interval of two chips from 0 to $2T$. For the binary PDM

$$S_1(t) = \begin{cases} a \sin(\omega t + \varphi_0) & 0 < t \le T \\ a \sin[\omega(t - T) + \varphi_0] & T < t \le 2T \end{cases}$$

$$S_2(t) = \begin{cases} a \sin(\omega t + \varphi_0) & 0 < t \le T \\ -a \sin[\omega(t - T) + \varphi_0] & T < t \le 2T \end{cases} \tag{1.11}$$

The signal $S_1(t)$ corresponds to the transmission of the phase difference $\Delta\varphi = 0$, the signal $S_2(t)$ to the phase difference $\Delta\varphi = \pi$. If an integer of periods of the carrier with frequency φ fills into a chip of T duration, the signals of (1.11) can be recorded in the interval $0 < t \le 2T$:

$$S_1(t) = a \sin(\omega t + \varphi_0)$$

$$S_2(t) = \text{sgn}\left(\sin\frac{\pi t}{T}\right) \sin(\omega t + \varphi_0) \tag{1.12}$$

Here the signal $S_1(t)$ does not have a phase hop in the interval $(0, 2T)$, but the signal $S_2(t)$ has a hop of π at the moment $t = T$, that is expressed in (1.12) through the signum function.

Note that the two-position PDM signals are antipodal within the one-chip interval as well as the PM signals, but within the interval of two chips they are intensely orthogonal as FM signals. The latter follows from the following equation:

$$\int_0^{2T} S_1(t) S_2(t) \, dt = 0$$

which is valid for (1.11) and (1.12) at any initial phases φ_0.

The generation of the PDM signals can be reduced to the generation of the PM signals by repeated encoding (or differential encoding) of the transmitted binary sequence. The algorithm of repeated encoding is simple: If we designate the information symbol having to be transmitted by nth chip as $J_n = \pm1$, the differentially encoded symbol J_n' in the nth chip is defined by the following recurrent rule:

$$J_n' = J_n J_{n-1}' \tag{1.13}$$

Now for the generation of a signal with two-position PDM it is enough to multiply the carrier by the signal of (1.13).

Figure 1.5 shows the scheme of a modulator and demodulator of signals with the two-position PDM. As can be seen from the comparison between this scheme and the scheme shown in Figure 1.4, the PDM modulator contains the same carrier $f_0(t)$ oscillator and multiplier as the PM modulator. However, here the repeatedly encoded information signal J_0' comes to the second input of a multiplier, and the differential encoder (enclosed by a dashed line) consists of a multiplier and a storage element that operate with binary symbols according to the algorithm (1.13). The demodulator of the two-position PDM signal contains the same phase detector as the demodulator of PM signals. The phase detector includes a multiplier and a low-pass filter; the reference oscillation $f_0'(t)$ coincides with a phase of one of the variants of the transmitted signals. The following calculation of a phase difference and the determination of the transmitted binary symbol are executed by multiplying the signals at the phase detector output delayed relative to each other for the duration of the chip. This is done by a differential decoder (enclosed by a dashed line).

The algorithms of modulation and demodulation of PDM signals are illustrated by time diagrams shown in Figure 1.6. Figure 1.6(a) shows the transmitted binary information; Figure 1.6(d) the repeatedly (differentially) encoded information signal; Figure 1.6(b) the carrier; and Figure 1.6(c,e) the signals with PM and PDM at the modulator output. In Figure 1.6(f) we can see that on the boundary between the third and fourth chips the phase hop of 180 degrees is specially entered into the reference oscillation $f_0'(t)$ of a demodulator. It enables us to illustrate the appearance of errors in systems with PM and PDM. The faulty symbols are marked in Figure 1.6(g,h) by asterisks. In the system with PM all the following symbols are faulty after the reference oscillation polarity has been changed. In this case the reception of binary symbols opposite to those transmitted (the so-called "antipodal operation") will take place up to the next phase hop of the reference oscillation or the input signal, that is, the so-called error propagation. At the same time, in the PDM system the hoplike changing of the reference oscillation results in a single error and has no other negative consequences.

Thus, the phase-difference modulation completely solves the problem of elimination of the uncertainty of the signal initial phase and the ambiguity of the

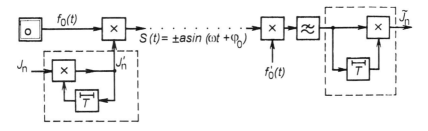

Figure 1.5 The simplest phase-difference modem.

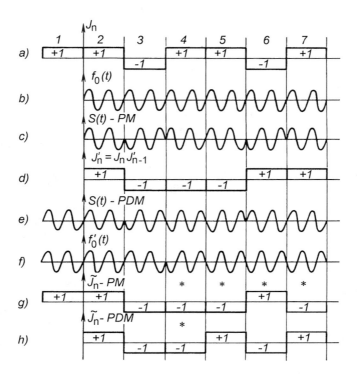

Figure 1.6 The time diagrams of signals in the PM and PDM modems.

demodulator operation inherent to transmission of messages by phase-modulated signals. However, we should note that the result cannot be reached "free of charge." In comparison with the perfect PM system, the PDM system has some defects. First, we are required to transmit one starting chip at the beginning of a communications session; second, the error probability is increased about two times; third, in a digital flow there are basically double errors, which complicates a codec slightly when using correcting codes; fourth, a PDM modem is a bit more complex than the PM. However, all these disadvantages do not reduce the basic advantages of PDM, which are especially obvious if we compare the PDM modem not with the perfect and infeasible PM modem but with those variants of PM that make the real alternative to PDM.

To realize the system with PM it is necessary at any rate to determine the exact conformity between signal variants at the demodulator input and transmitting discrete symbols. One possible method is the transmission of a special synchrosignal (marking signal) corresponding to the stipulated beforehand discrete symbol, for example, to binary zero. According to this signal we can, for example, set the required polarity of the reference oscillation $f_0'(t)$ in a demodulator of PM signals (see Fig. 1.4). To keep conformity set it is necessary to transmit a synchrosignal periodically because in the real communications channel and in the demodulator itself unpredictable changes and phase hops of a signal due to interference can appear.

So, the phase modulation and a periodically transmitted synchrosignal is one alternative to phase-difference modulation. Another way to realize the system with absolute PM is to use special codes or signals with redundancy that enables detection of errors such as all symbols inverting.

Thus, the implementation of the real system with PM is inevitably connected with certain losses—in power, rate, and equipment—in comparison with the perfect PM system. These losses appear to be more significant than the losses of PDM in comparison with that perfect PM. Therefore the choice between PM and PDM requires us to take into account not only the characteristics of appropriate modems but also the system situation as a whole. Some additional discussion on this topic is given in Chapter 3, which is devoted to the coherent reception of phase-modulated signals.

Comparing the opportunities of modems with PM and PDM, one should take into account that the reception of the PDM signal, in contrast to the PM one, can be fulfilled by incoherent methods and hence the possibility to choose the method of signal processing in the systems with PDM is considerably larger than in the systems with PM.

1.3 CLASSIFICATION OF METHODS OF PHASE-MODULATED SIGNAL RECEIVING

The elementary phase modems of digital transmission systems are built on the basis of coherent processing, which descends from the first attempts at the phase-modulated signals reception: synchronous detecting [14].

The invention of PDM has created new opportunities in this field, the main one of which is the possibility that incoherent processing can be applied to phase-modulated signals, something that has heretofore been considered inconceivable.

We consider the essence of various methods of a phase-modulated signal reception by presenting an example for the two-position PDM already discussed.

The phase difference of adjacent chips is the information parameter of a PDM signal:

$$S_{n-1}(t) = a \sin(\omega t + \varphi_{n-1}) \qquad (n-1)T < t \leq nT$$
$$S_n(t) = a \sin(\omega t + \varphi_n) \qquad nT < t \leq (n+1)T \qquad (1.14)$$

The receiver has to determine the phase difference transmitted by the oscillations (1.14):

$$\Delta_n\varphi = \varphi_n - \varphi_{n-1} \qquad (1.15)$$

Chips distorted by an interference come to the receiver input. If the interference is additive and has the values of $\xi_{n-1}(t)$ and $\xi_n(t)$ within intervals of $(n-1)$th and nth chips, the distorted chips can be written in the form of

$$x_{n-1}(t) = S_{n-1}(t) + \xi_{n-1}(t)$$
$$x_n(t) = S_n(t) + \xi_n(t) \tag{1.16}$$

For constituting the algorithms of demodulators operation it is convenient to represent chips $S_{n-1}(t)$ and $S_n(t)$ as vectors $\mathbf{S}_{n-1}(t)$ and $\mathbf{S}_n(t)$ of the functional signal space and chips $x_{n-1}(t)$ and $x_n(t)$ distorted by an interference as $\mathbf{x}_{n-1}(t)$ and $\mathbf{x}_n(t)$ vectors of the signal and interference functional space.

Vector algebra is expedient to use because the notions of a phase difference of two chips and the angle between the corresponding vectors of the functional signal space are identical. In vector algebra the angle $\Delta_n \varphi_\xi$ between any vectors $\mathbf{x}_{n-1}(t)$ and $\mathbf{x}_n(t)$, which are here vector sums of signals and interferences defined in the interval T, is determined from a general relation

$$\cos \Delta_n \varphi_\xi = \frac{(\mathbf{x}_n \mathbf{x}_{n-1})}{|\mathbf{x}_n| |\mathbf{x}_{n-1}|} \tag{1.17}$$

where[1]

$$(\mathbf{x}_n \mathbf{x}_{n-1}) = \int_0^T x_n(t)\, x_{n-1}(t)\, dt \tag{1.18}$$

is the scalar product of vectors $\mathbf{x}_{n-1}(t)$ and $\mathbf{x}_n(t)$, $|\mathbf{x}_{n-1}(t)|$ and $|\mathbf{x}_n(t)|$ are the length (norm) of the vectors $\mathbf{x}_{n-1}(t)$ and $\mathbf{x}_n(t)$:

$$|\mathbf{x}_{n-1}| = \sqrt{\int_0^T x_{n-1}^2(t)\, dt} \qquad |\mathbf{x}_n| = \sqrt{\int_0^T x_n^2(t)\, dt} \tag{1.19}$$

It is easy to determine that at absence of interferences the value of $\cos \Delta_n \varphi_\xi$ is equal to the cosine of the transmitted phase difference $\Delta_n \varphi$. In fact, under specified conditions we can derive from the formula (1.17):

---

1. Elementary signals $x_{n-1}(t)$ and $x_n(t)$ exist in different time intervals; therefore, the relation (1.18) should be understood as follows: For the integration the elementary signals are combined in time.

$$\cos \Delta_n \varphi_\xi = \frac{(S_{n-1} S_n)}{|S_{n-1}||S_n|} = \frac{\int_0^T a \sin(\omega t + \varphi_{n-1}) \, a \sin(\omega t + \varphi_n) \, dt}{\sqrt{\int_0^T a^2 \sin^2(\omega t + \varphi_{n-1}) \, dt} \sqrt{\int_0^T a^2 \sin^2(\omega t + \varphi_n) \, dt}}$$

$$= \frac{\dfrac{a^2 T}{2} \cos(\varphi_n - \varphi_{n-1})}{\dfrac{a\sqrt{T}}{\sqrt{2}} \dfrac{a\sqrt{T}}{\sqrt{2}}} = \cos \Delta_n \varphi$$

Thus, the transmitted information parameter $\Delta\varphi$ can be determined by (1.17) through the ratio of the scalar product of vectors of two adjacent chips to the lengths product of these vectors.

However, if there is interference, the value of $\cos \Delta_n \varphi_\xi$ calculated by (1.17) is not equal to the value of $\cos \Delta_n \varphi$. The task of the theory is to compose such algorithms of $\cos \Delta_n \varphi_\xi$ estimation that will give maximum probability of the correct identification of $\cos \Delta_n \varphi$ and $\cos \Delta_n \varphi_\xi$. The receiving methods corresponding to these algorithms will be optimum for the conditions for which they were synthesized.

For the reception, that is, the determination of a transmitted information symbol, we should find $\Delta_n \varphi_\xi$ by (1.17) and compare it to all possible values of $\Delta_n \varphi_i$. The phase difference that is considered to be transmitted is the one closest to the received value $\Delta_n \varphi_\xi$. At the two-position PDM there are two variants of transmitting phase differences: $\Delta_n \varphi_1 = 0$ and $\Delta_n \varphi_2 = \pi$, and the cosine of these differences accepts values ± 1. Thus, at this case the determination of the transmitted symbol is reduced to the determination of a sign of $\cos \Delta_n \varphi_\xi$ in (1.17). Mathematically this operation is represented as follows:

$$J_n = \text{sgn} \cos \Delta_n \varphi_\xi \tag{1.20}$$

which should be read: The information symbol J_n transmitted within the nth chip is equal to the sign of $\cos \Delta_n \varphi_\xi$.

Remembering that the product of vector norms is a positive value, then in accordance with (1.17) and (1.20) we get

$$J_n = \text{sgn}(\mathbf{x}_n \mathbf{x}_{n-1}) \tag{1.21}$$

Thus, to determine the transmitted information symbol we should find the sign of the scalar product of vectors $\mathbf{x}_{n-1}$ and $\mathbf{x}_n$.

Expression (1.21) is very important for the theory of phase-difference modulation. Using it we can get and classify all algorithms for PDM signal receiving.

In fact, the methods of PDM signal processing differ according to the means of calculation (algorithm) of the scalar product in (1.21). In its turn, the algorithm of the scalar product calculation depends on the information about the received signal parameters, namely, the dimension of the space of vectors $\mathbf{x}_{n-1}$ and $\mathbf{x}_n$.

The first and most obvious reception algorithm is called *autocorrelated* and is in direct realization of expression (1.21) taking into account (1.18):

$$J_n = \mathrm{sgn}\left[\int\limits_0^T x_n(t)\, x_{n-1}(t)\ dt\right] \tag{1.22}$$

The term *autocorrelated receiving method* is used because of the analogy between the mathematical expression for an autocorrelated function of a signal $x(t)$ and an expression in square brackets of the formula (1.22). In Chapter 5 we indicate that algorithm (1.22) corresponds to a rule of the extended maximum likelihood at receipt of unknown shaped signals in the additive white Gaussian noise (AWGN) channel and is optimum in this sense.

The demodulator realizing algorithm (1.22) is shown in Figure 1.7. After the input bandpass filter has selected the signal on a carrier frequency ω and limited the noise spectrum, the received signal comes to the autocorrelator consisting of a unit of signal delay by duration of the chip T, the multiplier and integrator. The integrator operates in the reset mode: The integration of an input signal is performed within the interval of every chip $(0, T)$, and then the integrator is forced to be set to zero. The sample of the output signal from the integrator at the moment T comes to the input of a sign discriminator (sgn). The discrete signal ± 1 received at its output corresponds to transmitted binary symbols. A time locker (TL) sets the time points for signal sampling and signal reset at the integrator output. The unit generates the appropriate time pulses by nonlinear conversion of an input signal.

The algorithm of autocorrelated processing according to (1.22) corresponds to the calculation of the scalar product of the vectors of the received signal in the infinite-dimensional functional space (Hilbert space). In real conditions the space dimension of the received signals is always limited and does not exceed the value

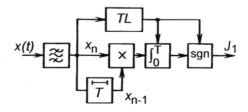

Figure 1.7 Autocorrelated demodulator of binary PDM signals.

B = 2FT where F is the bandpass width of the channel up to the demodulator input. (As shown in Figure 1.7, the parameter F is determined by the input bandpass filter.) For finite-dimensional signals the scalar product can be expressed by the projections of the corresponding vectors on coordinate (basic) function. If the signal space is given by the system of the orthonormal functions $f_i(t)$ the scalar product (1.18) can be expressed as a sum

$$(\mathbf{x}_n \mathbf{x}_{n-1}) = \int_0^T x_n(t)\, x_{n-1}(t)\; dt = \sum_{i=1}^{FT} a_{ni}\, a_{(n-1)i} + b_{ni}\, b_{(n-1)i} \tag{1.23}$$

where

$$a_{ni} = \int_0^T x_n(t)\, f_i(t)\; dt \qquad a_{(n-1)i} = \int_0^T x_{n-1}(t)\, f_i(t)\; dt$$

$$b_{ni} = \int_0^T x_n(t)\, f_i^*(t)\; dt \qquad b_{(n-1)i} = \int_0^T x_{n-1}(t)\, f_i^*(t)\; dt \tag{1.24}$$

(The Hilbert transformation is marked by an asterisk.) Thus, for the finite-dimensional case, general algorithm (1.21) can be written as

$$J_n = \mathrm{sgn} \sum_{i=1}^{FT} [\, a_{ni}\, a_{(n-1)i} + b_{ni}\, b_{(n-1)i}\,] \tag{1.25}$$

In fact, algorithm (1.25) is the modification of the receiving autocorrelated method. It is also a general record of incoherent processing algorithms of the PDM signals at which there is no information about initial phases of signal elements. If the elements of a composite signal coincide with the basic functions $f_i(t)$ but the phases of these elements are unknown and unequal the algorithm (1.25) is optimum in the AWGN channel and is called the *incoherent reception algorithm* with incoherent accumulation.

The further synthesis of particular algorithms revealing the general relation (1.21) is connected with the use of a priori information about the received signal to reduce the dimension of the considered vector space.

If we know that the received signal in the interval $(0, T)$ is a harmonic oscillation with the frequency ω, but the initial phase of this oscillation is unknown, the space dimension of the expected signals is becoming 2. Let us take two orthogonal oscillation as the basic functions:

$$f_1(t) = \sin \omega t$$
$$f_2(t) = \cos \omega t \tag{1.26}$$

In the two-dimensional space determined by basis (1.26) the scalar product (1.18) is

$$(\mathbf{x}_n \mathbf{x}_{n-1}) = \frac{X_n X_{n-1} + Y_n Y_{n-1}}{|\mathbf{f}_1||\mathbf{f}_2|} \tag{1.27}$$

where

$$X_n = \int_0^T x_n(t) \sin \omega t \, dt \qquad X_{n-1} = \int_0^T x_{n-1}(t) \sin \omega t \, dt$$

$$Y_n = \int_0^T x_n(t) \cos \omega t \, dt \qquad Y_{n-1} = \int_0^T x_{n-1}(t) \cos \omega t \, dt \tag{1.28}$$

$$|\mathbf{f}_1|^2 = |\mathbf{f}_2|^2 = \int_0^T \sin^2 \omega t \, dt = T/2$$

Substituting (1.27) and (1.28) into (1.21), we obtain

$$J_n = \operatorname{sgn}(X_n X_{n-1} + Y_n Y_{n-1}) \tag{1.29a}$$

or in the expanded form

$$J_n = \operatorname{sgn}\left[\int_0^T x_n(t) \sin \omega t \, dt \int_0^T x_{n-1}(t) \sin \omega t \, dt + \int_0^T x_n(t) \cos \omega t \, dt \int_0^T x_{n-1}(t) \cos \omega t \, dt \right] \tag{1.29b}$$

Processing algorithm (1.29) corresponds to one of the main (alongside the autocorrelated one) methods of PDM signal reception. It is called the *optimum incoherent method*. As introduced in Chapter 4, algorithm (1.29) provides the minimum error probability for the processing of two chips of a phase-modulated signal with unknown and uniformly distributed initial phase in the AWGN channel.

Figure 1.8 shows a scheme for the corresponding demodulator. Here the carrier oscillator (O) and phase shifter $\pi/2$ generate orthogonal oscillations sin

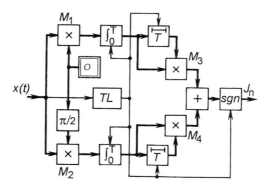

Figure 1.8 Optimum incoherent demodulator of binary PDM signals.

ωt and $\cos \omega t$, which multiply the received signal in the multipliers M_1 and M_2. The integrators operate in a reset mode. Then there follow the memory units T and multipliers of the integrators outputs M_3 and M_4. In the scheme only one sample of a signal should be stored in the memory unit instead of a whole signal as it is in the autocorrelated demodulator shown in Figure 1.7. In this case the time locker (TL) provides time intervals to read out and reset the integrators, to write and read out in memory units, and to give out the received binary symbol in a sign discriminator (sgn). The synchronizing connections are shown by thin lines in Figure 1.8.

Let's consider the initial algorithm (1.21). Different means of its realization have been obtained by transition from the infinite-dimensional space to the finite-dimensional one and then to two-dimensional vector space using more and more a priori information of the signal. The algorithm of calculation of the scalar product of vectors of the adjacent chips used in the optimum incoherent receiver is based on the assumption that the signal is known completely with the exception of its initial phase.

If we now assume that the initial (noninformation) phase of the signal φ_0 is also known, the dimension of an expected signal will become equal to 1 because its variants are fully determined by a coefficient at the unique coordinate function $f_1(t) = a \sin(\omega t + \varphi_0)$ coinciding with a transmitted signal up to a sign and scale. Therefore the scalar product (1.8) is equal to

$$(\mathbf{x}_n \mathbf{x}_{n-1}) = \frac{X_{n0} X_{(n-1)0}}{|f_1|^2} \qquad (1.30)$$

where

$$X_{n0} = \int_0^T x_n(t)\ \sin(\omega t + \varphi_0)\ dt$$

$$X_{(n-1)0} = \int_0^T x_{n-1}(t)\ \sin(\omega t + \varphi_0)\ dt$$

(1.31)

After the substitution of (1.30) into (1.21) we have

$$J_n = \mathrm{sgn}(X_{n0}\,X_{(n-1)0}) = \mathrm{sgn}\ X_{n0}\ \mathrm{sgn}\ X_{(n-1)0}$$

(1.32a)

or extended as

$$J_n = \mathrm{sgn}\left[\int_0^T x_n(t)\ \sin(\omega t + \varphi_0)\ dt\right]\mathrm{sgn}\left[\int_0^T x_{n-1}(t)\ \sin(\omega t + \varphi_0)\ dt\right]$$

(1.32b)

Algorithm (1.32) corresponds to a coherent method of PDM signals reception. As introduced in Chapter 3, this algorithm provides the minimum error probability at receiving the PDM signals in the AWGN channel.

The scheme of the corresponding demodulator was considered in the previous paragraph (see Fig. 1.5). We discuss it in more detail in Figure 1.9. Figure 1.9 shows the carrier recovery unit or the reference oscillation selector (ROS), which performs selection of the coherent reference oscillations. It is one of main parts of the coherent demodulator. The principles and schemes of these units are given in Chapter 3. We merely note here that the coherent reference oscillation is generated by means of an input signal of a demodulator as well as results of its processing in the demodulator. Figure 1.9 shows corresponding connections. The memory element and multiplier M_2 operate here only with binary symbols.

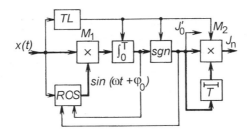

Figure 1.9 Coherent demodulator of binary PDM signals.

Note that the synthesized coherent demodulator of PDM signals includes the coherent demodulator of PM signals—that part of the circuit from the input up to the output of the unit sgn shown in Figure 1.9. The corresponding algorithm is easy to write:

$$J_n' = \text{sgn} \int_0^T x_n(t) \, \sin(\omega t + \varphi_0) \, dt \qquad (1.33)$$

In this case the binary symbols J_n' coincide with transmitted ones if by means of the corresponding synchronization the unique conformity between these symbols and the reference oscillations has been determined. At the same time the symbols J_n at the output of the PDM demodulator (see Fig. 1.9) coincide with the transmitted ones without determination of the specified conformity.

So, three of the mathematical algorithms indicated above, (1.18), (1.27), and (1.30), are used for calculating the *scalar product* of vectors of $(n-1)$ and nth signal chips $\mathbf{x}_{n-1}$ and $\mathbf{x}_n$. The scalar product is proportional to the cosine of the received phase difference, and the algorithms define three fundamental methods of reception of signals with phase-difference modulation:

- *Autocorrelated reception,* when the scalar product of vectors is calculated directly [see (1.22)];
- *Optimum incoherent reception,* when the scalar product is calculated by multiplying the projections of the received signal on the orthogonal coordinate functions sin ωt and cos ωt with arbitrary phase [see (1.29)];
- *Coherent reception,* when the scalar product is calculated by multiplying the projections of the received signal on the coordinate function $\sin(\omega t + \varphi_0)$ coinciding (coherent) with one of the signal variants [see (1.32)].

Analysis shows that in spite of the great variety of algorithms and schemes for PDM demodulators they all correspond to the classifications just given. If more complex algorithms are used, for example in receiving aggregates, they also are a combination of the three specified algorithms.

In these three algorithms the main operation during extraction of transmitted data from the received signal is calculation of the convolution of the received signal and a reference signal. The received signal with a certain time shift is used as the reference signal in the autocorrelated method; the orthogonal harmonic oscillations with an arbitrary initial phase are used with the optimum incoherent method; and in-phase (synchronous) with one of signal variants (or shifted relative to it by a certain angle) oscillation is used with the coherent method. These features permit us to determine to which of the three types of receivers a particular demodulator belongs.

The problem of the comparative noise immunity of the considered methods of PDM signal processing is very important (see Chapter 6). In terms of noise immunity reduction, the methods can be ordered as follows: coherent, optimum incoherent, and autocorrelated.

Such fluctuated interference as white noise forms multidimensional space that is uniformly filled by interference in the sense that the interference projections on any coordinate axis of the space have identical distributions. The interference space is always crossed by the signal space (otherwise the interference could be fully separated from a useful signal). It is clear that the smaller the dimension of the signals space, the smaller the amount of interference (in the sense of probability) that can distort the signal.

In particular, for coherent reception, the mixture of a signal and interference is projected onto the single coherent coordinate axis. Hence, the receiver extracts only one projection of an interference together with a useful signal.

For optimum incoherent reception, the mixture of a signal and interference is projected on two coordinate axes and, hence, two projections of the interference are extracted together with a useful signal; the signal-to-interference ratio at the receiver output decreases in comparison with the coherent reception.

Finally at the autocorrelated reception, when the scalar product of vectors $\mathbf{x}_{n-1}$ and $\mathbf{x}_n$ is calculated by means of the general formula of scalar products, a receiver together with a useful signal extracts the interference from the multidimensional noise space. The dimension of this space is, as a rule, more than 2; therefore, the noise immunity of the autocorrelated method of reception is less than for the correlation methods, that is, the coherent and optimum incoherent methods.

However, we should emphasize that our discussion is valid only when the optimization conditions for each of these methods of receiving are satisfied. Otherwise, for example, the incoherent demodulator can give less error probability than the coherent one operating under inadequate conditions. The conditions that make each different algorithm optimum are considered in Chapters 3, 4, and 5.

1.4 FIRST-ORDER AND SECOND-ORDER PHASE-DIFFERENCE MODULATION AND INVARIANCE PROPERTIES

All that has been said to this point about the signals with PDM and methods of receiving these signals were concerned with the special case of PDM called first-order phase-difference modulation, or PDM-1.

Note that here it is not a matter of the *modulation multiplicity*, determined by a number of discrete levels of a transmitted signal or of its information parameter, but of an *order of phase differences* used as the information parameter.

The concept of the order of phase differences of a signal is introduced as follows.

Let's deal with series transmissions. We have chips of a harmonic PM signal with the initial phases

$$\varphi_0, \ \varphi_1, \ \varphi_2, \ \ldots, \ \varphi_{n-1}, \ \varphi_n, \ \varphi_{n+1}, \ \ldots \qquad (1.34)$$

We can calculate phase differences between each pair of the adjacent chips as follows:

$$\Delta_1^1 \varphi = \varphi_1 - \varphi_0$$
$$\Delta_2^1 \varphi = \varphi_2 - \varphi_1$$
$$\vdots$$
$$\Delta_n^1 \varphi = \varphi_n - \varphi_{n-1}$$
$$\Delta_{n+1}^1 \varphi = \varphi_{n+1} - \varphi_n$$
$$\vdots$$

$$(1.35)$$

Phase differences (1.35) are called phase differences of the first order (or simply first phase differences) because they have been derived from the initial phase sequence (1.34) by single-step subtraction. It is marked by the top index 1 at the operator of difference calculation Δ. Thus, the designation $\Delta_n^1 \varphi$ shows that the matter is of the calculation of the first-order phase difference between the nth and $(n-1)$th chips of a signal. The sequence of the first-order phase differences and the sequence of the initial phases are expanded in time in accordance with chip transmission:

$$\Delta_0^1 \varphi, \ \Delta_1^1 \varphi, \ \Delta_2^1 \varphi, \ \ldots, \ \Delta_{n-1}^1 \varphi, \ \Delta_n^1 \varphi, \ \Delta_{n+1}^1 \varphi, \ \ldots \qquad (1.36)$$

Taking into account this sequence, we can calculate new differences under the same rule according to which (1.35) was made up of (1.34):

$$\Delta_1^2 \varphi = \Delta_1^1 \varphi - \Delta_0^1 \varphi$$
$$\Delta_2^2 \varphi = \Delta_2^1 \varphi - \Delta_1^1 \varphi$$
$$\vdots$$
$$\Delta_n^2 \varphi = \Delta_n^1 \varphi - \Delta_{n-1}^1 \varphi$$
$$\Delta_{n+1}^2 \varphi = \Delta_{n+1}^1 \varphi - \Delta_n^1 \varphi$$
$$\vdots$$

$$(1.37)$$

The phase differences of (1.37) are called the second-order phase differences (or simply second phase differences) because they have been received from the initial phase sequence (1.34) by double-step subtraction. It is marked by the top index 2 at the operator of subtraction Δ. Phase differences of the second-order make up a time sequence similar to the sequences of (1.34) and (1.36):

$$\Delta_0^2\varphi, \Delta_1^2\varphi, \Delta_2^2\varphi, \ldots, \Delta_{n-1}^2\varphi, \Delta_n^2\varphi, \Delta_{n+1}^2\varphi, \ldots \tag{1.38}$$

We can go on to calculate phase differences for third and higher orders. Table 1.1 shows the process of forming phase differences of a high order. Each component of this table is equal to the difference of two adjacent components in the preceding line. In this section we will concentrate on using the phase differences of the first and second orders. More detailed information about higher order modulation is given in the next chapter.

Now we can determine the PDM introduced earlier as follows: The method of forming a phase-modulated signal when the information is put into the values of first-order differences of the chip phases is called *first-order phase-difference modulation* (PDM-1).

The phase difference is an information parameter of a signal at PDM-1. It is determined by two chips of a signal

$$\Delta_n^1\varphi = \varphi_n - \varphi_{n-1} \tag{1.39}$$

Table 1.1
Forming the High-Order Phase Differences

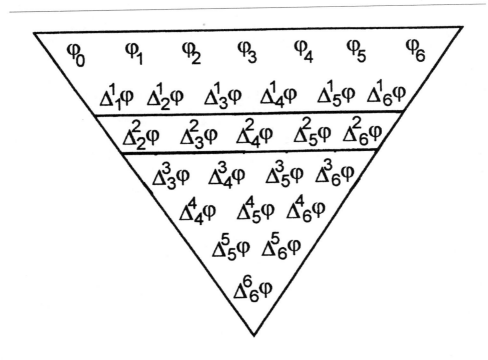

where the initial phase of the nth chip of a signal transmitted by a communications channel is

$$\varphi_n = \varphi_{n-1} + \Delta_n^1 \varphi \tag{1.40}$$

Expressions (1.40) and (1.39) determine the general algorithm for forming and processing the PDM-1. It is illustrated by the scheme shown in Figure 1.10(a).

At the transmitting site of a digital system, one of the admissible levels of the first phase difference $\Delta_n^1 \varphi$ conforms with the discrete information symbol J_n. Then the initial phase of the next nth chip of a transmitting signal is formed by a chip-time-delay element and by an adder according to (1.40). At the receiving site, when the initial phases of two adjacent chips of a signal have been measured, the information phase difference $\Delta_n^1 \varphi$ is calculated by a chip-time-delay element and by a subtractor. Then the information phase difference $\Delta_n^1 \varphi$ is identified with the transmitted discrete symbol i. The operation of phase meters and PDM-1 signal receivers was illustrated in the previous paragraph.

Let's determine the second-order phase-difference modulation (PDM-2). PDM-2 is a method of forming PM signals when the information is put into values of the second-order differences of the chip phases.

The information parameter of the PDM-2 signal is a difference between phase differences determined by three chips

$$\Delta_n^2 \varphi = \Delta_n^1 \varphi - \Delta_{n-1}^2 \varphi = (\varphi_n - \varphi_{n-1}) - (\varphi_{n-1} - \varphi_{n-2}) = \varphi_n - 2\varphi_{n-1} + \varphi_{n-2} \tag{1.41}$$

From (1.41), we obtain the initial phase of the next nth chip of a signal transmitted by a communications channel:

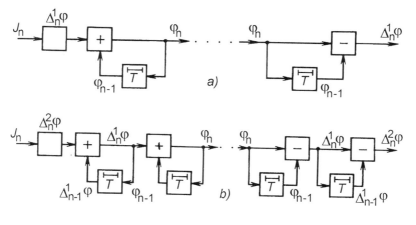

Figure 1.10 Principles of forming and processing the PDM signals: (a) PDM-1 and (b) PDM-2.

$$\varphi_n = \Delta_n^2\varphi + 2\varphi_{n-1} - \varphi_{n-2} \tag{1.42}$$

It can be presented similar to (1.40) as two recurrent expressions:

$$\begin{aligned} \varphi_n &= \Delta_{n-1} + \Delta_n^1\varphi \\ \Delta_n^1\varphi &= \Delta_{n-1}^1\varphi + \Delta_n^2\varphi \end{aligned} \tag{1.43}$$

A modulator of the PDM-2 signal, operating according to algorithm (1.43), consists of two PDM-1 modulators connected in series, and a receiving processor consists of two first phase difference calculators connected in series [Fig. 1.10(b)].

Let's discuss the main property of PDM-1 and PDM-2: the property of invariance. In engineering this term refers to the insensitivity of a system to any external interference or destabilizing factors. In this case some characteristic or parameter of a system is an invariant of the transformations caused by the interference or destabilizing factors.

As has already been noted, the PDM-1 method allows us to eliminate the ambiguity of the demodulator decision that results from the uncertainty of the initial phase of the received signal. This is explained by the fact that the first phase difference is the invariant of any signal transformation adding an arbitrary initial phase to the information chip phases. Really, if an arbitrary and unknown initial phase φ_0 is added to information phases φ_{n-1} and φ_n of the $(n-1)$th and nth chips, the phase difference of the $(n-1)$th and nth chips of a signal will not be changed because of such a conversion:

$$\Delta_n\varphi = (\varphi_n + \varphi_0) - (\varphi_{n-1} + \varphi_0) = \varphi_n - \varphi_{n-1} = \text{invar } \varphi_0 \tag{1.44}$$

Thus, PDM-1 is invariant to the initial phase of a signal. It is important for this property of PDM-1 to actually be saved at all methods of receiving. When the reception is optimum incoherent or autocorrelated, it does not require any information about the initial phase. When the reception is coherent, the initial phase should be known up to fixed shifts depending on modulation multiplicity; up to 180 degrees at two-position PDM-1 that is quite feasible. The elementary schemes of PDM-1 modems realizing the property of invariance to the signal initial phase were indicated in the previous section. It is possible to say that the considered incoherent demodulators of the PDM-1 signals are absolutely invariant, that is, they are entirely insensitive to the initial phase of a signal.

The transfer to PDM-2 permits us to achieve complete insensitivity—not only to any initial phase shift but also to any frequency shift. This is explained by the fact that the second phase difference is the invariant of any signal transformation leading to an arbitrary frequency shift of a carrier. Indeed, let us assume that the nominal carrier frequency $\omega = m2\pi/T$ (m is an integer) has been changed by arbitrary value $\Delta\omega$. Then if the phase of the $(n-1)$th chip is equal to

$(\varphi_{n-1} + \varphi_0)$, the phase of the nth chip will become equal to $(\varphi_n + \varphi_0 + \Delta\omega T)$ and the phase of the $(n + 1)$th chip will become $(\varphi_{n-1} + \varphi_n + 2\Delta\omega T)$. Now it is clear that the first phase differences

$$\Delta_{n+1}^1\varphi = \varphi_{n+1} - \varphi_n + \Delta\omega T$$

$$\Delta_n^1\varphi = \varphi_n - \varphi_{n-1} + \Delta\omega T$$

do not depend on the initial phase φ_0. However, they depend on the frequency shift $\Delta\omega$ while the second phase difference does not depend on either φ_0 or $\Delta\omega$:

$$\Delta_{n+1}^2\varphi = \Delta_{n+1}^1\varphi - \Delta_n^1\varphi = \varphi_{n+1} - 2\varphi_n + \varphi_{n-1} = \mathrm{invar}(\varphi_0, \Delta\omega) \qquad (1.45)$$

Thus, PDM-2 is invariant to the carrier frequency. This unique property of PDM-2 considerably enhances the ability to use digital communications systems with phase-modulated signals. At absolute PM, signal reception is possible only when the initial phase is known exactly. This extremely limits the practical application of absolute phase modulation. At PDM-1 it appears possible that we will be able to receive phase-modulated signals with an arbitrary initial phase. At PDM-2 there appears to be an extra possibility that we will be able to receive phase-modulated signals with an uncertain carrier frequency.

The PDM-2 signals as well as the PDM-1 signals can be received by the algorithms of coherent, optimum incoherent, and autocorrelated reception. When using the first two methods the property of invariance to the frequency will not be realized because the coherent and optimum incoherent methods of a reception achieve their potential opportunities only when the carrier frequency is known exactly, and they degrade when the frequency of a signal deviates from the frequency of a reference oscillation in demodulators. When the signal frequency is known exactly, the PDM-2 method has the same noise immunity as PDM-1 at the coherent reception, and the noise immunity of PDM-2 is higher than that of PDM-1 at the optimum incoherent reception. This is an important advantage of PDM-2.

The property of absolute invariance to carrier frequency is achieved by the autocorrelated reception of the PDM-2 signals. In the elementary case of the binary PDM-2 with the second phase differences $\Delta^2\varphi_1 = 0$, $\Delta^2\varphi_2 = \pi$ the operation algorithm of the appropriate demodulator can be made up by presenting the cosine of the second phase difference of the received signal as a scalar product of the adjacent chips, similar to how it was done in the previous section for PDM-1. It is obvious that in this case the transmitted information symbol J_n is determined by the sign of cosine of the second phase difference $\Delta_n^2\varphi_\xi$ on the nth chip:

$$J_n = \mathrm{sgn}\, \cos \Delta_n^2\varphi_\xi = \mathrm{sgn}(\cos \Delta_n^1\varphi_\xi \cos \Delta_{n-1}^1\varphi_\xi + \sin \Delta_n^1\varphi_\xi \sin \Delta_{n-1}^1\varphi_\xi)$$

$$(1.46)$$

Cosines and sines of the first phase differences being a part of (1.46) are presented by the formulas of vector algebra—similar to (1.17)—as follows:

$$\cos \Delta_n^1\varphi_\xi = \frac{(x_n x_{n-1})}{|x_n||x_{n-1}|} \qquad \sin \Delta_n^1\varphi_\xi = \frac{(x_n x^*_{n-1})}{|x_n||x_{n-1}|} \tag{1.47}$$

$$(x_n x_{n-1}) = \int_0^T x_n(t)\, x_{n-1}(t)\ dt \qquad (x_n x^*_{n-1}) = \int_0^T x_n(t)\, x^*_{n-1}(t)\ dt$$

where $x_n(t)$ and $x_{n-1}(t)$ are the nth and $(n-1)$th chips of a signal at the demodulator input. Having substituted (1.47) into (1.46) we receive the required algorithm of autocorrelated processing of the two-position PDM-2 signals, which is absolutely invariant to the carrier frequency.

Figure 1.11 shows the appropriate demodulator, which is a part of the elementary digital transmission system with PDM-2. There is also a PDM-2 signal modulator in this system. It consists of a carrier oscillator (O), multiplier, and differential encoder. The last unit contains a two-chip-delay binary element and a multiplier of binary symbols ±1. The operation algorithm of the differential encoder can be easily made up by relation (1.42). As long as at the binary PDM-2 all phases included in this relation are equal either to 0 or π their addition can be replaced by either modulo-2 addition of digits 0 and 1 or multiplication of digits +1 and −1. In the last case the algorithm of encoding the transmitted binary symbols J_n will be in the form of the following recurrent relation:

$$J'_n = J_n J'_{n-2} \tag{1.48}$$

The algorithm (1.48) is realized in the scheme shown in Figure 1.11.

The operation of the modem (see Fig. 1.11) is illustrated by the time diagrams in Figure 1.12. Figure 1.12(a) shows the input information signal J_n. Figure 1.12(b,c) shows the output signals of the modulators at PDM-1 and PDM-2: The first signal

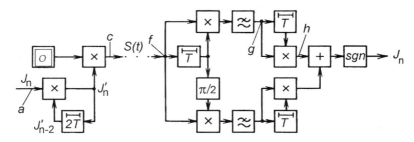

Figure 1.11 Autocorrelated PDM-2 modem providing invariance to the carrier frequency.

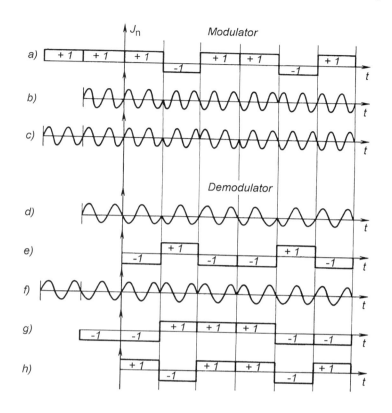

Figure 1.12 Time diagrams for PDM-1 and PDM-2 signal processing.

corresponds to the modulator in Figure 1.5 and the second one to the modulator in Figure 1.11.

Let's assume that in a channel the carrier frequency has changed such that the chip does not contain an integer of carrier cycles as was the case for the modulator output. Instead it carries an odd number of half-cycles. Then the signals with PDM-1 and PDM-2 at the appropriate demodulator inputs will be of the kind shown in Figure 1.12(d,f), respectively.

Figure 1.12(e) shows an output signal of the PDM-1 autocorrelated demodulator (see Fig. 1.7). As can be seen from the comparison of the diagrams in Figures 1.12(e) and 1.12(a), the PDM-1 autocorrelated demodulator produces binary symbols opposite to the transmitted ones at the given change of the carrier frequency. The demodulator operates as a coherent demodulator of signals with absolute PM when the phase of the received signal does not correspond the reference oscillation.

Now let's consider the PDM-2 autocorrelated demodulator. The signal at the top correlator output (point *g* in Fig. 1.11) is submitted in Figure 1.12(g). In this case the signal at the bottom correlator output is equal to zero because, at the chosen relation between the signal frequency and chip duration, the signals at the inputs of this correlator are orthogonal. From this it follows that the demodulator output signal results from multiplying the signs of the adjacent chips of a signal shown in Figure 1.12(g). This output signal is shown in Figure 1.12(h) and coincides with the transmitted signal.

This is a brief description of how signals are processed with PDM-2 demodulating them without errors at unknown carrier frequency shifts. The algorithms and schemes of PDM-2 signal processing are considered in Chapters 3 through 6 in more detail. We also illustrate that the noise immunity of the system with PDM-2 is not worse than of the system with PDM-1.

1.5 MULTIPOSITION PHASE-MODULATED SIGNALS

In the previous sections the various methods of generating and processing of phase-modulated signals corresponding to PM, PDM-1, and PDM-2 were illustrated primarily for the example of binary signals.

In this section we consider principles of forming m-ary (multiposition) phase-modulated signals and signals with combined methods of phase and amplitude modulation. As a rule, the difference between PM, PDM-1, or PDM-2 is not usually stipulated. If, for example, some set of signal vectors is considered, they can refer to any of these modulation methods and the variants of signal phases can be equally the variants of phase differences. The only difference is that in the first case the angles on the vector diagram are the transmitted absolute phases, and in the second case they are the phase differences of adjacent chips.

The principle of forming m-ary phase signals is rather simple: m signal vectors (points) are located on a circle with the radius depending on signal power (or chip energy) at equal distances with the angle interval $2\pi/m$ radian. Such sets of signal points/vectors for the cases when $m = 2, 3, 4,$ and 8 are shown in Figure 1.13.

The radii of all circles are equal to a square root of energy of an elementary signal: $R = \sqrt{E}$. If the harmonic oscillation with the parameters a, ω, and φ is transmitted in the T duration chip, this value will be equal to

$$R = \sqrt{E} = \sqrt{\int_0^T a^2 \sin^2(\omega t + \varphi) \, dt} = \frac{a\sqrt{T}}{\sqrt{2}} \tag{1.49}$$

The value of (1.49) coincides with the Euclidean distance between the center of the circle and any point located on it.

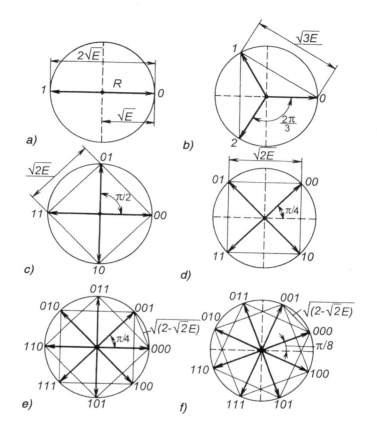

Figure 1.13 Multiple phase signals: (a) two phase, (b) three phase, (c,d) four phase, and (e,f) eight phase.

Figure 1.13(a) shows a two-phase (single-valued) signal with the phases 0 and π. Here the distance between signal points $d_2 = 2\sqrt{E}$ is the maximum possible distance between two points of the circle with radius $\sqrt{E}$. It fully defines the potential interference resistance of this two-position system. The distance $d_2 = 2\sqrt{E}$ is the starting value. The distance between signal variants in different multiposition systems is compared with this starting value.

The two-phase signal ís encoded quite simply: The symbol 0 conforms to one of its variants and the symbol 1 to the other [see Fig. 1.13(a)].

Figure 1.13(b) shows a three-phase signal with vectors standing off from each other at 120 degrees. This is the unique multiposition system in which the distances between all signal variants are identical (equidistant signals). We should note that the distance between two harmonic signals S_1 and S_2 with duration T and phase-difference φ is equal to

$$d(S_1,\ S_2) = \sqrt{\int_0^T [S_1(t) - S_2(t)]^2\ dt}$$

$$= \sqrt{\int_0^T [a\sin(\omega t + \varphi) - a\sin\omega t]^2\ dt}$$

$$= \sqrt{a^2 T - a^2 T \cos\varphi} = \sqrt{2E}\sqrt{1 - \cos\varphi} \qquad (1.50)$$

where $E = a^2 T/2$.

The distances d_m, calculated by formula (1.50), between the nearest signal variants in m-ary phase systems and the energy loss of m-ary systems in comparison with a two-phase system are indicated in Table 1.2.

In a general case the minimum distance between the signals is not the one-to-one characteristic of system noise immunity. Error probability in the perfect receiver depends on the whole signal set and not just on the distance between the nearest signal variants. However, for the given configuration of signal points the error probability is a steadily decreasing function of the minimum distance. This parameter is good for qualitative comparison of various signal systems.

In the case of a three-phase system [see Fig. 1.13(b)] the distance between signals is $d_3 = \sqrt{3E}$ and differs from d_2 by 1.25 dB. Therefore, at equal duration of

Table 1.2
Distances Between Nearest Signal Points and Energy Losses

Multiplicity of Modulation N	Number of Phases m	Minimum Phase Difference	Minimum Euclidean Distance Between Signals d_m	d_2/d_m (dB)
1	2	π	$2\sqrt{E}$	0
$\log_2 3$	3	$2\pi/3$	$\sqrt{3E} \approx 1.73\sqrt{E}$	1.25
2	4	$\pi/2$	$\sqrt{2E} \approx 1.41\sqrt{E}$	3.01
3	8	$\pi/4$	$\sqrt{(2-\sqrt{2})E} \approx 0.765\sqrt{E}$	8.34
4	16	$\pi/8$	$\sqrt{(2-\sqrt{2+\sqrt{2}})E} \approx 0.39\sqrt{E}$	14.2
5	32	$\pi/16$	$\sqrt{(2-\sqrt{2+\sqrt{2+\sqrt{2}}})E} \approx 0.196\sqrt{E}$	20.2

a chip, the three-phase system slightly yields to the two-phase one on the noise immunity but surpasses it on the transmission rate by $\log_2 3 = 1.58$ times. As indicated later, the three-phase system is optimum with respect to the energy per one information bit. Despite this, the system is used comparatively seldom because of the need for a special process to convert the binary code into ternary code.

Four-phase (double-valued) signals are the ones most used in practice [see Fig. 1.13(c,d)]. The double-valued system is submitted in two most popular variants: with phases (or phase differences) 0, $\pi/2$, π, and $3\pi/2$ [Fig. 1.13(c)] or $\pi/4$, $3\pi/4$, $5\pi/4$, and $7\pi/4$ [Fig. 1.13(d)]. In both cases the minimum distance between signal variants is $d_4 = \sqrt{2E}$, which corresponds to the distance between the orthogonal signals.

Such four-phase signals are the most noise-resistant signals among all two-dimensional 4-ary signals. Besides, these signals have a remarkable property: At equal bit rates they provide the same bit error rate as two-phase (antipodal) signals because the time needed to decrease the distance between signals is completely compensated by a two times increase in the chip duration.

Each chip of the 4-ary PM signal contains two information bits. Figure 1.13(c,d) shows a typical system of encoding for the four-phase signals by the two-digit binary combinations. This system is called the *Gray code*. The peculiarity of the code is that the code combinations corresponding to the nearest phases (or phase differences) differ by only one binary symbol. At 4-ary PM the Gray code is a unique keying code with such a property. Due to its use, a fault reception of a chip, as a rule, results in distortion of only one of two transmitted binary symbols.

Note that the signal systems shown in Figure 1.13(c,d) have the same potential noise immunity and differ only by the aspects of realization. As shown in Figure 1.13(d) the minimum phase shift in the system is equal to $\pi/4$. In each chip the system has a nonzero phase hop. In contrast, the system of Figure 1.13(c) has a zero phase hop at the transmission of a combination 00. It predetermines some features of modulator and demodulator realization. This problem is discussed later.

Figure 1.13(e,f) shows two variants of eight-phase (three-valued) signals. In the first option the phases or phase differences accept the values of $(i - 1)\pi/4$ where $i = 1, 2, \ldots, 8$, and in the second option they equal the values of $(2i - 1)\pi/8$. The first variant integrates the signals of two four-phase systems [see Fig. 1.13(c,d)] shifted relative to each other by $\pi/4$. The second variant does not contain a zero phase and has a constant shift relative to the first one by $\pi/8$. Every chip of an eight-phase signal contains three binary units of information. In a three-valued system, as well as in a double-valued system, there exists only one optimum keying code in which the three-digit binary combinations corresponding to the nearest signal vectors differ by only one binary symbol; this is the Gray code. The Gray code words are written down at the appropriate vectors (see Fig. 1.13). The distance between the nearest vectors in 8-ary signal systems shown in Figure 1.13 is

$$d_8 = \sqrt{(2 - \sqrt{2})E}$$

which is 8.3 dB less than in a two-phase system. This distance is not minimum as in the cases of two-, three- and four-phase signals considered earlier.

The uniform disposition of all signal points on the circle, that is, the use of signals with the same power and different phases, is optimum only for two-, three- and four-phase signals [m = 2, 3, 4; see Fig. 1.13(a–d)]. At $m > 4$ the optimum signals are the signals of nonequivalent power. They differ by phase as well as by amplitude and are located regularly inside the circle, the radius of which is determined by maximum admissible signal power. From the point of view of modulation theory, these signals refer to combined modulation methods at which some signal parameters change simultaneously. In the case considered, such parameters are the amplitude and phase or phase difference of signals: the amplitude-phase (APM) or amplitude-phase difference modulation (APDM), both of which are equivalent to quadrature amplitude modulation (QAM).

The elementary principle of forming signals with APM is to locate signal points on two concentric circles. However, it does not always lead to the optimum result. Figure 1.14 shows an example of such 8-ary signal with APM. The signal points are marked by crosses.

In Figure 1.14, four signals are located on the circle with the radius $R = \sqrt{E}$, and the other four are on the circle with radius $r < R$ and with the phase shifted by $\pi/4$. This set of signals is optimized (according to the criterion of choosing the maximum of the minimum distance between the signals) by choosing the radius ratio R/r. In fact, the more r at fixed R the more distance between the points of the internal circle, but the less distance between the points of internal and external circles. Therefore, the required maximum is achieved when these

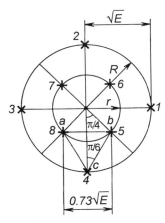

Figure 1.14 An 8-ary APM signal. Signal points are located on two concentric circles.

distances become equal, that is, when the triangle *abc* shown in Figure 1.14 is equilateral:

$$ab = bc = ca \qquad (1.51)$$

We need to find the ratio R/r at which condition (1.51) is satisfied. The following equations are obvious:

$$\frac{ab}{2} = r \sin \frac{\pi}{4}$$

$$\frac{bc}{\sin \pi/4} = \frac{R}{\sin(\pi - \pi/4 - \pi/6)} \qquad (1.52)$$

Having substituted (1.52) into (1.51) we obtain

$$R/r = 2 \cos \pi/12 \approx 1.932 \qquad (1.53)$$

Thus at the optimum ratio R/r the minimum distance between the signals is

$$d_8 = ab = \frac{\sqrt{E} \sin \pi/4}{\cos \pi/12} \approx 0.73\sqrt{E} \qquad (1.54)$$

This distance is a little bit shorter than the distance of a system with 8-ary PM signals located on one circle with the radius $R = \sqrt{E}$ [see Fig. 1.13(e,f) and Table 1.2]. Thus, in the case of the 8-ary system, the disposition of signal vectors on two concentric circles instead of one does not give any profit.

The elementary 8-ary system with APM appears to be optimum by the criterion of choosing the maximum of the minimum distance: Seven signals are located in regular intervals on the circle with the radius $R = \sqrt{E}$, and the eighth signal is equal to zero [15]. This set of signal points is shown in Figure 1.15.

The minimum distance of the system shown in Figure 1.15 is

$$d_8 = \frac{\sqrt{E} \sin 2\pi/7}{\cos \pi/7} \approx 0.86\sqrt{E} \qquad (1.55)$$

The signal constellation shown in Figure 1.15 has the same defect as all signals with APM, that is, unequal power of variants. This causes certain difficulties when these signals are transmitted over a communications channel (especially at nonlinear conversions or fading) and also in the realization of optimum processing.

Despite certain defects of APM, this modulation method is widely used because the performance of the phase-modulated equipower signals quickly degrades at increasing *m*.

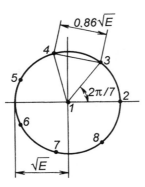

Figure 1.15 The best 8-ary signal in the two-dimensional space.

Let's consider 16-ary signals with PM and APM. Two systems of such signals are shown in Figure 1.16. The usual 16-ary PM signals are designated by points on a circle of the $R = \sqrt{E}$ radius. In this case the signal points are regularly located in intervals along the circle with a $\pi/8$ step. Usually they are encoded by four-digit binary words of the 16-ary Gray code. However, in this case the Gray code is not a unique optimum keying code (see Chapter 2). The distance between the nearest signal points in 16-ary systems with PM or PDM is

$$d_{16}(\text{PM}) = 0.39\sqrt{E}$$

which is a 14-dB energy loss in comparison with the binary PM system (see Table 1.2). It is obvious that the better system of signals with APM can be formed in the circle of this radius.

The 16-ary APM signal constellation is marked by crosses in Figure 1.16. The odd points of this constellation are regularly located on the circle of a greater radius with a $\pi/4$ interval, and they coincide with the odd signal points of the signals with PM. The even points of the APM signal are regularly located on the circle of a smaller radius $r < R$ with the same $\pi/4$ interval and with a common $\pi/8$ shift relative to the odd points.

The minimum distance between the signal points is maximized by the choice of such ratio R/r at which the distance between adjacent points on the circle of r radius is equal to the distance between the nearest points on two circles: $ab = bc = ca$. The same is applicable for (1.51), (1.52) and (1.53):

$$ab = 2r \sin \pi/8 \qquad (1.56)$$

$$bc = \frac{R \sin \pi/8}{\sin(\pi - \pi/8 - \pi/6)} \qquad (1.57)$$

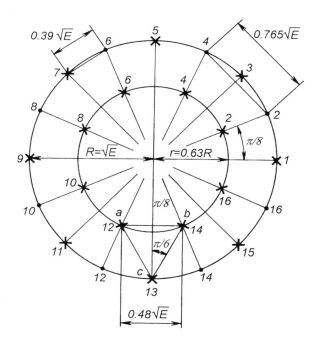

Figure 1.16 The 16-ary APM signals with signal points on two concentric circles.

Equating (1.56) to (1.57) we have

$$R/r = 2 \cos 5\pi/24 \approx 1.587 \qquad (1.58)$$

At the given optimum ratio R/r the minimum distance between signals is

$$d_{16}(\text{APM}) = ab = \frac{\sqrt{E} \sin \pi/8}{\cos 5\pi/24} \approx 0.482\sqrt{E} \qquad (1.59)$$

that is more than d_{16} (PM).

Another 16-ary APM signal constellation is shown in Figure 1.17. Here the signal points enumerated and marked by crosses are located in nodes of a square array. The size of the array is defined by the diagonal of a large square, which is equal to the diameter of the described circle: $2R = 2\sqrt{E}$. Sixteen of the usual PM signal points are located on the circle for comparison. In the example considered, both comparable signal sets have the same limit to maximum signal power, which is achieved in the system with APM at signal numbers 3, 7, 11, and 15. Under such conditions the minimum distance between signal points is

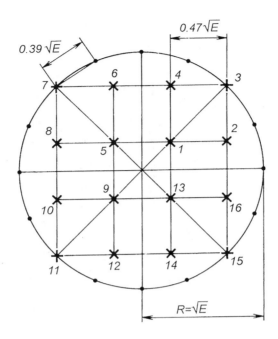

Figure 1.17 The 16-ary APM signal with signal points in the nodes of a square array at limited maximum power.

$$d_{16}(\text{APM}) = \sqrt{2E}/3 \approx 0.47\sqrt{E} \qquad (1.60)$$

This distance is more than in the system with 16-ary PM and almost the same as the distance in the system of signals with APM shown in Figure 1.16.

The transfer of PM to APM in systems with limited average power is mostly effective. In this case signal points 3, 7, 11, and 15, shown in Figure 1.17, are located on a circle the radius of which is more than $R = \sqrt{E}$ and, hence, the minimum distance will be more than $0.47\sqrt{E}$.

Figure 1.18 shows as an example the same 16-ary signals with PM and APM as in Figure 1.17 but for the case of limited average power. Sixteen signal points of the phase-modulated signal are shown on the circle of $R = \sqrt{E}$ radius. Sixteen signal points of a system with APM marked by crosses are located in the nodes of a square array, which in this case falls outside the limits of a circle.

To determine the comparative power parameters of signals, we should proceed from the equation of average power or average energy of signals with PM and APM. Signals with PM have equal average and maximum energy:

$$E_{\text{avr}}(\text{PM}) = E_{\text{max}}(\text{PM}) = E \qquad (1.61)$$

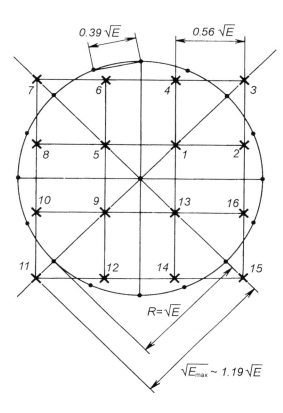

Figure 1.18 The 16-ary APM signal with signal points in the nodes of a square array at the limited average power.

The average power of signals with APM is expressed by the energy of variants

$$E_{avr}(APM) = \frac{1}{16}\sum_{i=1}^{16} E_i \qquad (1.62)$$

Signals 3, 7, 11, and 15 have the maximum energy

$$E_3 = E_7 = E_{11} = E_{15} = E_{max} \qquad (1.63a)$$

The energy of the rest of the signals is expressed in terms of E_{max}:

$$E_1 = E_5 = E_9 = E_{13} = E_{max}/3 \qquad (1.63b)$$

$$E_2 = E_4 = E_6 = \ldots = E_{16} = \sqrt{5}E_{max}/3 \qquad (1.63c)$$

Thus substituting (1.63) into (1.62) we obtain

$$E_{\mathrm{avr}}(\mathrm{APM}) = (2 + \sqrt{5})E_{\max}/6 \qquad (1.64)$$

Equating (1.61) with (1.64) we have

$$\sqrt{E_{\max}} = \sqrt{6/(2 + \sqrt{5})}\sqrt{E} \approx 1.19\sqrt{E} \qquad (1.65)$$

As can be seen, at limited average power the maximum length of an APM signal vector is about 20% more than the length of a PM signal vector. At the same time the minimum distance between vectors is increased from $0.47\sqrt{E}$ at the limited maximum power (see Fig. 1.17) up to

$$d_{16}(\mathrm{APM}) \approx 0.56\sqrt{E} \qquad (1.66)$$

at the limited average power (see Fig. 1.18).

Calculations similar to those indicated show that in the case of APM signals located on two concentric circles (see Fig. 1.16) and at limited average power, the maximum signal length and the minimum signal distance are equal to

$$\sqrt{E_{\max}} \approx 1.2\sqrt{E}$$
$$d_{16}(\mathrm{APM}) \approx 0.58\sqrt{E} \qquad (1.67)$$

Thus, the considered 16-ary APM signals (with one variant shown in Fig. 1.16 and the other in Figs. 1.17 and 1.18) are nearly equivalent in terms of the potential noise immunity when both maximum and average power is limited. These variants are not strictly optimum; however, they are close to optimum in terms of noise immunity [15]. These variants are convenient for realization and therefore are frequently used in practice.

The considered multiposition phase-modulated signals and multiposition signals with amplitude-phase (phase-difference) modulation are the most frequently used in digital communication systems. The two-, four-, and eight-phase signals with PM and PDM as well as signal-code structures based on these signals are almost exclusively applicable to radio and satellite channels. The 16-ary APM signals and m-ary APM signals at $m > 16$ are used in wire and radio-relay channels.

1.6 SIGNAL-CODE STRUCTURES BASED ON PHASE-MODULATED SIGNALS

Use of the multiphase signals permits an increase in the transmission rate almost without an increase in the occupied frequency band. As a result the specific message bit rate, that is, bit rate divided by the occupied frequency bandwidth (the major characteristic of digital transmission systems), can be considerably improved.

However, the opportunities of such transmission rate increases are limited because if an information bit is added at a chip interval, the number of phase or amplitude levels are twice increased, with the result being a significant degradation of the noise immunity.

One of the ways to increase performance is to apply error correcting codes that allow us to correct certain parts of errors. However, the error correcting codes require redundancy, which leads to the increase of modem operation rate. Besides, an increase of the modem operation rate results in the increase of the error probability at the decoder input, which is not always compensated for to a sufficient degree by code correctability. Certainly, it is also possible to increase code efficiency even at small redundancy at the expense of code length; however, in that case the complexity of a decoder grows sharply.

Thus, the bit rate increase is limited by noise immunity degradation, and the noise immunity, in its turn, is a function of a limited opportunity to achieve a high enough transmission rate. The solution to the problem is not an easy one. Actually the following problems are comparatively easy to solve: (1) the need to provide a high bit rate in channels with insignificant interferences and (2) the need to provide high noise immunity at a low bit rate or insignificant restrictions on the occupied frequency band [16].

At the same time the up-to-date digital communication systems are often designed under conditions of rather limited frequency and power resources. The problem becomes especially complicated when at these restrictions we still have the demand for a high transmission rate and low error probability simultaneously. This problem is connected to the classical problem of achieving an information transmission rate that is close to communication channel capacity. One of the prospective ways to solve the problem is to use phase signal-code structures in combination with the reception as a whole.

A set of sequences of modulated signal chips at the continuous channel input incorporated in a block by this or that code with redundancy is called a *signal-code structure*. Each sequence (block) of a signal-code structure consisting of a certain number of chips conforms to some segment (block) of transmitted messages. Thus, with the help of signal-code structures the discrete information sequences are converted into a continuous communication channel signal. Different elementary signals, in particular the parts of harmonic oscillations with amplitude, frequency, phase (phase-difference), or mixed kinds of modulation can be used as elements of signal-code structures. The phase signal-code structures differ in that their element is a chip of a harmonic signal with phase or phase-difference modulation.

How does a signal-code structure differ from a correcting code with redundancy? The correcting code is intended for a discrete communication channel (or, as they say, for discrete representation of a continuous channel); with the help of this code the information symbols are converted into discrete symbols coming to a modem input the parameters of which, as a rule, are not connected with a structure of a code.

A signal-code structure can be said to be a code for a continuous channel. It integrates encoding operations and modulation in one procedure where sometimes it is difficult to separate the encoding from modulation. The synthesis of signal-code structures does not reduce to the separate and independent choices of a code and a modulation method but is executed as a whole taking into account the interactivity of these procedures. It is also important to emphasize that the reception of signal-code structures is executed exclusively as a reception of a whole signal-code block. At the reception of signals encoded with redundancy the reception procedure, as a rule, consists of demodulation and decoding. At the reception of signal-code blocks, these operations are not separated and the decision is accepted not element by element but for the whole signal-code block.

Note that the effect of a simultaneous increase in the bit rate and decrease in the error bit rate, which signal-code structures potentially provide, is achieved at the expense of two factors: the integration of encoding and modulation in the common, optimum process and the realization of the optimum reception of signal-code blocks as a whole. The main principle of the system approach is realized in this case—optimization as a whole is preferable compared to optimization by parts.

The theory of forming and reception of signal-code structures is rather complicated and is a developing field in modern communication theory. Simple and concrete examples follow that explain the main features of how signal-code structures are formed and the main point of those positive effects which can be achieved with their help on the basis already considered earlier for phase-modulated signals.

As the first example we make up a signal-code structure that enables us to transmit binary information with the noise immunity superior to the noise immunity of the antipodal signals [17]. For this purpose we consider three-digit binary code combinations. There are $2^3 = 8$ such combinations; they are written in columns 1 through 3 of Table 1.3 and represent combinations of a without-redundancy binary code (3,3). The specified three-digit combinations can usually be transmitted by three chips of a two-phase signal compared to each binary symbol a phase or phase difference 0 or π, as shown in columns 4 through 6 of Table 1.3. Note that at this transmission method, the minimum code distance between the combinations of a without-redundancy binary code (3,3)—Hamming distance—is equal to 1, and the minimum Euclidean distance between the signals corresponding to these combinations in the interval of three chips is equal to

$$d_{\min} = 2\sqrt{E} \tag{1.68}$$

where E is the energy of duration T signal chip.

The same three-digit binary combinations can be encoded by two-digit ternary code with symbols 0, 1, 2. Because two-digit ternary combinations contain $3^2 = 9$ combinations, one combination remains superfluous and can be excluded. Encod-

Table 1.3
Six-Dimensional Signal-Code Structure on the Base of 3-Level PM

Binary Code (3,3)			Two-Phase Signal			Ternary Code (3,2)			Signal-Code Structure		
1	*2*	*3*	*4*	*5*	*6*	*7*	*8*	*9*	*10*	*11*	*12*
0	0	0	0	0	0	0	0	0	0	0	0
0	0	1	0	0	π	0	1	2	0	$2\pi/3$	$4\pi/3$
0	1	0	0	π	0	0	2	1	0	$4\pi/3$	$2\pi/3$
0	1	1	0	π	π	1	0	2	$2\pi/3$	0	$4\pi/3$
1	0	0	π	0	0	1	1	1	$2\pi/3$	$2\pi/3$	$2\pi/3$
1	0	1	π	0	π	1	2	0	$2\pi/3$	$4\pi/3$	0
1	1	0	π	π	0	2	0	1	$4\pi/3$	0	$2\pi/3$
1	1	1	π	π	π	2	1	0	$4\pi/3$	$2\pi/3$	0
—	—	—	—	—	—	2	2	2	$4\pi/3$	$4\pi/3$	$4\pi/3$

ing of eight initial three-digit binary combinations by two-digit ternary numbers is indicated in columns 7 and 8 of Table 1.3. Now to make up a required signal-code structure we need to add to each without-redundancy two-digit combination of a ternary code one more verifying ternary symbol. This symbol should be chosen according to the following rule: The sum (modulo 3) of all three symbols must equal zero. This three-digit ternary code (3,2) with one redundant symbol is indicated in columns 7 through 9 of Table 1.3.

Let's compare symbols of this code and chips of a PM signal with three permitted phases of 0, $2\pi/3$, and $4\pi/3$, that is, the three-phase signal [see Figure 1.13(b)]. As a result, we have a signal-code structure with the phases specified in columns 10 through 12 of Table 1.3. This structure is six dimensional because it contains three two-dimensional signals. The bit rate of this structure is more than the bit rate of the usual binary PM signal because in the interval of the same three chips it transfers $\log_2 9$ of information bits instead of 3 bits at the binary PM. If one combination, for example, phases $4\pi/3$, $4\pi/3$, $4\pi/3$, is not used, the transmission rates of these two systems are identical.

Now we calculate the minimum signal distance for the synthesized multidimensional signal and compare it with the distance of (1.68) for the binary PM signal. The code combinations of the considered ternary code (3,2) differ by not less than in two symbols, that is, the minimum Hamming distance in the given code is equal to 2. Hence, the phases at least two of three chips of every signal-code block are various. According to (1.50) the minimum Euclidean distance for the three-phase signal is equal to $\sqrt{3E}$; therefore, the desired minimum distance is

$$d''_{min} = \sqrt{6E} \approx 2.45\sqrt{E} \tag{1.69}$$

which is 1.76 dB more than the distance (1.68) for the antipodal signals.

We should emphasize that the indicated energy gain will be realized only in the case of the ideal coherent reception of three chips of the signal. To realize such reception as a whole, it is necessary to generate in a receiver eight reference oscillations coinciding to chip phases with variants of the signal-code block (columns 10 through 12 of Table 1.3). Then using phase detectors/correlators we should calculate convolutions (integrals of products) of the received signal and the specified reference oscillations. The decision about a transmitted three-digit combination is received by a maximum value of the calculated convolutions.

Forming and processing of the considered signal-code structure is illustrated by time diagrams (see Fig. 1.19). Figure 1.19(a) shows a sequence of transmitting binary symbols divided into three symbols blocks. Figure 1.19(b) shows a normal two-phase signal in vector representation. This signal represents the transmitting information by phase shift π according to Table 1.3. Figure 1.19(c) shows three-chips signals also in vector representation, corresponding to columns 10 through 12 of Table 1.3. Figure 1.19(d,e) depicts the output signals of a phase detector with a harmonic reference oscillation for the incoming signals shown in Figure 1.19(b,c), respectively. The shaded parts characterize the Euclidean distance between the signal, corresponding to a given code combination, and the signal,

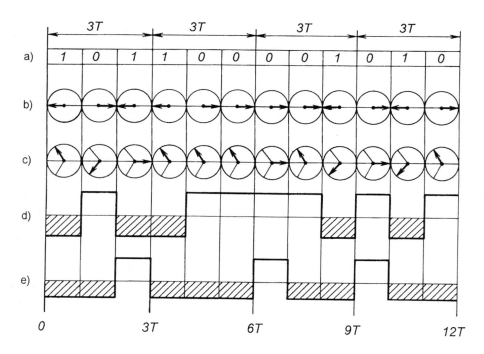

Figure 1.19 Time diagram of the formation and processing of the signal-code structure based on a three-phase signal.

corresponding to the zero combination. For half of all combinations, this distance appears to be large at the binary PM [e.g., for the combination in an interval $(0,3T)$ in Fig. 1.19], however, the minimum distance, which basically determines noise immunity, is more at the considered signal-code structure.

Note that in this case it is rather simple to realize the reception of signals with absolute PM at the expense of the unused code combination 2, 2, 2 (see the last line of Table 1.3). Let us assume that a synchrosignal was transmitted at the beginning of a communication session and this signal has provided one-to-one conformity between reference oscillation phases and transmitted three-digit combinations. After that any combination of permitted code words will not form at its boundary an unused combination 2, 2, 2 (see the ternary code entries in Table 1.3). Hence, the absence of combinations 2, 2, 2 at the receiver output indicates the correctness of phase synchronization. In contrast, the occurrence of the combination 2, 2, 2 is an attribute of incorrect conformity between reference oscillation phases and transmitting symbols. Using this attribute it is possible to repeat phase synchronization and restore the correct work of the receiver.

The considered elementary signal-code structure permits us to reduce the error probability having saved the same transmission rate as in the case of the binary PM.

At a more complex signal-code structure it is possible to increase the bit rate and decrease the error bit rate simultaneously. Usually at their synthesis the problem is put in the following way: to achieve a transmission rate that is more than that at the N-valued PM with the noise immunity appropriate to the $(N-1)$-valued PM.

For example, the popular system of digital transmission with QPSK, that is, four-phase (two-valued) PM or PDM, sometimes cannot meet the high requirements on the bit rate and noise immunity. The question is this: Can we find a signal-code structure on the base of PM signals that would permit us to obtain the higher bit rate and at the same time the lower error bit rate than at the conventional QPSK? A positive answer to this question has been given in [18].

Every signal-code block of the corresponding structure consists of 24 chips and is a 48-dimensional signal. If the usual four-phase signal (QPSK) was applied in the considered block, it would contain 48 information bits. However, let's apply the eight-phase signal at which the nth chip phase is determined by three binary symbols q_{1n}, q_{2n}, and q_{3n} and is written as

$$\varphi_n = q_{1n}\pi + q_{2n}\pi/2 + q_{3n}\pi/4 \tag{1.70}$$

The symbols q_{1n}, q_{2n}, q_{3n} take the values 0 or 1 and determine if there is a phase shift by π, $\pi/2$, or $\pi/4$, respectively. Thus, the whole signal-code block consisting of 24 chips will be determined by a rectangular matrix of three lines and 24 columns:

$$Q = \begin{Vmatrix} q_{11} & q_{12} & \cdots & q_{1.24} \\ q_{21} & q_{22} & \cdots & q_{2.24} \\ q_{31} & q_{32} & \cdots & q_{3.24} \end{Vmatrix} \tag{1.71}$$

The columns of matrix (1.71) according to (1.70) determine the phase of the appropriate signal chip, and the lines determine the sequence of phase changing by π (first line), by $\pi/2$ (second line), and by $\pi/4$ (third line). The lines of matrix (1.71) are code combinations of three codes (24, K_l, d_l) of identical length 24 but with different numbers of information symbols K_l and code distances d_l ($l = 1, 2, 3$).

The indicated codes, corresponding to the lines of matrix (1.71), should provide protection against interference influence for the most vulnerable phase shifts. Because the third line defines phase changing about the least value $\pi/4$, it should be coded by the most effective code with the most redundancy; the second line determines phase changing about $\pi/2$ (by the code with less redundancy) and the first line (by the without-redundancy code at all) because it determines phase changing about the biggest angle π. In the example considered, the without-redundancy code with $K_1 = 24$ and $d_1 = 1$ corresponds to the first line, the check on parity code with parameters $K_2 = 23$ and $d_2 = 2$ to the second line, and the extended Golay code with $K_3 = 12$ and $d_3 = 8$ to the third line.

Let's find rate and power parameters for the described multidimensional signal-code structure. The length of a signal-code block is equal to 24 chips, and total number of bits that can be transmitted in such a block is

$$K = K_1 + K_2 + K_3 = 24 + 23 + 12 = 59$$

that is, $59/24 = 2.46$ bits in average are transmitted in every chip. Thus, in this system the transmission rate is 23% more in comparison with the conventional QPSK system.

To analyze the potential noise immunity of the synthesized signal system, we have to determine the minimum Euclidean distance between different signal-code blocks (SCB). Let's assume that two SCBs differ from each other only by combinations of the third line of a forming matrix (1.71). Because the minimum Hamming distance of the Golay code is equal to eight, at least eight symbols q_{3n} of the compared SCBs will be different and, hence, in eight chips these SCBs will be different by about phase $\pi/4$. The chips differing by about phase $\pi/4$, as shown in Section 1.5, are apart from each other at the distance

$$d_{\pi/4} = \sqrt{(2 - \sqrt{2})E}$$

As long as the number of such chips is not less than eight, the distance between two SCBs with identical symbols q_{1n} and q_{2n} is not less than

$$d_{3\min} = \sqrt{8}\sqrt{(2 - \sqrt{2})E} \approx 2.16\sqrt{E} \tag{1.72}$$

which is 0.69 dB more than at the binary PM with antipodal signals.

Now let the compared SCBs have identical third lines but different second lines for the forming matrix (1.71). Then SCBs differ not less than in two chips ($d_2 = 2$), the phase shift between which is equal to $\pi/2$. As long as the chips with phase shift $\pi/2$ are apart from each other by $d_{\pi/2} = \sqrt{2E}$, the distance between SCBs with identical q_{1n} and q_{3n} will be not less than

$$d_{2\min} = \sqrt{2}\sqrt{2E} = 2\sqrt{E} \tag{1.73}$$

which is equal to the distance between signals of the binary PM with antipodal signals.

Last, if we are forming matrix (1.71) with identical second and third lines that correspond to two different SCBs, the first lines will certainly be different. In this case the minimum difference consists of a phase shift π in one of 24 chips that corresponds to the distance

$$d_{1\min} = 2\sqrt{E} \tag{1.74}$$

Integrating the results of (1.72), (1.73), and (1.74) we get that the minimum distance between any blocks (24 chips length) of the considered signal-code structure is not less than $2\sqrt{E}$, that is, not less than in the system with binary PM (BPSK) using the antipodal signals.

Thus, the considered signal-code structure provides 23% more transmission rate than the four-phase signals (QPSK) and the same noise immunity as two-phase signals (BPSK).

The principle of generating the appropriate signals is completely determined by relations (1.70) and (1.71). The transmitting unit should contain three eight-position phase modulators and two encoders—an elementary encoder for adding a symbol of check on parity and a Golay encoder. Binary information from the system input is divided into three blocks containing 24, 23, and 12 binary symbols, respectively. The first block comes directly to the input of the eight-phase modulator which performs 180-degree phase shifts of chips. The second 23-chip block comes to the encoder where the 24th parity check symbol is added to it, then it comes to the input of the modulator changing phases of chips by 90 degrees. The third block comes to the Golay encoder where 12 verifying symbols are added to it and then to the third input of the modulator shifting chips phases by 45 degrees.

To realize the potential noise immunity of the considered signal-code structure it is necessary to fulfill a coherent reception as a whole of every signal-code block containing 24 chips. The attempts of the frontal decision of this problem, that is, by convolution of the received signal with all possible variants, are hopeless because a number of variants of an eight-position signal in 24 chips is equal to 8^{24}. Therefore an economical procedure close to an optimum reception as a whole consisting of several stages has been developed in [18].

A suboptimum reception as a whole of the Golay code transmitted by the third subchannel of a three-valued system with PM is executed on the first stage. This is the most complex part of the signal processing procedure and results in the determination of a code combination $(q_{31}, q_{32}, \ldots, q_{324})$. On the second stage, the received signal block of 24 chips, digitally recorded in operative memory, is transformed by means of the received, on the first stage, combination $(q_{31}, q_{32}, \ldots, q_{324})$ to eliminate all $\pi/4$ phase shifts—this is the operation of $\pi/4$ keying removal. As a result the signal is transformed from an eight-phase signal into a four-phase one. Reception as a whole of the code combinations with the parity check is performed on the third stage.

The reception as a whole is executed according to the Vagner algorithm [1]: The binary symbol is corrected in that chip in which the calculated reliability is the least. As a result we have the code combination $(q_{21}, q_{22}, \ldots, q_{224})$. The $\pi/2$ keying removal is executed by the received combination on the fourth stage, and signal is transformed from a four-phase signal into a two-phase one. Element-by-element reception of the without-redundancy code combination $(q_{11}, q_{12}, \ldots, q_{124})$, transmitted by the first binary subchannel of the PM system, occurs on the last (fifth) stage. Then the received information symbols $q_{11}, q_{12}, \ldots, q_{124}, q_{21}, q_{22}, \ldots, q_{223}, q_{31}, q_{32}, \ldots, q_{312}$, that is, all 59 binary symbols, are given to the consumer.

This is the general principle used for creating the algorithm of a reception as a whole of the considered signal-code structure. Its detailed study requires the attraction of additional information concerning the decoding theory and reception as a whole as well as the theory on element-by-element processing of phase-modulated signals.

The examples given for forming signal-code structures, based on using phase-modulated signals, serve as an introduction to the new area of PM and PDM applications. This is one of the most promising fields in digital communication theory and engineering where qualitatively new achievements are possible. Among a considerable number of the proposals and developments in this field, the multi-dimensional hierarchical signal-code structures developed by Ginzburg are most noticeable due to their theoretical generality and depth [19–21].

References

[1] Fink, L. M., *Discrete Message Transmission Theory*, 2nd ed., Moscow: Sov. Radio, 1970.

[2] Fink, L. M., et al. *Signal Transmission Theory*, Moscow, Radio & Svyaz, 1986.

[3] Proakis, J. G., *Digital Communications*, New York: McGraw-Hill Book Company, 1983.

[4] Simon, M. K., S. M. Hinedi, and W. C. Lindsey, *Digital Communication Technique*, Englewood Cliffs, NJ: Prentice-Hall, 1995

[5] Viterbi, A. J., and J. K. Omura, *Principles of Digital Communication and Coding*, New York: McGraw-Hill Book Company, 1979.

[6] Wozencraft, J. M., and I. M. Jacobs, *Principles of Communication Engineering*, New York: John Wiley & Sons, 1965.

[7] Kolesnik, V. D., and G. Sh. Poltyrev, *Information Theory*, Moscow: Science Publishers, 1982.

[8] Shannon, C. E., "A Mathematical Theory of Communication," *Bell System Tech. J.*, Vol. 27, 1948.

[9] Sapozhkov, M. A., and V. G. Mihailov, *Vocoder Communications*, Moscow: Radio & Svyaz, 1983.

[10] Clark, G. C., and J. B. Cain, *Error-Correction Coding for Digital Communicatons*, New York: Plenum Press, 1987.

[11] McWilliams, F. J., and N. J. A. Sloane, *The Theory of Error-Correcting Codes*, Amsterdam: North-Holland Publishing Company, 1977.

[12] Peterson, W. W., and E. J. Weldon, *Error-Correcting Codes*, Cambrdige, MA: MIT Press, 1972.

[13] Korzhik, V. I., L. M. Fink, and K. N. Schelkunov, *Noise Immunity of Discrete Message Transmission Systems, Handbook*, Moscow: Radio & Svyaz, 1981.

[14] Momot, Ye. G., *The Problems and Methods of Synchronous Radio Reception*, Svyazizdat, 1961.

[15] Banket, V. L., and V. M. Dorofeev, *Digital Methods in Satellite Communications*, Moscow: Radio & Svyaz, 1988.

[16] Zuko, A. G., et al. *Noise Immunity and Efficiency of Information Transmission Systems*, Moscow: Radio & Svyaz, 1985.

[17] Petrovich, N. T., and O. N. Porohov, "Three-Level Keying in Communications Systems," *Radiotechnika*, No. 7, 1985, pp. 3–8.

[18] Antonov, G. V., et al., "Multidimensional Signals Modem for the Satellite Communication System," *Electrosvyaz*, No. 5, 1988.

[19] Ginzburg, V. V., "Procedures of Receiving as a Whole the Signals with Redundancy," *Radiotechnika*, No. 11, 1978, pp. 20–33.

[20] Ginzburg, V. V., "Multidimentional Signals for Continuous Channels," *Problems of Infor. Transmission*, Vol. 20, No. 1, 1984, pp. 28–46.

[21] Ginzburg, V. V., "Algorithms of Multidimentional Signals Processing," *Radiotechnika*, No. 5, 1988, pp. 42–52.

CHAPTER 2

General Algorithms for PDM Signal Formation and Processing

2.1 INVARIANCE PROPERTY OF SIGNAL DISCRETE-DIFFERENCE TRANSFORMATION

This chapter is dedicated to advancing our knowledge of the formation and processing of multiposition PDM signals and will provide more information about the difference modulation methods.

We have already said that PDM-1 permits us to remove the harmful influence of an initial phase uncertainty, and PDM-2 permits us to get rid of the influence of carrier frequency instability. These properties of PDM, called the *invariance properties* in terms of the corresponding interfering effects, are actually the consequence of fundamental properties of finite differences of functions, representing the real signals. Let's discuss some statements of the theory of finite differences of functions, from which these properties follow [1–3].

The calculation of finite differences and the theory of difference equations are the important areas of study in modern mathematics in that they are a junction between the classical results of differential calculation and differential equation theory, on the one hand, and the constructive applications of these results in a number of the engineering fields, on the other hand. The methods, based on the use of finite differences of functions, have found their application in robotics, computers, radio engineering, and communications for solving the problems of signal discrete approximation, construction of specialized computer algorithms, development of adaptive systems of data processing, and systems of data transmission through communications channels.

Finite differences of a function $f(t)$, expressed by the samples at the moments $t_1, t_2, \ldots, t_n$, are usually designated by the symbol $\Delta_n^k f$. The top index specifies the order of a finite difference, and the bottom one specifies the number of a difference that is appropriate to its time position. If the result is concerned with any time position of the kth order difference, only the top index $\Delta^k f$ is saved at the finite difference symbol; if the result is concerned with any time position and any order difference, this is represented by the symbol Δf without indexes.

The representation of finite differences by means of a recurrent equation is mostly usable:

$$\Delta_n^k f = \Delta_n^{k-1} f - \Delta_{n-1}^{k-1} f \qquad (2.1)$$

Expression (2.1) along with the formula for calculation of the first-order difference

$$\Delta_n^1 f = f(t_n) - f(t_{n-1}) \qquad (2.2)$$

completely determines the algorithm for calculating all values of any order of finite difference. It is also possible to use representation of finite differences directly through function samples:

$$\Delta_n^k f = \sum_{i=0}^{k} (-1)^{k-i} C_k^i f(t_{n-k+i}) \qquad (2.3)$$

where C_k^i are the binomial coefficients.

The $k + 1$ function samples participate in forming each of the values of the kth-order finite difference. In particular, for the differences of the second and third orders we have

$$\Delta_n^2 f = f(t_n) - 2f(t_{n-1}) + f(t_{n-2}) \qquad (2.4)$$

$$\Delta_n^3 f = f(t_n) - 3f(t_{n-1}) + 3f(t_{n-2}) - f(t_{n-3}) \qquad (2.5)$$

In the constructive theory of functions the method of finite differences is used in the tasks of interpolation and approximation. In particular, for the function $f(t)$, given by its samples in $N + 1$ points at intervals δt, the interpolating polynomial is

$$\tilde{f}(t) = f(t_0) + \sum_{k=1}^{N} \frac{\Delta^k f}{k!(\delta t)^k}(t - t_0)(t - t_0 - \delta t) \dots [t - t_0 - (k-1)\delta t] \qquad (2.6)$$

and the error of representation by polynomial (2.6) is equal to

$$\tilde{f}(t) - f(t) = \left(\frac{\Delta^k f}{\Delta t^k}\right)\frac{(t - t_0)(t - t_0 - \delta t) \dots (t - t_0 - N\delta t)}{(N+1)!} \qquad (2.7)$$

where $(\Delta^k f / \Delta t^k)$ is the conditional designation of the so-called kth-order-divided finite difference being the analog of the kth derivative.

The divided difference is not explicitly expressed through the appropriate derivative, but it is known that within the interpolation interval (t_0, t_N) there will be a point τ in which the following equation is valid:

$$\left(\frac{\Delta^k f}{\Delta t^k} \right) = \frac{1}{k!} \frac{d^k f(t)}{dt^k} \bigg|_{t=\tau} \tag{2.8}$$

At the same time the expression of the divided difference, represented through the finite difference of the same order, is known as

$$\left(\frac{\Delta^k f}{\Delta t^k} \right) = \frac{\Delta^k f}{k! \, (\delta t)^k} \tag{2.9}$$

from which the next important relation follows:

$$\Delta^k f = (\delta t)^k \frac{d^k f(t)}{dt^k} \bigg|_{t=\tau} \tag{2.10}$$

which means that if the kth derivative of a signal is identically equal to zero within some time interval, the kth-order finite differences, calculated within this interval (at any quantization step), are also equal to zero.

Let's call the discrete-difference transformation of a signal $S(t)$ or its parameter $\gamma(t)$ the operation at which the sequence of signal samples $S(t)$ or signal's parameter $\gamma(t)$ at equally separated moments t_i is one-to-one transformed into the sequence of the high-order finite differences $\Delta_i^k S$ or $\Delta_i^k \gamma$.

The submitted relations permit us to formulate two fundamental properties of the discrete-difference transformation:

1. The property of linearity following from (2.3):

$$\Delta^k (S_1 + S_2) = \Delta^k S_1 + \Delta^k S_2 \tag{2.11}$$

2. The property of invariance following from (2.10):

$$\Delta^k S \equiv 0 \quad \text{if } d^k S / dt^k \equiv 0 \tag{2.12}$$

In its turn, the property of invariance to certain interfering effects follows from (2.11) and (2.12). If there is an additive mixture of a useful signal and an interference

$$x(t) = S(t) + \xi(t)$$

and the latter is such that

$$\frac{d^k \xi(t)}{dt^k} \equiv 0 \qquad (2.13)$$

then we have according to (2.11) and (2.12)

$$\Delta^k x = \Delta^k(S + \xi) = \Delta^k S + \Delta^k \xi = \text{invar } \xi \qquad (2.14)$$

The last notation means that the kth difference of a signal S is the invariant of transformation, consisting of adding the interference ξ, which satisfies condition (2.13), to the signal.

Thus, with the help of discrete-difference transformation of the signal and interference mixture it is possible to be released from the influence of certain interfering effects or to reduce this influence. As long as both the interference and the signal are subjected to the transformation, it is natural to assume that the kth difference of the useful signal is not equal to zero:

$$d^k S(t)/dt^k \neq 0$$
$$\Delta^k S \neq 0 \qquad (2.15)$$

Inequality (2.15) alongside equality (2.13) determine the conditions for achieving the invariance to interfering effect $\xi(t)$ when transmitting the useful signal $S(t)$ and using the discrete-difference transformation of the kth order.

From here the idea of using the finite differences of signal parameters as modulated ones follows. A new kind of signal modulation that appeared can be called the *finite difference modulation* or the *difference modulation of the high order*; in the latter case we are emphasizing the opportunity and expediency of using the various order differences depending on properties of interfering effects with which we have to struggle.

The property of the invariance of systems with the high-order difference modulation concerns both the additive and nonadditive interferences. In the former case it is illustrated by relation (2.14). In the last case the system invariance to nonadditive interfering effects can be achieved if this effect leads to signal parameter deviations satisfying (2.13).

If $\gamma(t)$ is the transmitted signal parameter and

$$\gamma_\xi(t) = \gamma(t) + \xi(t)$$

is the received (distorted) signal parameter, with the interfering effect $\xi(t)$ satisfying condition (2.13), and in addition we can satisfy the condition $\Delta^k \gamma \neq 0$, then

$$\Delta^k \gamma_\xi = \Delta^k(\gamma + \xi) = \Delta^k \gamma + \Delta^k \xi = \Delta^k \gamma = \text{invar } \xi \qquad (2.16)$$

Any signal parameter can be used as the parameter γ; for example, in the case of a simple harmonic signal, amplitude-difference, frequency-difference, or phase-difference modulation of the high order are possible.

In a general case, when using kth-order difference modulation, the information is put into the kth-order differences of a signal parameter that is subjected to distortions. With that end in view, the transmitter calculates this signal parameter using its kth-order difference, determined by information symbols of a message source. The reverse operation is executed in the receiver—the kth difference of the parameter is calculated by the values of the received signal parameter.

The potential opportunities of the systems with difference modulation, in the sense of achieving the properties of absolute or relative invariance, can be determined by revealing the class of interfering effects relative to which the system has specified properties. Revealing such classes of effects presents an independent problem in each particular communication system. However, it is not difficult to make some general conclusions about the classes of interfering effects with respect to which it is possible to achieve the invariance in the systems with difference modulation of this or that order. The following statement is true.

The system with the kth-order difference modulation possesses the potential opportunity to achieve absolute invariance to all interfering effects, causing the information parameter variations by the value $\xi_1(t)$ presented as the power polynomial of the $(k-1)$th degree with random coefficient

$$\xi_1(t) = \sum_{i=0}^{k-1} \alpha_i t^i \qquad (2.17)$$

As long as the kth derivative of function (2.17) is identically equal to zero, the information parameter has the property of absolute invariance (2.12).

In robotics and communications the term *absolute invariance* designates the property of complete (strict) tolerance to the interfering effect [4], which is stipulated in the case considered by strict equality to a zero of the kth finite differences of the interfering effect. At the same time, this statement declares only the potential opportunity to achieve the absolute invariance. To actually achieve system absolute invariance, the kth derivative of the useful signal (or its information parameter) should not be equal to zero; otherwise the kth differences of the mixture of the useful signal and interference will be also equal to zero. In the case considered, the useful signal should be presented, for example, as a power polynomial of the degree not less than k:

$$S_1(t) = \sum_{i=0}^{k} \beta_i t^i \qquad (2.18)$$

Then for all kth differences of the sum

$$x(t) = S_1(t) + \xi_1(t)$$

we have

$$\Delta^k x = \Delta^k S_1 + \Delta^k \xi_1 = \Delta^k S \neq 0 \qquad (2.19)$$

from which it follows that the signal kth differences can serve as an information parameter that is absolutely invariant to the interfering effect.

At harmonic interfering effects, the achievement of absolute invariance, strictly speaking, is impossible, because not a single one from the derivatives of these effects is identically equal to zero. At the same time, using (2.10) it is not difficult to establish, that for the harmonic signal

$$\xi_2(t) = a \sin \omega t$$

the kth differences of its sample difference, taken at intervals T, do not exceed the value

$$\max|\Delta^k \xi_2(t)| \leq T^k a\omega^k \qquad (2.20)$$

Thus, in the system with difference modulation the relative invariance can be provided to the harmonic interfering effect, because according to (2.20) the kth-order difference can be made reasonably small by reducing the quantization interval T.

Under the relative invariance we understand the small sensitivity of a system to interfering effects, when the information signal or parameter is changed under the influence of the interfering effect by no more than some value ϵ given beforehand.

Table 2.1 indicates, as an example, maximum values of the finite differences of the first six orders for the harmonic interfering effect

$$\xi(t) = \sin 2\pi t / \tau$$

at various relations between the period of this oscillation τ and the interval of the discrete-difference transformation T. As shown in the table, we can decrease the interfering effect to as small a value as desired by decreasing the interval of the discrete-difference transformation, that is, the elementary signal duration T, and/or by increasing the order of finite differences. If we thus modulate the appropriate finite differences of a useful signal by information symbols, it is possible to ensure the required degree of relative invariance.

Table 2.1
Finite Differences of a Sinusoidal Function

τ/T	$\Delta^1\xi$	$\Delta^2\xi$	$\Delta^3\xi$	$\Delta^4\xi$	$\Delta^5\xi$	$\Delta^6\xi$
72	0.087	0.008	0.0008	0.0004	0.0002	0.0001
36	0.174	0.031	0.0052	0.001	0.0004	0.0001
18	0.342	0.119	0.041	0.014	0.004	0.002
9	0.684	0.460	0.320	0.215	0.114	0.088

This conclusion can be extended to any interfering effects with a limited spectrum. As long as the maximum value of the kth derivative of the function $\xi_3(t)$ with a spectrum, limited overhead by frequency ω, satisfies the inequality [3]

$$\max\left|\frac{d^k\xi_3(t)}{dt^k}\right| \le \omega^k \max|\xi_3(t)| \qquad (2.21)$$

then according to (2.10) we have

$$|\Delta^k\xi_3(t)| \le T^k\omega^k \max|\xi_3(t)| \qquad (2.22)$$

Because the real interference and signals are always power limited, from (2.22) it follows that the kth finite difference of the interfering effect can be made as small as desired by the appropriate choice of the parameters T and k. It proves the principal possibility that we can achieve at least the relative invariance to the interfering effect with a limited power and spectrum. To realize this opportunity, we need the kth signal differences to be more than the kth interference differences. In the case considered, this means that the signal spectrum should be wider than the interference spectrum.

Further discussion of the condition of how to realize the invariance property of the discrete-difference transformation requires a more detailed definition of the problem. Such a concrete definition will be presented later for high-order PDM, when the interfering effects cause the carrier phase and frequency deviations. But before discussing high-order PDM, we consider general algorithms of signal formation and signal processing for the difference modulation of an arbitrary parameter.

2.2 GENERAL ALGORITHMS FOR SIGNAL FORMATION AND PROCESSING AT A HIGH-ORDER DIFFERENCE MODULATION

Let's formulate the considered algorithms for an arbitrary signal or an arbitrary signal parameter, the samples of which in discrete moments $t_i = iT$ we shall designate as $\gamma_i = \gamma(iT)$. Here T is the interval of time quantization or the duration of an elementary signal. If the kth-order differences $\Delta^k \gamma$ of the parameter γ belong to a discrete set, a digital system takes place, but if the differences belong to a continuous set, then an analog system with discrete time takes place.

The algorithms for signal formation and processing in such systems are based on the general formulas for the calculation of final differences [see (2.1), (2.2), and (2.3)]. The first (recurrent) algorithm of signal forming at the kth-order difference modulation is based on the following recurrent formula:

$$\Delta_n^{k-1} \gamma = \Delta_{n-1}^{k-1} \gamma + \Delta_n^k \gamma \qquad (2.23)$$

A sequence of the kth differences of the parameter γ conforms (in the case of discrete symbols on the basis of keying codes) with the transmitted sequence of information symbols (discrete or continuous):

$$\Delta_1^k \gamma, \Delta_2^k \gamma, \ldots, \Delta_n^k \gamma \qquad (2.24)$$

Then using (2.24) according to formula (2.23), the finite differences of $(k-1)$th, $(k-2)$th order, etc. (up to the differences of the zero order, the values of the parameter γ) are sequentially calculated:

$$\gamma_1 = \gamma_0 + \Delta_1^1 \gamma$$
$$\gamma_2 = \gamma_1 + \Delta_2^1 \gamma$$
$$\vdots$$
$$\gamma_n = \gamma_{n-1} + \Delta_n^1 \gamma \qquad (2.25)$$

Sequence (2.25) defines the transmitted signal chips

$$S(t, \gamma_1), S(t, \gamma_2), \ldots, S(t, \gamma_n) \qquad (2.26)$$

When processing a signal with the kth-order difference modulation, the elementary signals (2.26) are converted into the parameter γ in the operation of measuring (estimation) the parameter γ:

$$\tilde{\gamma}_1, \tilde{\gamma}_2, \ldots, \tilde{\gamma}_n, \ldots \qquad (2.27)$$

At digital transmission the estimates (2.27) and the transmitted parameters (2.25) belong to the same discrete set and are determined by a demodulator. Then on the basis of the received sequence (2.27), using the recurrent formula

$$\Delta_n^k \tilde{\gamma} = \Delta_n^{k-1} \tilde{\gamma} - \Delta_{n-1}^{k-1} \tilde{\gamma} \qquad (2.28)$$

we can sequentially calculate the first, second, and other differences of the parameter γ, up to the kth-order differences:

$$\Delta_2^1 \tilde{\gamma} = \gamma_2 - \gamma_1; \qquad \Delta_3^1 \tilde{\gamma} = \gamma_3 - \gamma_2; \qquad \ldots;$$
$$\Delta_3^2 \tilde{\gamma} = \Delta_3^1 \tilde{\gamma} - \Delta_2^1 \gamma; \qquad \Delta_4^2 \tilde{\gamma} = \Delta_4^1 \tilde{\gamma} - \Delta_3^1 \tilde{\gamma}; \qquad \ldots;$$
$$\vdots \qquad\qquad\qquad\qquad\qquad\qquad\qquad\qquad\qquad (2.29)$$
$$\Delta_{k+1}^k \tilde{\gamma} = \Delta_{k+1}^{k-1} \tilde{\gamma} - \Delta_k^{k-1} \tilde{\gamma}; \qquad \Delta_{k+2}^k \tilde{\gamma} = \Delta_{k+2}^{k-1} \tilde{\gamma} - \Delta_{k+1}^{k-1} \tilde{\gamma}; \qquad \ldots$$

As we can see from (2.29), the first actually received information symbol is the $(k + 1)$th transmitted one. It is explained by the fact that to calculate the first value of the kth difference of the parameter it is necessary to have $k + 1$ of its samples. Thus, the first k information symbols are considered to be the "references" for calculating the following ones and are lost during the transmission.

The structural scheme of the transmitting part of a system with the kth-order difference modulation [Fig. 2.1(a)] corresponds to recurrent algorithm (2.23). Using the encoder the transmitted information J will be converted into the values

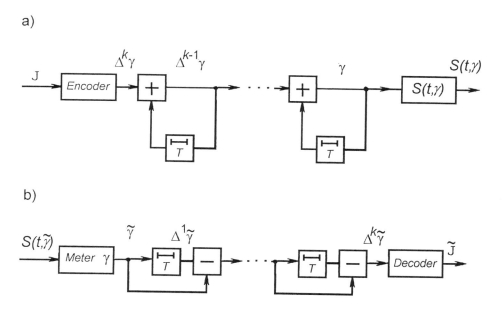

Figure 2.1 (a) Formation and (b) processing of signals with kth-order difference modulation.

of the kth difference $\Delta^k \gamma$ of the parameter γ. Then k converters connected in series, including the adder and memory element for the time period T, calculate the values of the parameter γ; and a signal generator $S(t, \gamma)$ generates the transmitted signal chips corresponding to these values. Note that when using a digital system the encoder is the converter of binary code combinations into discrete values of the kth difference of the modulated parameter, and the signal generator is the modulator of the appropriate carrier parameter.

The block diagram of the receiving part of the system [Fig. 2.1(b)], corresponding to the recurrent algorithm (2.28), contains the γ parameter meter and k identical converters (a memory element for the time period T and a subtractor), connected in series, calculating the finite difference from the first up to the kth order. The scheme ends with the decoder, which transforms the calculated difference $\Delta^k \gamma$ to the information symbol J. In the digital system, the decoder converts the received difference into a binary code combination, and the meter is a demodulator of the parameter γ.

The second algorithm for formation and processing the signals with kth-order difference modulation is based on (2.3), which in this case is the following

$$\Delta_n^k \gamma = \sum_{i=0}^{k} (-1)^{k-i} C_k^i \gamma_{n-k+i} \qquad (2.30)$$

The given algorithm is a peculiar one because the kth difference is calculated here directly by $k + 1$ samples of the parameter γ, omitting the calculation of the lower order differences.

Similar to (2.30) the algorithm for calculating the current value of parameter γ by its previous values and current value of the kth difference can be found, having presented (2.30) as

$$\Delta_n^k \gamma = \sum_{i=0}^{k-1} (-1)^{k-i} C_k^i \gamma_{n-k+i} + \gamma_n$$

whence

$$\gamma = \Delta_n^k \gamma - \sum_{i=0}^{k-i} (-1)^{k-i} C_k^i \gamma_{n-k+i} \qquad (2.31)$$

Formulas (2.31) and (2.30) determine algorithms of the transmitting and receiving parts of a system with the kth-order difference modulation. The schemes for signal formation and processing units according to these algorithms are shown in Figures 2.2 and 2.3.

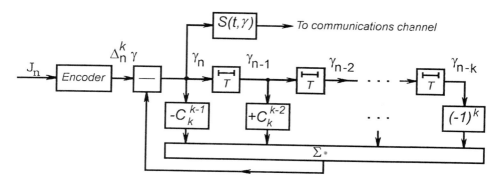

Figure 2.2 Formation of signals with the *k*th-order difference modulation.

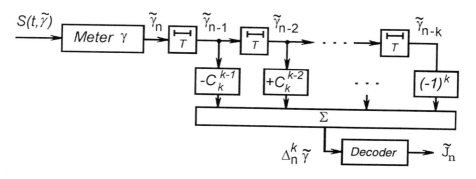

Figure 2.3 Processing of signals with the *k*th-order difference modulation.

These schemes, besides the encoder, decoder, γ parameter meter, and other elements, which are performing the same functions as those of the schemes shown in Figure 2.1, contain multipliers by constant numbers of a kind

$$(-1)^{k-i} C_k^i$$

The submitted schemes illustrate the general concept of how to construct the systems with finite difference modulation. The particular features of such systems are defined by many factors. The structure of a signal generator, encoder, and decoder depends on a kind of transmitted information, used keying code and a carrier. The structure of a parameter meter depends on communication channel statistics and a priori information about the signal. In many cases the optimum algorithms result in measurement of some functions of parameters. For example, in the systems with phase-difference modulation, the optimum meters determine trigonometric functions of a signal phase.

2.3 HIGH-ORDER PHASE-DIFFERENCE MODULATION

Phase-difference modulation is practically the most important type of difference modulation. Having the same advantages as classical phase modulation, namely, high specific bit rate and low bit error rate, PDM allows us to achieve the transmission system invariance to the changes of a signal phase and signal frequency. Frequency-difference modulation (FDM) has the last property, however with other conditions being equal, PDM provides a higher specific bit rate of information transmission and higher noise immunity.

At the kth-order PDM the information is put into the sequence of the kth differences of carrier phases, which can accept a finite number of values:

$$\Delta^k \varphi_1; \quad \Delta^k \varphi_2; \quad \ldots; \quad \Delta^k \varphi_m$$

The appropriate system is referred to as an m-ary system with PDM of the kth order or, at $m = 2^N$, the system with N-valued PDM of the kth order. In the N-valued systems the variants of the transmitted kth differences of phases take the following values:

$$\Delta^k \varphi_i = \Delta^k \varphi_0 + (i - 1) \pi / 2^{N-1} \quad \text{for } i = 1, 2, \ldots, 2^N$$

where $\Delta^k \varphi_0$ is the initial value of the kth phase difference, which is chosen out of convenience to realize the operations of modulation and demodulation.

The choice of the PDM or FDM order in this or that system is defined by characteristics of the signal frequency and phase instability. We consider this question in more detail.

Let the signal be the following oscillation:

$$S(t) = a \sin \psi(t) \tag{2.32}$$

From the argument of the given trigonometric function we allocate the linear component $\omega_0 t$, which does not result in the changes of an initial phase in the moments, divisible by the duration of the elementary signal T, that is, obviously, true at $\omega_0 = 2\pi l / T$, where l is the integer. Then the difference $\psi(t) - \omega_0 t = \varphi(t)$ will characterize the changes of the signal initial phase in the specified moment, and signal (2.32) can be presented as

$$S(t) = a \sin[\omega_0 t + \varphi(t)] \tag{2.33}$$

At digital transmission by phase-modulated signals, the function $\varphi(t)$ represents both the transmitted information and disturbing effects. The last ones are caused by interference as well as by nonlinear and parametrical conversions of a signal in a communications channel. These conversions, as a rule, can be presented within a finite interval by the polynomial model with random coefficients φ_0, φ_1, $\varphi_2, \ldots$, describing the constant, linear, squared, and other changes of the signal initial phase, namely,

$$\varphi(t) = \sum_{i=0}^{k-1} \varphi_i t^i + \varphi_{in}(t) \tag{2.34}$$

where $\varphi_{in}(t)$ is the signal information phase. As far as the signal frequency ω is equal to the derivative of phase (2.34), we have

$$\omega(t) = \frac{d\varphi}{dt} = \sum_{i=1}^{k-1} i\varphi_i t^{i-1} + \omega_{in}(t) \tag{2.35}$$

where ω_{in} is the information deviation of the signal frequency. From (2.34) and (2.35) we finally have

$$\frac{d^k[\varphi(t) - \varphi_{in}(t)]}{dt^k} \equiv 0 \qquad \frac{d^{k-1}[\omega(t) - \omega_{in}(t)]}{dt^{k-1}} \equiv 0 \tag{2.36}$$

The given identities of (2.36) in combination with the previously proven property of invariance of the discrete-difference transformation demonstrate that the kth differences of the signal phase and $(k - 1)$th differences of the frequency do not depend on the polynomial-type disturbing effects described by function (2.34).

Thus, if the spurious variations of a signal phase within the interval $(k + 1)T$ can be presented as a polynomial of the $(k - 1)$th order, the absolute system invariance to these variations can be provided by PDM of the kth order or by FDM of the $(k - 1)$th order.

Let's consider the most typical special cases of the signal phase and frequency variations.

Uncertain, but constant initial phase:

$$\varphi(t) = \varphi_{in}(t) + \varphi_0$$
$$\omega(t) = \omega_{in}(t) \tag{2.37}$$

This situation can appear for these reasons:

- The incoherence of the reference oscillators in the communications line;
- The received signal and the reference signal in a demodulator are not strictly coherent (coherence with an accuracy up to discrete shift).

As for this case

$$\frac{d[\varphi(t) - \varphi_{in}(t)]}{dt} \equiv 0$$

$$\omega(t) - \omega_{in}(t) \equiv 0$$

we can conclude that

$$\Delta^1 \varphi = \Delta^1 \varphi_{in} = invar \; \varphi_0$$
$$\Delta^0 \omega = \omega_{in} = invar \; \varphi_0$$

(2.38)

which means that absolute invariance to constant deviations of a signal phase is provided by the first-order PDM and FDM of the zero (i.e., FM) order.

Uncertain and linearly varying initial phase (uncertain, but constant frequency):

$$\varphi(t) = \varphi_{in}(t) + \varphi_0 + \varphi_1 t \qquad \omega(t) = \omega_{in}(t) + \varphi_1$$

(2.39)

This situation is caused by these conditions:

- Instable and nonstandard reference oscillators in the communications line;
- The Doppler effect at communication with mobile or nonstationary object, when a transmitter and/or a receiver move with constant speed relative to each other.

As for this case

$$d^2[\varphi(t) - \varphi_{in}(t)]/dt^2 \equiv 0$$
$$d[\omega(t) - \omega_{in}(t)]/dt \equiv 0$$

we can conclude that

$$\Delta^2 \varphi = \Delta^2 \varphi_{in} = invar(\varphi_0, \; \varphi_1)$$
$$\Delta^1 \omega = \Delta^1 \omega_{in} = invar(\varphi_0, \; \varphi_1)$$

(2.40)

which means that the absolute invariance to linear deviations of a signal phase can be provided by the second-order PDM and the first-order FDM.

Uncertain and squarely varying initial phase (uncertain and linearly varying frequency):

$$\varphi(t) = \varphi_{in}(t) + \varphi_0 + \varphi_1 t + \varphi_2 t^2$$
$$\varphi(t) = \varphi_{in}(t) + \varphi_1 + 2\varphi_2 t \tag{2.41}$$

This situation is caused by these conditions:

- Instability of the reference oscillators in the communications line;
- The Doppler effect at the linearly accelerated movement of a transmitter and a receiver relative to each other.

As for this case

$$d^3[\varphi(t) - \varphi_{in}(t)]/dt^3 \equiv 0$$
$$d^2[\omega(t) - \omega_{in}(t)]/dt^2 \equiv 0$$

we can conclude that

$$\Delta^3 \varphi = \Delta^3 \varphi_{in} = \text{invar}(\varphi_0, \varphi_1, \varphi_2)$$
$$\Delta^2 \omega = \Delta^2 \omega_{in} = \text{invar}(\varphi_0, \varphi_1, \varphi_2) \tag{2.42}$$

which means that the absolute invariance to square deviations of the signal phase (linear deviations of the frequency) can be provided by the third-order PDM and the second-order FDM.

Some additional remarks should be made on conditions and means of achieving the invariance in considered situations.

Concerning the conditions we are going to discuss, note that the approximate model of phase and frequency variations, set by relations (2.34) and (2.35), should be correct for the interval equal to the duration of $k + 1$ chips. If, for example, PDM-1 is used, to achieve the absolute invariance to the initial phase, the last one should be kept constant within the interval of two chips. If PDM-2 is used, then to achieve absolute invariance to the carrier frequency, this frequency should be constant within the interval of three chips. If the specified conditions are not executed, and the phase variations are not expressed precisely by the polynomial of the $(k - 1)$th degree within $k + 1$ chips, the application of PDM-k can provide only the relative invariance to the appropriate spurious changes of the signal phase and frequency.

The means to achieve invariance to signal frequency and phase variations are through adequate algorithms for formation (modulation) and processing (demodulation) of the appropriate signals.

Implementation of the PDM-k modulators does not contain obstacles. From common algorithms (2.24), (2.25), and (2.31) it is easy to get the appropriate algorithms for PDM-k. The general schemes for formation of the signals with difference modulation, shown in Figures 2.1(a) and 2.2, also contain the comprehensive information, necessary for making up functional schemes for the PDM-k modulators.

As an example, Figure 2.4 shows the scheme of signal modulators with PDM-2 and PDM-3. They consist of coders, phase modulators (PM), and units converting phase differences of the second or third order into the current signal phase. The last units are usually called as differential encoders (drawn out by dashed lines); their algorithms are considered in the following section.

A more complex problem is the realization of PDM-k signal demodulators. The general schemes of processing the signals with difference modulation [see Figs. 2.1(b) and 2.3] contain a meter of an information parameter, the algorithm of which is quite difficult to make up in a general case. At PDM this is the signal demodulator, which is implemented differently depending on a priori information about the received signal and the order of phase differences. So, at PDM-1, as mentioned earlier, the property of the absolute invariance to the uncertain initial signal phase is achieved by any incoherent signal demodulator with PDM-1, including the optimum incoherent and autocorrelated demodulators. The operation algorithms and the schemes of such demodulators are considered in detail later. At PDM-2 the property of the absolute invariance to the carrier frequency is reached only by using the autocorrelated demodulators, which are also studied in detail

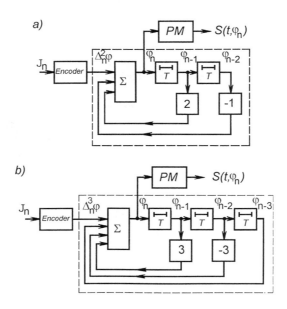

Figure 2.4 (a) Second- and (b) third-order PDM modulators.

later. At PDM-3 the autocorrelated demodulator, as will be shown, provides only the relative invariance to linear changes of the frequency, and to achieve the absolute invariance we have to synthesize a special "not two-dimensional" meter of the current signal phase. In the general case, the phase meter in systems with high-order PDM is the unit for optimum evaluation of a signal phase at a priori information, corresponding to that uncertainty, which the system should overcome.

2.4 OPTIMUM KEYING CODES FOR MULTIPHASE PM AND PDM SIGNALS

Modulation, in a broad sense, is understood as the conversion of one or several symbols into an elementary signal. As shown in the previous section, the algorithms in systems with difference modulation include three main operations: the transformation of the information symbol into a finite difference of the modulated parameter, the calculation of the modulated parameter according to its kth difference, and elementary signal formation. In systems with PDM, the first operation is called *keying encoding*, the second is *differential* or *repeat encoding*, and the third is the modulation itself. According to this algorithm, a modulator of the PDM system contains an encoder, differential encoder, and phase modulator.

On transmission of the digital binary information, the encoder associates a transmitted phase difference with the binary code combinations of a non-redundant code, which is referred to as a *keying code*. Examples of keying binary codes were discussed in Section 1.5 when describing multiposition PM signals (see Fig. 1.13).

Let's consider the problem of non-redundant encoding in digital systems with PDM in more detail. In the specified systems, the sequence of 2^N phase differences

$$\Delta\varphi_1, \Delta\varphi_2, \ldots, \Delta\varphi_2^N \qquad (2.43)$$

arranged in ascending order, is put in unequivocal conformity to the sequence of N-digit binary numbers:

$$Z_1, Z_2, \ldots, Z_{2^N}, Z_i = (z_{i1}, z_{i2}, \ldots, z_{iN}) \qquad (2.44)$$

The table of conformity between (2.43) and (2.44), containing 2^N lines and N columns (Table 2.2), serves as a keying code. The columns of this table correspond to binary subchannels of a multiphase PDM system. The problem is to synthesize the keying codes with the highest noise immunity, taking into account the nonuniformity of the error distribution among the binary subchannels and the complexity of realization.

There are various numerical characteristics of noise immunity of the m-ary systems with PDM: the probability of a faulty reception of an m-ary symbol, the

Table 2.2
N-Valued Keying Code

Transmitted Difference of Phases	Symbol of a Keying Code for a Subchannel			
	1	*2*	...	*N*
$\Delta\varphi_1$	0	1	...	0
$\Delta\varphi_2$	1	1	...	0
⋮	⋮	⋮	...	⋮
$\Delta\varphi_{2^N}$	1	0	...	1

error probabilities in binary subchannels, and the average error probability of binary subchannels. All of these probabilities are the functions of the statistical characteristics of interferences. As a rule, the probability of receiving some certain phase instead of the other one is a decreasing function of the phase difference. Therefore, at the transmission of binary combinations by multiphase signals it is necessary to associate the phase differences that are more remote from each other with the binary code combinations that have more Hamming distance. It is especially important to make up the keying code so that all pairs of code combinations, appropriate to the nearest pairs of phase differences, have the minimum Hamming distance; that is, they should differ only in one binary symbol.

We call the optimum binary keying code the keying code, at which the Hamming distance between any two adjacent code combinations is equal to 1.

The main property of the corresponding *N*-valued optimum code is that at a faulty reception instead of the transmitted phase (or phase difference) of one of two nearest to it there appears an error only in one binary symbol of the *N*-digit binary combination. Besides, as long as the probability of being able to receive instead of transmit the phase (or phase difference) nearest to it is much more than the probability of receiving any other more remote phase, the given code, as a rule, provides the minimum, or rather close to minimum, average error probability of binary channels.

The most widely used optimum keying code is the Gray code [5]. Other optimum keying codes exist including the codes generalizing the Gray code. Before we pass on to a discussion of those codes, we should consider general properties of the optimum binary keying codes.

1. Let the information phase differences be a priori equiprobable and regularly located in the interval $(0, 2\pi)$ with a step $2\pi/2^N$, and let it be possible to neglect the probability of the phase difference error $\pi/2^{N-1}$ in comparison with the probability of the error $\pi/2^N$.

Then the error probability in a binary subchannel is directly proportional to a number of changes of binary symbols in the corresponding code column.

Following from this property is the thought that to have identical error probabilities in binary subchannels it is necessary to synthesize the keying code with the same number of symbol changes in the columns. As shown later, the synthesis of such a code is possible only for the system, in which the multiplicity of modulation is equal to the integer degree of number 2. In the other cases, the use of the optimum codes results in an unequal noise immunity of binary subchannels. The Gray code provides especially large nonuniformity of the error bit rates in binary subchannels.

2. If n_1, n_2, . . . , n_N are the numbers of symbol changes in the corresponding code columns, then, first, all n_i are even and, second,

$$\sum_{i=1}^{N} n_i = 2^N \tag{2.45}$$

3. The optimum code is invariant to the rearrangement of the columns. The rearrangement of the columns does not change either the common number of symbol changes or their distribution between columns. Hence, the codes, received one from the other by the rearrangement of the columns, are equivalent in the sense of the noise immunity.
4. Two symbol changes in succession in any column of the optimum code are not allowed.
5. At $N \neq 2^l$, where l is the natural number, optimum codes do not exist that have an identical number of symbol changes in all columns.

According to property 2, the numbers n_1, n_2, . . . , n_N cannot be equal, because 2^N in this case is not evenly divided by N.

The important consequence of properties 1 and 5 consists of the fact that it is impossible to have the same noise immunity in all three binary subchannels of the three-valued (eight-phase) PDM system when using the optimum keying code.

One of the general methods of synthesizing the optimum keying codes [6] consists of bringing the synthesis of the N-valued optimum code K_N to the synthesis of the $(N-1)$-valued optimum code K_{N-1}. According to this method, the N-valued code K_N is a combination of two $(N-1)$-valued codes K_{N-1}^1 and K_{N-1}^0, and its structure is the following:

$$
K_N = \begin{cases}
\begin{array}{l}
0 \\
0 \\
\vdots \\
0
\end{array} \quad \boxed{K_{N-1}^0} \\[2em]
\begin{array}{l}
1 \\
1 \\
\vdots \\
1
\end{array} \quad \boxed{K_{N-1}^1}
\end{cases}
\tag{2.46}
$$

As can be seen, the first column of the optimum code with the structure of (2.46) always has two symbols changes: $n_1^N = 2$. The columns from the second up to the Nth one form *two* $(N-1)$-valued optimum codes. If the numbers of symbol changes in the columns of the $(N-1)$-valued code are equal to

$$
n_2^{N-1}, \ n_3^{N-1}, \ \ldots, \ n_N^{N-1}
$$

then the numbers of symbol changes in the second through Nth columns of the N-valued code are equal to one of the following numbers:

$$
n^N = n_i^{N-1} + n_j^{N-1}
$$
$$
n^N = n_i^{N-1} + n_j^{N-1} - 2
\tag{2.47}
$$

where $i, j = 2, 3, \ldots, N$. The second equality from (2.47) is executed only in one code column, in which the symbol change occurs at transition from the last symbol of the $(N-1)$-valued code to the first one.

Let's use this general approach for development of the optimum keying codes. As the basis for this development we take the optimum four-level (two-valued) keying Gray code:

$$
\begin{array}{cc}
0 & 0 \\
0 & 1 \\[1em]
1 & 1 \\
1 & 0
\end{array}
\tag{2.48}
$$

This code is a unique optimum two-valued code. Uniqueness is understood in the sense that all other optimum two-valued codes are equivalent to the code of (2.48), that is, they can be obtained from it by means of column rearrangement and by the cyclic shift of the lines. It is impossible to construct the optimum code, which cannot be obtained from (2.48) by the specified conversions. In both columns of the code of (2.48) there are two symbol changes, that is, the binary subchannels of a four-phase system with PM or PDM have the same potential noise immunity.

Using two codes (2.48) and the general structure (2.46), it is not difficult to construct an optimum eight-level (three-valued) keying code:

$$
\begin{array}{c|cc}
0 & 0 & 0 \\
0 & 0 & 1 \\
0 & 1 & 1 \\
0 & 1 & 0 \\
\end{array}
$$

$$
\begin{array}{c|cc}
1 & 1 & 0 \\
1 & 1 & 1 \\
1 & 0 & 1 \\
1 & 0 & 0 \\
\end{array}
$$
(2.49)

This code is also the Gray code. Having used (2.47), we know that in three-valued code columns two or four symbol changes are possible, that is,

$$
n_i^3 = \begin{cases} 2 \\ 4 \end{cases}
$$
(2.50)

According to (2.45)

$$
\sum_{i=1}^{3} n_i^3 = 8
$$
(2.51)

there is possible only one combination of symbol changes in three-valued optimum code columns: $2 + 2 + 4$, that is observed in the code of (2.49).

The last statement proves that the Gray code (2.49) is the unique three-valued optimum keying code. When using this code, two first subchannels of the three-valued system have the same error probability, and in the third subchannel the error probability is up to two times greater than in the first two.

Using the three-valued Gray code, we can make up optimum four-valued codes. First of all let's define what combinations of symbol changes are possible here. Because in the three-valued code these numbers are equal to 2 and 4, then according to (2.47) in the four-valued code they are equal to

$$
n_i^4 = 2, 4, 6, 8
$$
(2.52)

As far as

$$\sum_{i=1}^{4} n_i^4 = 16 \qquad (2.53)$$

the following four combinations of symbol change numbers are possible in the columns of the four-valued optimum code:

$$2 + 2 + 4 + 8 \qquad 2 + 4 + 4 + 6$$
$$2 + 2 + 6 + 6 \qquad 4 + 4 + 4 + 4 \qquad (2.54)$$

The optimum keying codes, corresponding to the combinations of (2.54), are submitted in Table 2.3. The numbers of symbol changes in the columns of these codes are specified on the last line of the table.

The first three codes are created according to the described technique. The first of them—the four-valued Gray code—is obtained by direct substitution of the three-valued Gray code (2.49) into the structure (2.46). The second and the third codes are obtained by substituting into (2.46) two codes (2.49) with column rearrangement occurring in all of them [6]. The fourth of the submitted codes was obtained by another method [7].

Table 2.3
Four-Valued Optimum Keying Codes

Phase or Phase Difference	Four-Valued Optimum Keying Codes			
	Gray Code	Okunev Codes		Khvorostenko Code
0	0 0 0 0	0 0 0 0	0 1 0 0	0 0 0 0
$\pi/8$	0 0 0 1	0 0 0 1	0 0 0 0	0 0 0 1
$2\pi/8$	0 0 1 0	0 0 1 1	0 0 0 1	0 1 0 1
$3\pi/8$	0 0 1 1	0 0 1 0	0 0 1 1	1 1 0 1
$4\pi/8$	0 1 1 0	0 1 1 0	0 0 1 0	1 1 0 0
$5\pi/8$	0 1 1 1	0 1 1 1	0 1 1 0	0 1 0 0
$6\pi/8$	0 1 0 1	0 1 0 1	0 1 1 1	0 1 1 0
$7\pi/8$	0 1 0 0	0 1 0 0	0 1 0 1	1 1 1 0
π	1 1 0 0	1 1 0 0	1 1 0 1	1 0 1 0
$9\pi/8$	1 1 0 1	1 1 1 0	1 1 1 1	1 0 0 0
$10\pi/8$	1 1 1 1	1 1 1 1	1 1 1 0	1 0 0 1
$11\pi/8$	1 1 1 0	1 1 0 1	1 0 1 0	1 0 1 1
$12\pi/8$	1 0 1 0	1 0 0 1	1 0 1 1	1 1 1 1
$13\pi/8$	1 0 1 1	1 0 1 1	1 0 0 1	0 1 1 1
$14\pi/8$	1 0 0 1	1 0 1 0	1 0 0 0	0 0 1 1
$15\pi/8$	1 0 0 0	1 0 0 0	1 1 0 0	0 0 1 0
Numbers of Symbol Changes	2 2 4 8	2 2 6 6	2 4 4 6	4 4 4 4

Thus, there are four nonequivalent four-valued optimum keying codes for 16-ary PM and PDM signals. They differ by the number of symbol changes in the columns and, hence, by the noise immunity of binary subchannels. The Gray code has the most nonuniformity of the noise immunity—in the fourth subchannel the error probability can be four times more than in the first two subchannels. The Khvorostenko code has the same noise immunity in subchannels. The Okunev codes take, according to this attribute, an intermediate position between the Gray code and the Khvorostenko code.

Now using the four-valued codes we obtained, it is possible to synthesize optimum five-valued keying codes. The number of nonequivalent five-valued codes increases considerably. As in the four-valued code columns, there are 2, 4, 6, or 8 symbol changes, then according to (2.45) and (2.47) in a five-valued code the numbers of symbol changes satisfy the following two conditions:

$$n_i^5 = 2,\ 4,\ 6,\ 8,\ 10,\ 12,\ 14,\ 16$$

$$\sum_{i=1}^{5} n_i^5 = 32 \qquad (2.55)$$

Thus, there are 16 nonequivalent optimum five-valued codes with the structure of (2.46). The combinations of symbol changes in the columns of these codes in correspondence with (2.55) are as follows:

(1)	$2 + 2 + 4 + 8 + 16$	(9)	$2 + 4 + 4 + 10 + 12$
(2)	$2 + 2 + 6 + 6 + 16$	(10)	$2 + 2 + 8 + 8 + 12$
(3)	$2 + 4 + 4 + 6 + 16$	(11)	$2 + 4 + 6 + 8 + 12$
(4)	$2 + 2 + 4 + 10 + 14$	(12)	$2 + 2 + 8 + 10 + 10$
(5)	$2 + 2 + 6 + 8 + 14$	(13)	$2 + 4 + 6 + 10 + 10$
(6)	$2 + 4 + 4 + 8 + 14$	(14)	$2 + 4 + 8 + 8 + 10$
(7)	$2 + 2 + 4 + 12 + 12$	(15)	$2 + 6 + 6 + 8 + 10$
(8)	$2 + 2 + 6 + 10 + 12$	(16)	$2 + 6 + 8 + 8 + 8$

(2.56)

Code (1) in (2.56) is the Gray code and has the greatest (up to eight times) noise immunity nonuniformity of the binary subchannels, and code (16) has the least (not more than four times) nonuniformity. These codes have the following representation:

(1)	00000	(16)	00000	
	00001		00001	
	00011		00011	
	00010		00010	
	00110		00110	
	00111		00111	
	00101		00101	
	00100		01101	
	01100		01111	
	01101		01110	
	01111		01010	
	01110		01011	
	01010		01001	
	01011		01000	
	01001		01100	
	01000		00100	
	11000		10100	(2.57)
	11001		10101	
	11011		10001	
	11010		11001	
	11110		11011	
	11111		10011	
	11101		10111	
	11100		11111	
	10100		11101	
	10101		11100	
	10111		11110	
	10110		10110	
	10010		10010	
	10011		11010	
	10001		11000	
	10000		10000	

Though there are many nonequivalent optimum five-valued codes, among them there is no code with an identical number of symbol changes in columns.

In spite of the fact that the Gray code has the greatest nonuniformity of the number of symbol changes along the columns, it is used most often in practice. This use can be explained. First, the Gray code is the unique optimum keying code for the most used four- and eight-phase signals. Second, it has a standard structure at any number of phases. Last, when using the Gray code the operation of decoding the multiphase PM and PDM signals is most simply realized. It consists in definition of the transmitted binary symbols according to trigonometric functions of the received phase difference. This last property of the Gray code follows from general algorithms of demodulating the PDM signals, which are considered in the next section.

2.5 GENERAL ALGORITHMS OF DEMODULATION AND DECODING OF MULTIPHASE PDM SIGNALS

As shown earlier, the general algorithms of signal demodulation in the system with high-order difference modulation include three operations: measuring the determining signal parameter γ; calculating the information parameter, that is, the kth difference of the determining parameter $\Delta^k \gamma$; and decoding of the kth difference.

At PDM these operations correspond to the measurement of a phase cosine and sine, the calculation of a cosine and sine of the kth phase difference, and the calculation of the transmitted discrete symbols (decoding). The general scheme of a PDM signal demodulator is shown in Figure 2.5.

The methods of measuring (evaluation) the trigonometric functions of a phase and its finite differences depend on the statistical characteristics of interferences and a priori information about signals at the receiving site. These methods, including three main classes of algorithms (coherent, optimum incoherent, and autocorrelated), are studies in detail in the following chapters.

Here let's discuss the algorithms for converting the trigonometric functions of the received phase difference into the appropriate elements of the transmitted message. These algorithms are called the general algorithms of PDM signal demodulation and decoding because they are true with any methods of measuring (evaluation) of the received signal phase difference. The elements of the transmitting

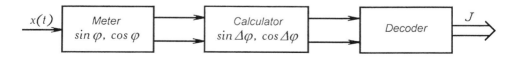

Figure 2.5 Demodulator of PDM signals.

message are more often presented as m-ary symbols or binary code combinations. In the latter case, the decoding algorithm depends on a keying code.

The demodulation problem at an arbitrary keying code can be formulated as follows: It is necessary to find the optimum algorithm, according to which every pair of values $\sin \Delta\varphi_\xi$ and $\cos \Delta\varphi_\xi$, where φ_ξ is the received phase difference from the continuous set, will be transformed into a phase difference $\Delta\varphi_i$ from the admissible discrete set of the phase differences $\Delta\varphi_1, \Delta\varphi_2, \ldots, \Delta\varphi_m$.

The given situation is characterized by conditional (a posteriori) probabilities that the phase difference $\Delta\varphi_i$ has been transmitted, if the phase difference $\Delta\varphi_\xi$ is received. Let's designate these conditional probabilities as $p(\Delta\varphi_i/\Delta\varphi_\xi)$, $i = 1, 2, \ldots, m$. Obviously, the phase difference $\Delta\varphi_i$ has been transmitted, if at all $j \neq i$ we have

$$p(\Delta\varphi_i/\Delta\varphi_\xi) > p(\Delta\varphi_j/\Delta\varphi_\xi) \tag{2.58}$$

When we have a priori equiprobable transmitted phase differences, the general algorithm (2.58) can be submitted by the likelihood function—the conditional probability density $W(\Delta\varphi_\xi/\Delta\varphi_i)$ of receiving the phase difference $\Delta\varphi_\xi$ at the transmission of the phase difference $\Delta\varphi_i$:

$$W(\Delta\varphi_\xi/\Delta\varphi_i) > W(\Delta\varphi_\xi/\Delta\varphi_j) \tag{2.59}$$

or, just the same, by the probability density $W(|\Delta\varphi_\xi - \Delta\varphi_i|)$ of the difference module $|\Delta\varphi_\xi - \Delta\varphi_i|$, that is,

$$W(|\Delta\varphi_\xi - \Delta\varphi_i|) > W(|\Delta\varphi_\xi - \Delta\varphi_j|) \tag{2.60}$$

If the function $W(\Delta\varphi_\xi - \Delta\varphi_i)$ is monotonically decreasing, (2.60) can be replaced by the rule in which any even and monotonically decreasing, in particular, cosine function of the difference $\Delta\varphi_\xi - \Delta\varphi_i$ is used instead of W. Thus, we obtain the following rule for choosing the transmitted phase difference:

$$\cos(\Delta\varphi_\xi - \Delta\varphi_i) > \cos(\Delta\varphi_\xi - \Delta\varphi_j) \tag{2.61}$$

Removing the parentheses in (2.61) we find the following general algorithm for demodulating the m-ary PDM signals:

$$\cos \Delta\varphi_\xi \cos \Delta\varphi_i + \sin \Delta\varphi_\xi \sin \Delta\varphi_i > \cos \Delta\varphi_\xi \cos \Delta\varphi_j + \sin \Delta\varphi_\xi \sin \Delta\varphi_j \tag{2.62}$$

for $i, j = 1, 2, \ldots, m$. This algorithm, being reasonably general, is not always convenient for practical applications because, as a rule, binary codes are used in the digital transmission systems, and hence, it is desirable for the transmitted binary symbols to become the demodulator output. The appropriate algorithms for the particular keying code can be derived from (2.62).

As an example of such a derivation let's find the demodulating algorithm for the four-phase signal when using the code corresponding to that of Table 2.4. Let us denote λ_i and calculate the right part of inequality (2.62) at all possible values of j:

$$\lambda_1 = \cos \Delta\varphi_\xi$$
$$\lambda_2 = \sin \Delta\varphi_\xi$$
$$\lambda_3 = -\cos \Delta\varphi_\xi \qquad (2.63)$$
$$\lambda_4 = -\sin \Delta\varphi_\xi$$

Comparing the data of the Table 2.4 with (2.62) and (2.63), it follows that at transmission of the sign "+" by the first subchannel, it should take the place of the inequality

$$\lambda_1 + \lambda_2 > \lambda_3 + \lambda_4 \qquad (2.64)$$

and at transmitting the "–" sign, the inverse inequality. Substituting (2.63) into (2.64), we get

$$\cos \Delta\varphi_\xi + \sin \Delta\varphi_\xi > -(\cos \Delta\varphi_\xi + \sin \Delta\varphi_\xi)$$

that is,

$$J_1 = \text{sgn}(\cos \Delta\varphi_\xi + \sin \Delta\varphi_\xi) \qquad (2.65)$$

At transmission of the plus sign by the second subchannel, it should take the place of the inequality

$$\lambda_1 + \lambda_4 > \lambda_2 + \lambda_3 \qquad \text{or} \qquad \cos \Delta\varphi_\xi - \sin \Delta\varphi_\xi > - (\cos \Delta\varphi_\xi - \sin \Delta\varphi_\xi)$$

Table 2.4
Four-Phase Code

j	$\Delta\varphi_j$	J_1	J_2
1	0	+	+
2	$\pi/2$	+	–
3	π	–	–
4	$3\pi/2$	–	+

Thus,

$$J_2 = \text{sgn}(\cos \Delta\varphi_\xi - \sin \Delta\varphi_\xi) \tag{2.66}$$

Relations (2.65) and (2.66) define the required algorithm of the four-phase (two-valued) PDM signal decoding.

For the multiphase systems when using a Gray keying code the similar (2.65) and (2.66) algorithms can be synthesized, making use of the peculiarities of this code structure.

The five-valued (32-phase) Gray code is given as an example in Table 2.5. In this code the 0 is submitted by a plus sign and the 1 by a minus sign. Notice that in every Gray code column the shift of a symbol sign occurs periodically, and the frequency of this shift is increased twice at the transition to every next column, starting from the second one. This permits us to express the binary symbols, transmitted in Gray code columns, directly by sines and cosines of the corresponding phase differences. To synthesize the appropriate algorithms, we enter the concept of a reduced phase difference system, to which we will reduce any arbitrary systems.

The sequence of 2^N increasing phase differences $\Delta_{\text{red}}\varphi_0, \Delta_{\text{red}}\varphi_1, \ldots, \Delta_{\text{red}}\varphi_{2^N-1}$ regularly located within the interval $(0, 2\pi)$, with a step $2\pi/2^N$ and with the minimum value of the phase difference $\Delta_{\text{red}}\varphi_0 = \pi/2^N$, is called the reduced N-valued system of the phase differences, that is,

$$\Delta_{\text{red}}\varphi_i = \pi/2^N + i\ \pi/2^{N-1}$$

$$i = 0, 1, 2, \ldots, 2^N - 1 \tag{2.67}$$

Table 2.5 shows an example (see the first column) of the reduced five-valued system in which the minimum phase difference is equal to $\Delta\varphi_0 = \pi/32$.

Any N-valued system of phase differences $\Delta\varphi_0, \Delta\varphi_1, \ldots, \Delta\varphi_{2^N-1}$ with a step $2\pi/2^N$ and with an arbitrary minimum shift $\Delta\varphi_0$ can be converted into the reduced system (2.67) using the following simple operation:

$$\Delta_{\text{red}}\varphi_i = \Delta\varphi_i - \Delta\varphi_0 + \pi/2^N \tag{2.68}$$

In general, if some arbitrary phase difference $\Delta\varphi_\xi$ is assumed, its reduced value in the N-valued system is

$$\Delta_{\text{red}}\varphi_\xi = \Delta\varphi_\xi - \Delta\varphi_0 + \pi/2^N \tag{2.69}$$

Now let us consider the keying Gray code (see Table 2.5) and compare binary symbols in its columns with the signs of the trigonometric functions of the reduced phase difference. It is not difficult to notice that the binary symbols transmitted in the first subchannel coincide with the sine sign of the reduced phase difference:

Table 2.5
Reduced Phase System and 32-Phase Gray Code

$\Delta\varphi_i$	J_1	J_2	J_3	J_4	J_5
$\pi/32$	+	+	+	+	+
$3\pi/32$	+	+	+	+	−
$5\pi/32$	+	+	+	−	−
$7\pi/32$	+	+	+	−	+
$9\pi/32$	+	+	−	−	+
$11\pi/32$	+	+	−	−	−
$13\pi/32$	+	+	−	+	−
$15\pi/32$	+	+	−	+	+
$17\pi/32$	+	−	−	+	+
$19\pi/32$	+	−	−	+	−
$21\pi/32$	+	−	−	−	−
$23\pi/32$	+	−	−	−	+
$25\pi/32$	+	−	+	−	+
$27\pi/32$	+	−	+	−	−
$29\pi/32$	+	−	+	+	−
$31\pi/32$	+	−	+	+	+
$33\pi/32$	−	−	+	+	+
$35\pi/32$	−	−	+	+	−
$37\pi/32$	−	−	+	−	−
$39\pi/32$	−	−	+	−	+
$41\pi/32$	−	−	−	−	+
$43\pi/32$	−	−	−	−	−
$45\pi/32$	−	−	−	+	−
$47\pi/32$	−	−	−	+	+
$49\pi/32$	−	+	−	+	+
$51\pi/32$	−	+	−	+	−
$53\pi/32$	−	+	−	−	−
$55\pi/32$	−	+	−	−	+
$57\pi/32$	−	+	+	−	+
$59\pi/32$	−	+	+	−	−
$61\pi/32$	−	+	+	+	−
$63\pi/32$	−	+	+	+	+

$$J_1 = \operatorname{sgn} \sin \Delta_{\mathrm{red}}\varphi_\xi \qquad (2.70)$$

Thus, the general algorithm of decoding the first binary subchannel of the N-valued system, taking into account (2.69), is the following:

$$J_1 = \operatorname{sgn} \sin(\Delta\varphi_\xi + \pi/2^N - \Delta\varphi_0) \qquad (2.71)$$

where, remember, $\Delta\varphi_\xi$ is the received phase difference, and $\Delta\varphi_0$ is the minimum value of the permitted phase differences.

Similarly, we can find that the binary symbols of the second subchannel (see Table 2.5) coincide with the cosine sign of the reduced phase difference; that is, the general algorithm of decoding the second binary subchannel of the N-valued system is

$$J_2 = \text{sgn} \cos(\Delta\varphi_\xi + \pi/2^N - \Delta\varphi_0) \tag{2.72}$$

or, by the sine of the reduced phase difference,

$$J_2 = \text{sgn} \sin(\Delta\varphi_\xi + \pi/2^N - \Delta\varphi_0 + \pi/2) \tag{2.73}$$

Because the structure of Gray code is regular, the symbol signs in every next column are alternated with the frequency, twice greater, than in the previous one. Therefore, the symbols of the third binary subchannel coincide with the cosine sign of the double reduced phase difference; the symbols of the fourth subchannel coincide with the cosine sign of the four times more reduced phase difference; the symbols of the fifth subchannel coincide with the cosine sign of the eight times more phase difference, and so on.

Thus, the general algorithm for decoding the third subchannel of the N-valued system is

$$J_3 = \text{sgn} \cos 2(\Delta\varphi_\xi + \pi/2^N - \Delta\varphi_0) \tag{2.74}$$

or, by the sine,

$$J_3 = \text{sgn} \sin 2(\Delta\varphi_\xi + \pi/2^N - \Delta\varphi_0 + \pi/4) \tag{2.75}$$

The similar general algorithms for decoding the fourth and fifth subchannels of the N-valued system can be presented as follows:

$$J_4 = \text{sgn} \cos 4(\Delta\varphi_\xi + \pi/2^N - \Delta\varphi_0) = \text{sgn} \sin 4(\Delta\varphi_\xi + \pi/2^N - \Delta\varphi_0 + \pi/8) \tag{2.76}$$

$$J_5 = \text{sgn} \cos 8(\Delta\varphi_\xi + \pi/2^N - \Delta\varphi_0) = \text{sgn} \sin 8(\Delta\varphi_\xi + \pi/2^N - \Delta\varphi_0 + \pi/16) \tag{2.77}$$

Integrating the results of (2.71), (2.73), (2.75), (2.76), and (2.77), we can derive the following general algorithm for decoding the ith binary channel of the N-valued system with PDM using the Gray code:

$$J_i = \begin{cases} \text{sgn} \sin(\Delta\varphi_\xi + \pi/2^N - \Delta\varphi_0) & i = 1 \\ \text{sgn} \sin[2^{i-2}(\Delta\varphi_\xi + \pi/2^N - \Delta\varphi_0 + \pi/2^{i-1})] & i > 1 \end{cases} \tag{2.78}$$

Expression (2.78), remarkable by its content, is the most general record of the algorithms for calculating the transmitted binary symbols J_i by the received phase difference $\Delta\varphi_\xi$.

As long as the results of measuring the phase difference are usually presented by $\sin\Delta\varphi_\xi$ and $\cos\Delta\varphi_\xi$, the algorithms of decoding the multiphase signals should be expressed by these functions.

From (2.78) we can obtain such algorithms for the systems with single-, two-, three- and four-valued PDM.

For single-valued ($N = 1$), that is, two-phase, PDM we have

$$J_1 = \text{sgn}(\cos\Delta\varphi_0 \cos\Delta\varphi_\xi + \sin\Delta\varphi_0 \sin\Delta\varphi_\xi) \tag{2.79}$$

In two-phase systems we usually use $\Delta\varphi_0 = 0$, which is why, as a rule, the following simple algorithm, coming from (2.79), is used

$$J_1 = \text{sgn}\cos\Delta\varphi_\xi \tag{2.80}$$

For two-valued ($N = 2$), that is, four-phase, PDM we have

$$J_1 = \text{sgn}[\cos(\pi/4 - \Delta\varphi_0)\sin\Delta\varphi_\xi + \sin(\pi/4 - \Delta\varphi_0)\cos\Delta\varphi_\xi]$$
$$J_2 = \text{sgn}[\cos(\pi/4 - \Delta\varphi_0)\cos\Delta\varphi_\xi - \sin(\pi/4 - \Delta\varphi_0)\sin\Delta\varphi_\xi] \tag{2.81}$$

In two-valued systems we usually apply two variants of the minimum permitted phase difference: $\Delta\varphi_0 = 0$ and $\Delta\varphi_0 = \pi/4$. At $\Delta\varphi_0 = 0$ we get

$$J_1 = \text{sgn}(\sin\Delta\varphi_\xi + \cos\Delta\varphi_\xi)$$
$$J_2 = \text{sgn}(\cos\Delta\varphi_\xi - \sin\Delta\varphi_\xi) \tag{2.82}$$

At $\Delta\varphi_0 = \pi/4$ we have

$$J_1 = \text{sgn}\sin\Delta\varphi_\xi$$
$$J_2 = \text{sgn}\cos\Delta\varphi_\xi \tag{2.83}$$

For three-valued ($N = 3$), that is, eight-phase, PDM from (2.78) after simple conversions we have

$$J_1 = \text{sgn}[\cos(\pi/8 - \Delta\varphi_0)\sin\Delta\varphi_\xi + \sin(\pi/8 - \Delta\varphi_0)\cos\Delta\varphi_\xi]$$
$$J_2 = \text{sgn}[\cos(\pi/8 - \Delta\varphi_0)\cos\Delta\varphi_\xi - \sin(\pi/8 - \Delta\varphi_0)\sin\Delta\varphi_\xi]$$
$$J_3 = \text{sgn}[\cos(3\pi/8 - \Delta\varphi_0)\sin\Delta\varphi_\xi + \sin(3\pi/8 - \Delta\varphi_0)\cos\Delta\varphi_\xi]$$
$$\times \text{sgn}[\cos(3\pi/8 - \Delta\varphi_0)\cos\Delta\varphi_\xi - \sin(3\pi/8 - \Delta\varphi_0)\sin\Delta\varphi_\xi] \tag{2.84}$$

In three-valued systems, the signals with $\Delta\varphi_0 = 0$ or $\Delta\varphi_0 = \pi/8$ are usually used. In the latter case the algorithm is simplified:

$$J_1 = \text{sgn} \sin \Delta\varphi_\xi$$
$$J_2 = \text{sgn} \cos \Delta\varphi_\xi$$
$$J_3 = \text{sgn}(\sin \Delta\varphi_\xi + \cos \Delta\varphi_\xi) \, \text{sgn}(\cos \Delta\varphi_\xi - \sin \Delta\varphi_\xi) \tag{2.85}$$

For four-valued ($N = 4$), that is, 16-phase, PDM from (2.78) by simple conversion it is possible to obtain the following algorithm of decoding at an arbitrary $\Delta\varphi_0$:

$$J_1 = \text{sgn}[\cos(\pi/16 - \Delta\varphi_0) \sin \Delta\varphi_\xi + \sin(\pi/16 - \Delta\varphi_0) \cos \Delta\varphi_\xi]$$
$$J_2 = \text{sgn}[\cos(\pi/16 - \Delta\varphi_0) \cos \Delta\varphi_\xi - \sin(\pi/16 - \Delta\varphi_0) \sin \Delta\varphi_\xi]$$
$$J_3 = \text{sgn}[\cos(5\pi/16 - \Delta\varphi_0) \sin \Delta\varphi_\xi + \sin(5\pi/16 - \Delta\varphi_0) \cos \Delta\varphi_\xi]$$
$$\times \text{sgn}[\cos(5\pi/16 - \Delta\varphi_0) \cos \Delta\varphi_\xi - \sin(5\pi/16 - \Delta\varphi_0) \sin \Delta\varphi_\xi]$$
$$J_4 = \text{sgn} \sin(\Delta\varphi_\xi + 3\pi/16 - \Delta\varphi_0) \, \text{sgn} \cos(\Delta\varphi_\xi + 3\pi/16 - \Delta\varphi_0)$$
$$\times \text{sgn} \cos(\Delta\varphi_\xi + 7\pi/16 - \Delta\varphi_0) \, \text{sgn} \sin(\Delta\varphi_\xi + 7\pi/16 - \Delta\varphi_0) \tag{2.86}$$

The common algorithms we have obtained for multiphase PDM signal demodulation and decoding are the basis of many real demodulators considered in the next chapters. The distinctions between the algorithms for the optimum incoherent and autocorrelated demodulators consist of the methods used to calculate the cosines and sines of the received first phase difference, included in the algorithms obtained earlier.

The cosine and sine of the first phase differences at the autocorrelated reception are presented by the elements of the received signal $x_n(t)$ and $x_{n-1}(t)$ as follows:

$$\cos \Delta^1\varphi_\xi = A\int_0^T x_n(t) x_{n-1}(t) \; dt$$
$$\sin \Delta^1\varphi_\xi = A\int_0^T x_n(t) x_{n-1}^*(t) \; dt \tag{2.87}$$

and at the optimum incoherent reception as

$$\cos \Delta^1\varphi_\xi = A(X_n X_{n-1} + Y_n Y_{n-1})$$
$$\sin \Delta^1\varphi_\xi = A(Y_n X_{n-1} - X_n Y_{n-1}) \tag{2.88}$$

where

$$X_i = \int_0^T x_i(t) \sin \omega t \; dt$$
$$Y_i = \int_0^T x_i(t) \cos \omega t \; dt$$

Having substituted (2.87) and (2.88) into (2.79) through (2.86), we can obtain the algorithms of the optimum incoherent and autocorrelated demodulators of the multiphase PDM signals.

References

[1] Gelfond, A. O., *Finite Differences Calculation*, Moscow: Gostechizdat, 1952.

[2] Bernshtein, C. N., *Collected Works, Vol. I, Constructive Theory of Functions*, Moscow: USSR Academy of Science Publishers, 1952.

[3] Mysovski, I. P., *Lectures on Calculations Methods*, Moscow: Phyzmatgiz, 1962.

[4] Okunev, Yu. B., *Telecommunications Systems with the Invariant Performance*, Moscow: Svyaz., 1973.

[5] Gray, F., "Pulse Code Modulation," U.S. Patent 2632058, 1953.

[6] Okunev, Yu. B., *Theory of Phase-Difference Modulation*, Moscow: Svyaz, 1979.

[7] Khvorostenko, N. P., *Statistical Theory of Discrete Signal Demodulation*, Moscow: Svyaz, 1968.

CHAPTER 3

Coherent Processing of PM and PDM Signals

3.1 ESSENCE AND APPLICATION OF COHERENT PROCESSING

Algorithms and the units of coherent reception have a central place in the theory and engineering of optimum signal processing. Historically they were developed before other algorithms and signal receiving units, and the permanent interest in coherent demodulators is caused by the fact that under certain conditions they provide the maximum possible noise immunity, which has not been exceeded by any other methods.

A strictly coherent demodulator is often called a perfect or ideal receiver, which serves as the standard for real demodulators. The error probability of the perfect coherent demodulator in the additive white Gaussian noise (AWGN) channel plays the same role in communication theory as, for example, the light speed in physics—it is possible to approach it and get as close as desired, but not to exceed it. Therefore the noise immunity curve of a coherent demodulator, that is, the theoretical dependence of error probability on the signal-to-noise ratio, is the basis for evaluating the quality of a demodulator operation in real conditions. As a rule, we use the energy loss of a real demodulator, showing how many decibels are necessary to increase the signal energy at the real demodulator input in comparison with ideal coherent one, to provide the same error probability at the given noise level.

Despite what we have just mentioned, we should not consider the potential noise immunity, provided by a coherent demodulator, to be an uncompromising limit. The fact is that this limit depends on the system of signals including a type of signal modulation, signal multiplicity, encoding method, and so on. In some cases, when receiving the signals with redundancy, the coherent demodulator performance depends on the processing interval, which is dependent, in its turn, on admissible delay of message delivery, admissible complexity of a receiver, and other factors. Therefore, having improved the signal system or having increased the processing interval, it is possible basically to reduce the error probability. This circumstance is sometimes incorrectly perceived as an opportunity to exceed the potential noise immunity. In reality, however, for the given signal system and at the given processing interval (the decision-making interval) there is a minimum error probability that can be achieved by the ideal coherent receiver.

The function of a demodulator usually includes making decisions about every signal element separately—the so-called process of "element-by-element reception." In this case the signal element energy and the spectral density of Gaussian white noise power completely define the potential noise immunity, that is, that minimum error probability that can be achieved at the demodulator output.

The error probability can be reduced if we receive signals with redundancy as a whole, that is, if we make a decision about several signal elements, suitably integrated, rather than one by one. This, however, does not mean that we have managed to exceed the potential noise immunity; instead it means that we have managed to achieve another level of the potential noise immunity that corresponds to a new signal set.

In this book, we do not discuss this problem further. It is necessary only to warn the specialists of extremes—which can result in a too shallow or fatal attitude to the concept of potential noise immunity.

For subsequent discussions, it is important that the coherent demodulators open the optimum demodulator hierarchy and under certain conditions provide minimum possible error probability during element-by-element reception.

The essence of coherent processing is that the received mixture of a signal and noise is compared with ideal samples of the transmitted signal, and then the sample is chosen that is most close to the received mixture (in the sense of Euclidean distance).

Strictly coherent reception can be realized only in that case, when all possible variants of the transmitted signal are precisely known at the demodulator input or, as they say, the receiving signal is a priori known with precision up to a certain information parameter. When the signal variants are segments of harmonic oscillations with these or those initial phases, as PM or PDM occurs, the ideal receiver–demodulator should store the samples of these oscillations, which coincide strictly with the receiving signals on frequency and phase. That is, the reference oscillations, coherent with the received signal, should be formed in the receiver. From here we get the names *coherent receiver* and *coherent demodulator.*

The generation of coherent reference oscillations, a precondition for the realization of coherent processing, is not a simple problem. The modern methods of high stable harmonic oscillation generation, for example, by quantum standards of frequency, basically permit us to generate, in the points with space diversity that correspond to the position of transmitter and receiver, two coherent oscillations during quite a long time interval. This, however, does not solve the problem of coherent reception. In wire and radio communication lines, the signal is subjected to several frequency conversions with respective alterations of its initial phase; besides, the signal phase at the receiving point depends on the time interval of electromagnetic oscillation propagation in the communication channel, and this time interval is changed because of both the change of the signal propagation path and the change of properties and parameters of the propagation media.

Thus, the initial phase, and in some cases the frequency of a signal at the demodulator input appear to be random, a priori unknown and unpredictable values. That is why the realization of purely coherent reception remains an abstraction, and so-called "quasicoherent" demodulators are actually realized in communication engineering. (We also use this term when it is necessary to emphasize that we are not dealing with pure coherence.)

The main point of all without exception quasicoherent demodulators is that the very precise samples of signal variants are generated by means of nonlinear (or parametrical) conversion and narrowband filtering of the received signal. Such generation is possible only in cases for which the initial phase of a carrier is changed much slower, under the influence of the above-mentioned factors, than the information phase or phase difference. At fulfillment of this condition and when interference is not too powerful, we manage to be able to predict the initial phase of a carrier with sufficient accuracy and to come nearer to an ideal coherent demodulator.

The units, carrying out the functions of generating the coherent signal samples in quasicoherent receivers, are called *reference oscillation selectors* (ROSs) or *carrier recovery units.*

In conventional ROSs the proper oscillation, coherent with the received signal, is formed by a narrowband passive filter or by a controlled oscillator, which is included in the phase-frequency-adjustment loop. The methods of reference oscillation selection and ROS schemes are considered in detail in this chapter.

During the last several years, so-called "algorithmic methods" of quasicoherent processing have been developed at which the proper reference oscillations in a demodulator are not generated, the processing is fulfilled using quadrature projections of the received signal, and signal samples are obtained by averaging the projections of the received signal, which is converted according to the results of demodulation. Controlled oscillations, coherent convolution units, and narrowband filters are not used in the appropriate demodulators, and their algorithms include basically arithmetic operations that are oriented to digital realization. The specified methods are considered in detail later as we study the reception of signals with PM, PDM, and APM.

Quasicoherent demodulators, based on the considered methods and the units of ROS and forming the signal samples, are reasonably simple to realize, and, mainly, so slightly conceded to an ideal coherent demodulator on noise immunity, that their further perfection in this direction loses any practical sense.

When considering the conditions necessary for applying coherent processing, it is important to notice the principal ambiguity of the results of forming signal samples using the information signal. The ambiguity is that, although the oscillation at the ROS output coincides with one of the signal variants, we do not know which one. This circumstance has predetermined the wide use of PDM, in which the ambiguity of the obtained signal samples does not influence the correctness of signal demodulation.

The other point is absolute PM, in which the right decision in a demodulator can be made only when one-to-one conformity exists between the sets of the received signals and signal samples. To establish this conformity, we must either use a special synchronization signal, usually transmitted before the information signal, or special encoding with redundancy, in which the code combinations are able to reveal the present ambiguity and to obtain the correct information. In either case, the use of PM entails additional power expenditures (for synchronization signals or protective encoding).

In this regard, the choice between PM or PDM always requires the comparative analysis in the context of a general system situation. PM in ideal conditions has a noise immunity that is slightly higher than PDM (about two times with respect to error probability), however, inevitably because of the additional power expenditures required in PM for the synchronization signal or the protective encoding, its advantages are practically eliminated. The modem used with PM is a little easier to realize than a PDM modem, however, again, the introduction of the synchronization signal or protective encoding inevitably results in a more complex system with PM when compared to PDM.

In accordance with usual practice, PDM is preferably used in systems of continuous transmission, and PM is frequently applied in systems of pulse communications or at the package operation mode (for example, in satellite systems with TDMA), in which it is necessary to use a synchronization signal for call access.

PM signals can be received using only the coherent method, whereas PDM signals can be processed by various incoherent methods (optimum incoherent, autocorrelated).

The comparison of coherent and incoherent methods of PDM signal reception on noise immunity are made in Chapter 6. Here we merely note that the advantage of using coherent reception in a channel with constant or slowly varying parameters grows in accordance with an increase in the signal multiplicity.

When choosing a number of phase levels in PM or PDM systems, we must take into account the following. The general tendency is that as the transmission rate improves as a result of increasing the number of phase (phase difference) levels, the noise immunity rapidly decreases. However, at the same information rate the transition from a two-phase signal to a four-phase one permits the use of a narrower frequency band (two times narrower) without noise immunity decreases, and even with it increasing, taking into account the intersymbol interference. The transition to the eight-phase signal, keeping the same information rate, results in some noise immunity losses, because the reduction of the minimum phase shift is not compensated by the lengthening of an elementary signal. However, reduction of the influence of intersymbol interference, which is caused by the elementary signal lengthening, can justify the transition to the eight-phase system. At further increases in the number of signal levels to achieve higher transmission rates, we must use combined amplitude-phase (or amplitude-phase-difference) methods of modulation (QAM) or multidimensional signal-code structures based, as a rule, on a combination of PM, PDM, and AM signals.

Coherent reception is widely used in wire communication channels and in relatively "quiet" radio channels. We can speak, for example, about the data transmission and multichannel transmission by wire telephone and wideband channels, in which PDM and APDM signals are used in a combination with coherent reception. Coherent PM and PDM signal reception is used in radio channels, at remote space communication, and in satellite communication links, in particular in digital multichannel systems with code, frequency, and time division. The coherent reception of multiposition PDM and APDM signals is also used in radio relay communication links. PM and PDM signal reception is beginning to see use in optic and fiber optic channels.

In all cases, for effective realization of coherent processing, steadiness or reasonably slow changing of the signal initial phase in a channel is required such that it can be evaluated more precisely, at least by the order, than the information phase.

3.2 COHERENT DEMODULATORS OF PM SIGNALS

When synthesizing algorithms and schemes for coherent demodulators of PM and PDM signals, we proceed from the basic results of the theory of optimum signal processing in the AWGN channel.

The algorithm for ideal coherent signal processing is formulated in this theory as follows: From a set of the known equally powerful signals $S_1(t)$, $S_2(t)$, . . . , $S_m(t)$, the signal $S_i(t)$ is considered transmitted, if

$$\int_0^T x(t)\, S_i(t)\ dt > \int_0^T x(t)\, S_j(t)\ dt \qquad \text{for } j = 1, 2, \ldots, m \quad j \neq i \qquad (3.1)$$

where $x(t)$ is the received signal, and T is the elementary signal (chip) duration.

In the case where simple signals are received with absolute phase modulation, the reference signals S_j included in (3.1) are the harmonic oscillations with the corresponding initial phases

$$S_j = a \sin(\omega t + \varphi_j) \qquad \text{for } j = 1, 2, \ldots, m \qquad (3.2)$$

At PM, as a rule, a number of discrete phases at the elementary signal interval T and, consequently, a number of reference signals (3.2) are equal to a number of information phase variants. At PDM, as is shown later, a number of discrete phase values and a number of reference oscillations can be more than a number of information phase difference variants.

A general scheme for a coherent demodulator of PM signals for an arbitrary set of information phases is shown in Figure 3.1(a). It contains m correlators and a unit of comparing and selecting the maximum correlator output (max). It is assumed here that the reference oscillations at correlators inputs are known.

Usually in PM systems the coherent reference oscillations are formed from a special synchronization signal, transmitted either simultaneously with the information signal or before it. In the latter case, the structure of a signal coming to the demodulator has the view shown in Figure 3.1(b). The *synchrosignal* S_{lock} (also called the *training signal* or *preamble*) is transmitted periodically in the beginning of every information batch, providing phase coherence during the whole batch (if the tuning is not made at the information elementary signal reception) and eliminating the ambiguity between a signal and its samples (ambiguity arises because of random phase hops in the communication channel and the demodulator).

As a rule, the number of reference oscillations and correlators in a PM signal demodulator is lower than the number of discrete phases. For example, if among the transmitted signals there are pairs of counterphase, that is, distinguished only by a signal's sign, then one correlator and one reference oscillation are sufficient for such a pair. The number of reference oscillations and correlators in a multiphase signal demodulator can be reduced to two when an appropriate calculator is applied.

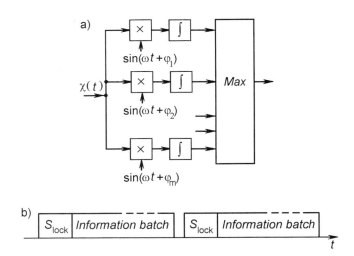

Figure 3.1 Coherent reception of PM signals: (a) demodulator scheme and (b) signal structure.

In fact, let there be convolutions of the received signal $x(t)$ and quadrature reference oscillations with a random initial phase φ_0, that is,

$$X_0 = \int_0^T x(t)\ \sin(\omega t + \varphi_0)\ dt$$

$$Y_0 = \int_0^T x(t)\ \cos(\omega t + \varphi_0)\ dt$$

(3.3)

Then any of the integrals included in algorithm (3.1) can be presented by the values from (3.3):

$$V_j = \int_0^T x(t)\ \sin(\omega t + \varphi_j)\ dt = \int_0^T x(t)\ \sin[(\omega t + \varphi_0) + (\varphi_j - \varphi_0)]\ dt$$

$$= X_0 \cos(\varphi_j - \varphi_0) + Y_0 \sin(\varphi_j - \varphi_0)$$

(3.4)

Thus, the general scheme of a coherent demodulator of multiphase signals can be submitted in the form shown in Figure 3.2. In this scheme the autonomous oscillator (O) and a phase shifter about $\pi/2$ produce quadrature reference oscillations with a random initial phase φ_0; the received signal projections onto these reference oscillations are calculated in two correlators, the values V_j are calculated in the V_j calculator under formula (3.4), and then the maximum value is defined of them.

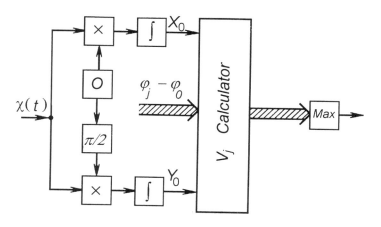

Figure 3.2 Coherent demodulator of multiposition PM signals.

For the scheme to operate, we must know the exact values of differences $\varphi_j - \varphi_0$ between phases of the received signal variants and the reference oscillation phase in the correlators. After finding these differences (for example, by measuring the phase shift between the synchrosignal and the oscillation of the reference oscillator), they are introduced into a calculator.

We emphasize that $\cos(\varphi_j - \varphi_0)$ and $\sin(\varphi_j - \varphi_0)$ included in algorithm (3.4) can be evaluated directly by the values of (3.3), calculated for the synchrosignal. Let us show how that can be done [1].

Let X_{0n} and Y_{0n} be the projections of the received synchrosignal, corresponding, for example, to the first variant of a multiphase signal with an initial phase φ_1, onto the reference oscillations with a random initial phase φ_0, calculated on the interval of the nth chip of the synchrosignal, that is,

$$X_{0n} = \int_{(n-1)T}^{nT} x_{syn}(t)\,\sin(\omega t + \varphi_0)\,dt$$

$$Y_{0n} = \int_{(n-1)T}^{nT} x_{syn}(t)\,\cos(\omega t + \varphi_0)\,dt$$

(3.5)

In an AWGN channel the values X_{0n} and Y_{0n} from (3.5) are normal random values with mathematical expectations, equal to the required projections of the synchrosignal undistorted by a noise. If these projections are constant during the observation interval, the arithmetic mean of (3.5) is the maximum likely estimate of the specified mathematical expectations [2–4].

Therefore, if the synchronization signal lasts during N elementary signals, the required estimates of the first variant of the PM signal are equal to

$$\tilde{X}_1 = \frac{1}{N}\sum_{n=1}^{N} X_{0n}; \qquad \tilde{Y}_1 = \frac{1}{N}\sum_{n=1}^{N} Y_{0n}$$

(3.6)

Note that in a channel with constant parameters the estimates of (3.6) are the maximum likely, unbiased, and effective ones [2–4].

By means of (3.6) and simple trigonometric transformations it is not difficult to find the estimates of projections of all other PM signal variants onto the orthogonal reference oscillations with a random phase φ_0:

$$\tilde{X}_j = \tilde{X}_1 \cos \Delta\varphi_j - \tilde{Y}_1 \sin \Delta\varphi_j$$

$$\tilde{Y}_j = \tilde{Y}_1 \cos \Delta\varphi_j + \tilde{X}_1 \sin \Delta\varphi_j$$

(3.7)

where $\Delta\varphi_j$ is the phase difference between the jth and the first variants of the transmitting PM signal for $j = 1, 2, \ldots, m$.

Now let's return to expression (3.4) and represent the trigonometric functions included in it by the obtained projection estimates of (3.7). As is obvious from comparing (3.4) and (3.7), in the first one there is a cosine and sine of the phase difference between the *j*th variant of PM signal and the reference oscillation, and in the second one we see the projections of the *j*th variant of a signal onto orthogonal components of the same reference oscillation. Hence,

$$\cos(\varphi_j - \varphi_0) = \tilde{X}_j/A \qquad \sin(\varphi_j - \varphi_0) = \tilde{Y}_j/A \qquad (3.8)$$

where $A = \sqrt{\tilde{X}_j^2 + \tilde{Y}_j^2}$ is the signal amplitude.

Having substituted (3.8) into (3.4) and excluded common multiplier for all V_j, we obtain the following expression for calculating the convolution of the received signal with a coherent reference oscillation, corresponding to the *j*th variant of the PM signal:

$$V_j = X_0(\tilde{X}_1 \cos \Delta\varphi_j - \tilde{Y}_1 \sin \Delta\varphi_j) + Y_0(\tilde{Y}_1 \cos \Delta\varphi_j + \tilde{X}_1 \sin \Delta\varphi_j) \text{ for } j = 1, 2, \ldots, m \qquad (3.9)$$

Remember that in (3.9) $\cos \Delta\varphi_j$ and $\sin \Delta\varphi_j$ are the constant coefficients, determined by a set of *m* transmitted signals, with $\Delta\varphi_j$ being a phase difference between the *j*th and the first variants of this signal set; X_0 and Y_0 are calculated by (3.3) for every current chip of an information signal; $\tilde{X}_1$ and $\tilde{Y}_1$ are calculated by (3.6) and (3.5) within the interval of the synchronization signal, which is the replica of the first signal variant.

Therefore, relation (3.9) along with (3.5), (3.6), and (3.3) represent the complete algorithm of coherent processing of an arbitrary multiphase signal with PM, preceded by synchronization signal transmutation.

The task of the demodulator is also to find the number with the maximum value from (3.9), which determines the transmitted information symbol.

In compact form the given demodulation algorithm can be recorded as follows:

$$i = \arg \max V_j \qquad \text{for } j = 1, 2, \ldots, m \qquad (3.10)$$

This algorithm is convenient to realize in digital or mixed analog–digital form; in the latter case convolutions (3.3) and (3.5) are realized by analog correlators, and the calculations in accordance with (3.6) and (3.9) are realized by a digital processor with a memory.

The appropriate scheme for the demodulator is shown in Figure 3.3. The demodulator operates in the following way. The input analog part, consisting of two correlators, a reference oscillator with a phase shifter, and two ADCs, calculates the convolutions of the received signal chips and the reference oscillation in the interval of the synchrosignal transmission, as well as at information chip transmis-

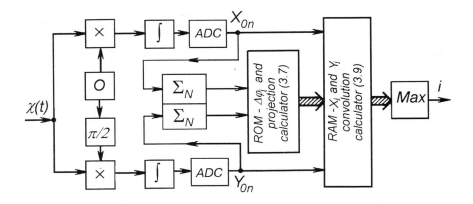

Figure 3.3 Analog–digital coherent demodulator of signals with multiphase modulation.

sion. Within the synchronization signal interval, the adders–accumulators Σ_N calculate the estimates of (3.6). The projections calculator, in which the values $\Delta\varphi_j$ are stored in read-only memory (ROM), determines $2m$ values (3.7) and directs them into random-access memory (RAM). Then, when the information signal has arrived, the convolution calculator determines the m values V_j of formula (3.9), compares them, and determines the maximum value according to (3.10).

The considered algorithm does not assume the "tuning" of the projections $\tilde{X}_j$ and $\tilde{Y}_j$ by the information signal and, hence, the correction of their values in the random-access memory. In this case the training signal should be transmitted quite frequently such that in the interval between two training signals the initial phase deviation of the received signal would not leave the admissible limits. However, it is not difficult to update the given algorithm in such a way that the tuning would be executed constantly—not only by the synchronizing signal, but by an information signal as well.

To make it up, let us come back to the above-described technique of estimating the parameters of the received signal, which is based on the averaging of the synchronizing signal projections [see (3.5) and (3.6)]. At the information chip reception, such averaging will not give the desired result as far as the estimated parameters becoming random values. It is possible, however, to convert the projections of the received signal, whatever it is, into the projections of any fixed variant, for example, the first variant, if we use the demodulation result, that is, the decision of what signal variant has been received.

This conversion of the received signal projections, using the decision feedback, is called the *projection reduction*; and the resulting new projections are called the *reduced projections*.

The reduction operation can be explained by Figure 3.4. Let us assume the signal vector **a** is received with the projections X_{0n} and Y_{0n}. Further, let this vector

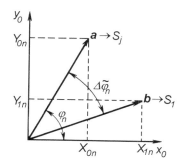

Figure 3.4 Determination of the reduced projections of a signal.

be identified in the demodulator with some signal variant S_j, which differs by a phase from the variant number 1 by $\Delta\tilde{\varphi}_n$. Then the reduction operation consists of converting projections X_{0n} and Y_{0n} of vector **a** into projections X_{1n} and Y_{1n} of vector **b**, corresponding to the first signal variant S_1 by using the known angle $\Delta\tilde{\varphi}_n$. The reduced projections can be averaged in time as well as the synchronizing signal projections, because at the correct demodulation their mathematical expectations do not depend on the transmitted information symbol.

We designate in terms of X_{0n} and Y_{0n} the received signal projections within the nth information chip

$$X_{0n} = \int_{(n-1)T}^{nT} x_{\text{inf}}(t) \, \sin(\omega t + \varphi_0) \, dt$$

$$Y_{0n} = \int_{(n-1)T}^{nT} x_{\text{inf}}(t) \, \cos(\omega t + \varphi_0) \, dt$$

(3.11)

and in terms of X_{1n} and Y_{1n} the received signal projections reduced to the first signal variant (see Fig. 3.4).

Also let φ_n be the phase of the received nth signal chip relative to the reference oscillation and $\Delta\tilde{\varphi}_n$ be the phase difference between the signal variant, for the benefit of which the decision has been made by the demodulator in the nth chip, and the first signal variant. In the latter case the wavy line shows that $\Delta\tilde{\varphi}_n$ is the estimation resulting from the nth chip demodulation.

From the obvious relations

$$X_{1n} = A \cos(\varphi_n - \Delta\tilde{\varphi}_n) \qquad Y_{1n} = A \sin(\varphi_n - \Delta\tilde{\varphi}_n)$$
$$\cos \varphi_n = X_{0n}/A \qquad\qquad \sin \varphi_n = Y_{0n}/A$$

(3.12)

where A is the amplitude of the received signal, we obtain

$$X_{1n} = X_{0n} \cos \Delta\tilde{\varphi}_n + Y_{0n} \sin \Delta\tilde{\varphi}_n$$
$$Y_{1n} = Y_{0n} \cos \Delta\tilde{\varphi}_n - X_{0n} \sin \Delta\tilde{\varphi}_n \tag{3.13}$$

The obtained reduced projections (3.13) are being averaged further similarly to (3.6) within the interval of N chips:

$$\tilde{X}_1 = \frac{1}{N}\sum_{n=1}^{N} X_{0n} \cos \Delta\tilde{\varphi}_n + Y_{0n} \sin \Delta\tilde{\varphi}_n$$
$$\tilde{Y}_1 = \frac{1}{N}\sum_{n=1}^{N} Y_{0n} \cos \Delta\tilde{\varphi}_n - X_{0n} \sin \Delta\tilde{\varphi}_n \tag{3.14}$$

The further algorithm of signal processing does not differ from that considered before: the estimates of (3.14) are substituted into (3.7), and the obtained values $\tilde{X}_j$ and $\tilde{Y}_j$ are substituted into (3.9); the maximum value of values V_j determines the number of the transmitted signal. The difference of the given algorithm of coherent reception is that the maximum value of V_j obtained, according to (3.10), defines the estimate of phase difference $\Delta\varphi_n$, which is substituted into (3.14) at every next step of accumulation: the so-called "decision feedback."

The demodulator (Fig. 3.5) that realizes algorithms (3.14), (3.7), and (3.9) differs from the demodulator shown in Figure 3.4 first of all by the fact that in this scheme between ADC and Σ_n there is a calculator of the reduced projection (3.13) that contains ROM of the phase differences $\Delta\tilde{\varphi}_n$ similar to ROM of the calculator of projections (3.7), and the choice of the required value $\Delta\tilde{\varphi}_n$ is executed according to the demodulation results (feedback from the scheme output).

Another difference is that in Figure 3.5 the processor calculates the averaged projections not only on receiving the synchronization signal, but on receiving the information signal as well. In this case the synchronization signal is needed only

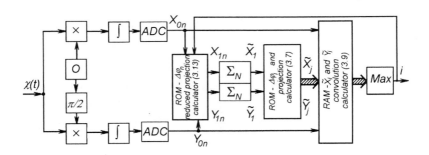

Figure 3.5 Analog–digital coherent demodulator of the PM signals with decision feedback.

for the initial installation of conformity between the receiving signal and signal samples, and then this conformity and the processing coherence are supported automatically by the information signal. However, here also we cannot fully abandon synchronization signal repeating, because in the transmitting process random hops of the signal initial phase are possible, which at absolute phase modulation leads to information distortion (the phenomenon of the so-called "reverse operation").

During projection tuning, according to the information chips the averaging of the reduced projections is fulfilled by N chips nearest to the demodulated element. The averaging interval of N chips in length is shifted to one step at the transition to demodulate every next chip—we average with a "floating window."

The scheme of such an averaging unit, illustrating the operation algorithm of the considered coherent demodulator, is shown in Figure 3.6. A branch of calculation $\tilde{X}_1$ by formula (3.14) is shown in the scheme. Projections X_{0n} and Y_{0n} from the correlator outputs and the values $\sin \Delta \tilde{\varphi}_n$ and $\cos \Delta \tilde{\varphi}_n$, selected from the ROM of the demodulator decision unit, come to the scheme input (see Fig. 3.5). As a result, one component of the sum (3.14) is formed at the adder output. Then the obtained reduced projection comes to the input of register R, consisting of N memory cells (MC) connected in series, in which the reduced projections of the signal received are stored. The content of the cells rewrites from one cell into another for every chip. The accumulator consists of an adder, a subtracting unit, and a memory cell MC_0. When the N-chip synchronization signal comes, the register begins to fill; in this case the nulls comes from the register output to the subtracting unit and consequently the sum of N reduced projections is accumulated in MC_0. Then, when the information signal comes [beginning with the $(N + 1)$th chip], one reduced projection of the previous chip is added to the content of MC_0 on every step and one projection of the chip, considered apart from the given chip by N steps, is subtracted. The tracking for the initial phase of the receiving signal is achieved in such a way.

The choice of the averaging interval, that is, a number of chips N in (3.14), is defined by the system situation: by the signal-to-noise ratio in a channel, by

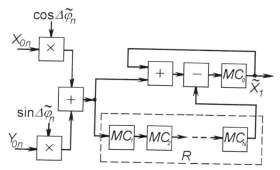

Figure 3.6 Reduced projections averaging unit.

demand for accuracy, by the number of phases, by the speed of phase change in a channel, and by other factors. This parameter is equivalent to the inertia of the carrier recovery unit.

In Chapter 6, when discussing the noise immunity of the given method of quasicoherent processing, we will show that in the AWGN channel it is sufficient to have three to five averaging intervals ($N = 3$–5) to achieve practically the potential noise immunity of the coherent reception in the systems with two- and four-phase PM.

We should emphasize that the approach developed for the synthesis of coherent demodulators leads to optimum schemes for PM signal quasicoherent demodulators, namely, the developed demodulators provide the minimum error probability at every step of demodulation, beginning with the first element. This is explained by the fact that at the first element of the received signal the synthesized algorithms actually realize the optimum incoherent processing, for there is no a priori information about the signal phase at this step. Then the maximum likely estimation with minimum possible dispersion is realized at every other step, which leads to most possible fast approximation to the ideal coherent processing. The proof of this fact is considered in Chapter 6, which discusses the noise immunity of the coherent methods of reception.

The general algorithms of quasicoherent demodulators of arbitrary PM signals were considered earlier. When using the specific systems of signals with PM, the algorithms and schemes of the demodulators are appreciably simplified.

First, let us consider a binary system with phases $\varphi_1 = 0$, $\varphi_2 = \pi$. In this case the estimates of phase differences $\Delta\tilde{\varphi}_n$ in (3.14) can take the values 0 and π, and the sign of $\cos \Delta\tilde{\varphi}_n$ coincides with the binary information symbol received at the demodulator output at the nth chip, that is,

$$\text{sgn} \cos \Delta\tilde{\varphi}_n = J_n \tag{3.15}$$

Therefore, (3.14) is transformed as

$$\tilde{X}_1 = \frac{1}{N}\sum_{n=1}^{N} J_n X_{0n}$$
$$\tilde{Y}_1 = \frac{1}{N}\sum_{n=1}^{N} J_n Y_{0n} \tag{3.16}$$

Similarly, we obtain from (3.9)

$$V_1 = \tilde{X}_1 X_0 + \tilde{Y}_1 Y_0$$
$$V_2 = -\tilde{X}_1 X_0 - \tilde{Y}_1 Y_0 \tag{3.17}$$

that is,

$$J_n = \mathrm{sgn}(\tilde{X}_1 X_{0n} + \tilde{Y}_1 Y_{0n}) \qquad (3.18)$$

Expression (3.18) coincides by its structure with the algorithm for the optimum incoherent reception of binary PDM signals—compare (3.18) with (4.10) in Chapter 4. Moreover, (3.18) transforms into (4.10) if in sums (3.16) there is only one component corresponding to the results of processing the received signal within the previous chip. Therefore, on the one hand, the algorithm for optimum incoherent reception is a special case of the algorithm of quasicoherent processing synthesized here, but, on the other hand, the last algorithm is the development of the first one on the basis of averaging the incoherent processing results.

A very elegant scheme for the coherent demodulator of signals with binary PM that can realize algorithm (3.18) and (3.16) is shown in Figure 3.7. The one-chip delay elements T are introduced into the estimation units for the received projections X_{0n} and Y_{0n} to arrive at the multipliers simultaneously with the received binary symbol J_n corresponding to them. The sign Σ_n designates the adders–accumulators for N chips with a "floating window," realizing algorithms (3.16). The scheme of such an accumulator is illustrated in Figure 3.8. (The description of its operation was given earlier when we discussed the scheme shown in Figure 3.6.)

The coherent demodulator shown in Figure 3.7 can be realized both in analog and in digital forms. The digital realization is preferable if it can be implemented using high-speed digital processors. In some cases the input part of the scheme, including the correlators, is implemented expediently in analog form, and all the rest in digital form.

In the scheme we obtained for a quasicoherent demodulator (see Fig. 3.7), the reference signals are not actually coherent with the receiving signal, and the

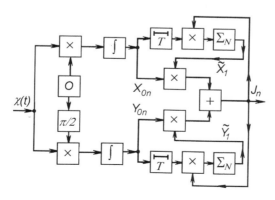

Figure 3.7 Coherent demodulator of the binary PM signals with incoherent reference signals.

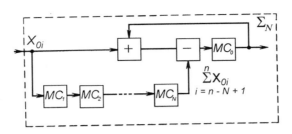

Figure 3.8 Adder–accumulator of the current signal chips.

autonomous reference oscillator generates the reference signal with an arbitrary initial phase. This oscillator should have reasonably high accuracy of setting the frequency, so that within the averaging interval the difference between the signal initial phase and the reference oscillation phase would be changed by a small fraction of the minimum information phase hop.

If an opportunity arises to form the coherent reference oscillation directly with the help of the received synchronization signal, the scheme of the coherent demodulator of binary PM signals takes the form shown in Figure 3.9.

This scheme, following from the general algorithm of (3.1) and (3.2) at $\varphi_1 = 0$ and $\varphi_2 = \pi$, looks considerably more simple than the schemes shown in Figure 3.7. However it should be taken into account that the formation of the coherent oscillation $S_1(t) = \sin(\omega t + \varphi_1)$, coinciding with the signal variant, to which the synchronization signal corresponds, requires an additional unit for selecting the reference oscillation, which is more complex than the given part of the demodulator scheme. That is why in some cases, especially when considering the digital realization, the scheme of Figure 3.7 can appear to be more preferable.

Let us now consider in more detail the coherent reception of four-phase signals with the initial phase variants $\varphi_1 = 0$, $\varphi_2 = \pi/2$, $\varphi_3 = \pi$, and $\varphi_4 = 3\pi/2$ using the keying Gray code [Fig. 3.10(a)].

As follows from the general algorithms of PM signal demodulation (2.71) and (2.72), in this case binary symbols in the subchannels are expressed by the signs of sine and cosine of the phase difference between the received signal $x(t)$ and the reference oscillation $S(t)$, shifted by an angle $\pi/4$ relative to the signal with a zero phase, that is,

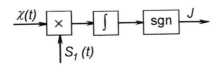

Figure 3.9 Coherent demodulator of binary PM signals with coherent reference oscillation.

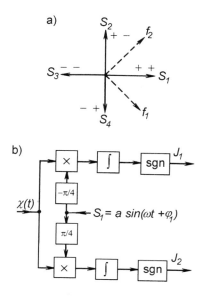

Figure 3.10 Coherent reception of four-phase PM signals: (a) reference oscillations and signals and (b) demodulator block diagram.

$$J_1 = \text{sgn} \int\limits_{0}^{T} x(t) \, \cos(\omega t + \varphi_1 - \pi/4) \, dt$$

$$ (3.19) $$

$$J_2 = \text{sgn} \int\limits_{0}^{T} x(t) \, \sin(\omega t + \varphi_1 - \pi/4) \, dt$$

The appropriate scheme is shown in Figure 3.10(b). In this demodulator, various methods and units for forming coherent reference oscillations can be used, including the methods based on filtering the synchronization signal or transforming the information signal. Note that because we are speaking here of absolute PM, the reference oscillation arriving at the correlator inputs should correspond to the strictly determined and assigned in advance signal variant, for example, the first variant, as shown in Figure 3.10(b).

As was done in the demodulator of Figure 3.7, it is possible to synthesize the scheme of a coherent demodulator of signals with four-phase PM without the coherent reference oscillations. We make up the appropriate algorithm, using the method discussed earlier.

The integrals included in (3.19) can be presented in the following form:

$$\int_0^T x(t) \cos(\omega t + \varphi_1 - \pi/4)\, dt = Y_0 \cos(\varphi_1 - \varphi_0 - \pi/4) - X_0 \sin(\varphi_1 - \varphi_0 - \pi/4)$$

$$\int_0^T x(t) \sin(\omega t + \varphi_1 - \pi/4)\, dt = Y_0 \sin(\varphi_1 - \varphi_0 - \pi/4) + X_0 \cos(\varphi_1 - \varphi_0 - \pi/4)$$

$$(3.20)$$

where X_0 and Y_0 are the projections of the received signal $x(t)$ onto the orthogonal reference oscillations with a random initial phase φ_0, calculated by (3.3). In turn, the sines and cosines included in (3.20) can be expressed by the projection estimates $\tilde{X}_1$ and $\tilde{Y}_1$ of the signal variant with initial phase φ_1 onto the reference oscillations with a phase φ_0 using the following formulas:

$$\cos(\varphi_1 - \varphi_0 - \pi/4) = \frac{1}{\sqrt{2}A}(\tilde{Y}_1 + \tilde{X}_1)$$

$$\sin(\varphi_1 - \varphi_0 - \pi/4) = \frac{1}{\sqrt{2}A}(\tilde{Y}_1 - \tilde{X}_1)$$

$$(3.21)$$

where $A = \sqrt{\tilde{X}_1^2 + \tilde{Y}_1^2}$.

Thus, from (3.21), (3.20), and (3.19), we obtain the following algorithm of demodulation for the four-phase signals, in which binary symbols are expressed by the quadrature projections of the received signal:

$$J_{1n} = \operatorname{sgn}[Y_{0n}(\tilde{Y}_1 + \tilde{X}_1) - X_{0n}(\tilde{Y}_1 - \tilde{X}_1)]$$

$$J_{2n} = \operatorname{sgn}[Y_{0n}(\tilde{Y}_1 - \tilde{X}_1) + X_{0n}(\tilde{Y}_1 + \tilde{X}_1)]$$

$$(3.22)$$

Let us now find expressions for the estimates $\tilde{X}_1$ and $\tilde{Y}_1$ by means of projections X_{0n} and Y_{0n}. For this purpose we first return to formulas (3.13) to obtain current values for the indicated projections X_{1n} and Y_{1n}. Because in this case $\Delta\tilde{\varphi}_n$ is equal to 0, $\pi/2$, π, or $3\pi/2$, then X_{1n} and Y_{1n} take the values specified in columns 2 and 3 of Table 3.1. Columns 4 and 5 specify the values of binary symbols in the first and second subchannels, corresponding to the phases indicated in the first column. From the Table 3.1 the following relations can be derived:

$$X_{1n} = \frac{J_{1n} + J_{2n}}{2}X_{0n} + \frac{J_{1n} - J_{2n}}{2}Y_{0n}$$

$$Y_{1n} = \frac{J_{1n} + J_{2n}}{2}Y_{0n} - \frac{J_{1n} - J_{2n}}{2}X_{0n}$$

$$(3.23)$$

Table 3.1
Reduced Projections in the Four-Phase System

1	2	3	4	5
$\Delta\bar{\varphi}_n$	X_{1n}	Y_{1n}	J_{1n}	J_{2n}
0	X_{0n}	Y_{0n}	$+1$	$+1$
$\pi/2$	Y_{0n}	$-X_{0n}$	$+1$	-1
π	$-X_{0n}$	$-Y_{0n}$	-1	-1
$3\pi/2$	$-Y_{0n}$	X_{0n}	-1	$+1$

Further the estimates $\tilde{X}_1$ and $\tilde{Y}_1$ are formed, as before, by averaging the values (3.23)

$$\tilde{X}_1 = \frac{1}{N}\sum_{n=1}^{N} X_{1n}$$
$$\tilde{Y}_1 = \frac{1}{N}\sum_{n=1}^{N} Y_{1n}$$

(3.24)

Expressions (3.22), (3.23), and (3.24) determine the required algorithm of coherent processing of signals with the four-level PM. The scheme for the appropriate demodulator is shown in Figure 3.11.

The purpose of elements and connections in this scheme is obvious from their comparison with formulas (3.22), (3.23), and (3.24). The adders–accumulators that perform the operation of (3.24) can be realized using the scheme shown in Figure 3.8. The part of the scheme after the correlator outputs is easily realized in digital form.

Let us proceed to the consideration of coherent demodulators of eight-phase signals with the initial phases $\varphi_i = (i - 1)\pi/4$, where $i = 1, \ldots, 8$. The schemes of demodulators in this case essentially depend on how many reference oscillations are used and how they are oriented relative to signal vectors.

If four reference oscillations are generated, coinciding with the four first variants of a signal [Fig. 3.12(a)], the demodulator can be constructed according to algorithm (3.1) and the scheme depicted in Figure 3.1(a). In this case four projections of the received signal onto reference oscillations f_1, f_2, f_3, and f_4 and the same projections with opposite signs come to the comparison unit. According to the maximum of these values, the demodulator determines the number of the transmitted phase and the corresponding three-digit binary combination.

Another variant of the coherent demodulator can be obtained if the reference oscillations are four signals with the phase shift by $-\pi/8$, as has been shown in Figure 3.12(b). In this case the transmitted binary code combinations are completely

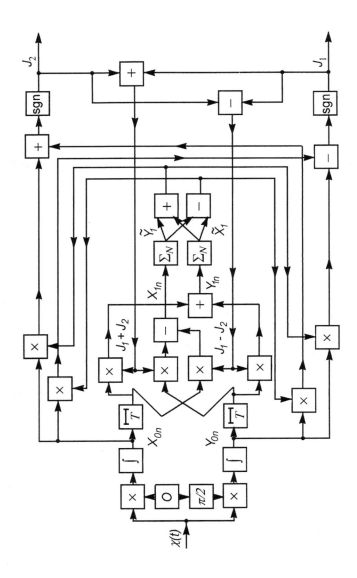

Figure 3.11 Coherent demodulator for four-phase PM signals.

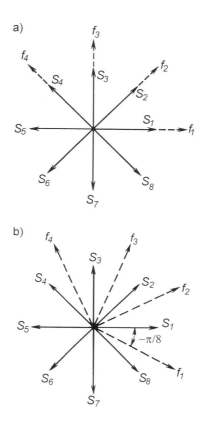

Figure 3.12 Reference signals in the system with eight-level PM.

defined by the signs of the projections of the received signal onto the specified reference oscillations:

$$\text{sgn } X_j = \text{sgn} \int_0^T x(t) f_j(t) \ dt \qquad \text{for } j = 1, 2, 3, 4 \qquad (3.25)$$

In Table 3.2 the projection signs for all of the transmitted phases are submitted in columns 3 through 6. In columns 7, 8, and 9 the binary code combinations corresponding to the keying Gray code are indicated.

The algorithm illustrated by Table 3.2 can be realized directly with the help of a ROM of columns 3 through 9 or by using logic schemes. In the latter case, the table serves as a truth table, describing the operation of logic unit with four binary inputs and three binary outputs. Using the rules of Boolean algebra, it is not difficult to make up the algorithm of such unit at any keying code.

Table 3.2
Projection Signs for Transmitted Signals

1	2	3	4	5	6	7	8	9
i	φ_i	sgn X_1	sgn X_2	sgn X_3	sgn X_4	J_1	J_2	J_3
1	0	+1	+1	+1	−1	+1	+1	+1
2	$\pi/4$	+1	+1	+1	+1	+1	+1	−1
3	$\pi/2$	−1	+1	+1	+1	+1	−1	−1
4	$3\pi/4$	−1	−1	+1	+1	+1	−1	+1
5	π	−1	−1	−1	+1	−1	−1	+1
6	$5\pi/4$	−1	−1	−1	−1	−1	−1	−1
7	$3\pi/2$	+1	−1	−1	−1	−1	+1	−1
8	$7\pi/4$	+1	+1	−1	−1	−1	+1	+1

At the same time, when using the Gray code as shown in Table 3.2, a simple conformity exists between the projection signs and code symbols: The symbols of columns 7 and 8 coincide with the symbols of columns 5 and 3, respectively, and the symbols of column 9 are equal to the product of the corresponding symbols of columns 4 and 6 with the opposite sign. Thus, we obtain:

$$J_1 = \text{sgn } X_3 \qquad J_2 = \text{sgn } X_1 \qquad J_3 = -\text{sgn } X_2 \text{ sgn } X_4 \qquad (3.26)$$

Figure 3.13 shows the structural scheme of a demodulator that realizes the algorithm described in (3.25) and (3.26), with reference oscillations allocated as shown in Figure 3.12(b).

We emphasize that any four oscillations, shifted relative to the received signals by the angle $\pi/8$, can be used as reference signals, and only the output part of the

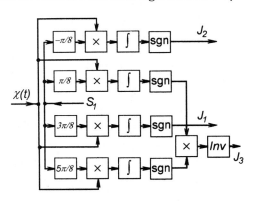

Figure 3.13 Coherent demodulator of eight-phase PM signals.

demodulator will be changed. However after choosing the reference signals, the oscillation, corresponding precisely to a definite received signal, has to come to the input of the phase shifter. The $S_1 = a \sin(\omega t + \varphi_1)$ is such an oscillation when we have the reference oscillations as shown in Figure 3.12(b).

Because in the two-dimensional space all vectors can be presented by a linear combination of two orthogonal vectors, the scheme shown in Figure 3.13 can be transformed into a scheme with only two correlators. Such a scheme for an arbitrary keying code is illustrated in Figure 3.14. The output part of the demodulator is the logic circuit or ROM circuit, the algorithms of which are synthesized by the same truth table as Table 3.2.

A similar (3.26) algorithm for a coherent demodulator for eight-phase signals with the Gray code can be obtained if we use the general algorithms for decoding the multiposition PM signals, as indicated in Section 2.5. These algorithms provide the relationship between binary symbols and the received phase difference.

At absolute phase modulation, the received phase difference is equal to the phase difference of the received signal chip and the reference oscillation, coinciding with the first variant of a signal, the phase of which is taken for zero. Then, as follows from algorithm (2.78), the symbols of the first two subchannels are the signs of the sine and cosine of the received phase φ relative to the reference oscillation with an additional shift by $-\pi/8$ and the symbol of the third subchannel is the product of the sine and cosine signs of the phase φ of the received signal relative to the reference oscillation with an additional shift by $(-\pi/8 - \pi/4)$, that is,

$$J_1 = \operatorname{sgn} \sin(\varphi - \pi/8) = \operatorname{sgn} \int_0^T x(t) \cos(\omega t + \varphi_1 - \pi/8) \, dt$$

$$J_2 = \operatorname{sgn} \cos(\varphi - \pi/8) = \operatorname{sgn} \int_0^T x(t) \sin(\omega t + \varphi_1 - \pi/8) \, dt$$

$$J_3 = \operatorname{sgn}[\sin(\varphi - \pi/8) - \cos(\varphi - \pi/8)] \operatorname{sgn}[\cos(\varphi - \pi/8) + \sin(\varphi - \pi/8)]$$

$$= \operatorname{sgn}\left[\int_0^T x(t) \cos(\omega t + \varphi_1 - \pi/8) \, dt + \int_0^T x(t) \sin(\omega t + \varphi_1 - \pi/8) \, dt\right]$$

$$\times \operatorname{sgn}\left[\int_0^T x(t) \cos(\omega t + \varphi_1 - \pi/8) \, dt - \int_0^T x(t) \sin(\omega t + \varphi_1 - \pi/8) \, dt\right] \qquad (3.27)$$

where φ_1 is the initial phase of the first signal variant. The appropriate (3.27) demodulator is shown in Figure 3.15 and does not require any additional explanation.

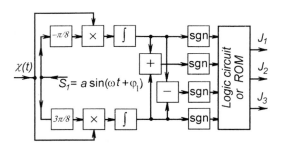

Figure 3.14 Coherent demodulator of eight-position PM signals with two correlators.

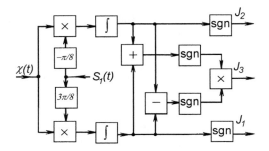

Figure 3.15 Modification of the coherent demodulator of eight-phase PM signals.

Finally, the realization of coherent reception of the eight-position PM signal by reference oscillations with an arbitrary phase is of interest. The integrals included in (3.27) can be presented similarly to (3.20) in the following form:

$$\int_0^T x(t)\,\cos(\omega t + \varphi_1 - \pi/8)\,dt =$$

$$Y_0 \cos(\varphi_1 - \varphi_0 - \pi/8) - X_0 \sin(\varphi_1 - \varphi_0 - \pi/8)$$

$$\int_0^T x(t)\sin(\omega t + \varphi_1 - \pi/8)\,dt =$$

$$Y_0 \sin(\varphi_1 - \varphi_0 - \pi/8) + X_0 \cos(\varphi_1 - \varphi_0 - \pi/8)$$

(3.28)

where X_0 and Y_0 are the projections of the received signal onto the orthogonal reference oscillations with an arbitrary initial phase φ_0 [see (3.3)]. The sines and cosines, included in (3.28), can be expressed by the estimates $\tilde{X}_1$ and $\tilde{Y}_1$ of the first signal variant projections (with the initial phase φ_1) onto the reference oscillations with the phase φ_0 under the following formulas:

$$\cos(\varphi_1 - \varphi_0 - \pi/8) = \frac{1}{A}(\tilde{Y}_1 \sin \pi/8 + \tilde{X}_1 \cos \pi/8)$$

$$\sin(\varphi_1 - \varphi_0 - \pi/8) = \frac{1}{A}(\tilde{Y}_1 \cos \pi/8 - \tilde{X}_1 \sin \pi/8)$$

(3.29)

where $A = \sqrt{\tilde{X}_1^2 + \tilde{Y}_1^2}$.

From (3.28), (3.29), and (3.27) we obtain the following algorithm of demodulation of signals with eight-position PM:

$$J_{1n} = \text{sgn}[\, Y_{0n}(\tilde{Y}_1 \sin \pi/8 + \tilde{X}_1 \cos \pi/8) - X_{0n}(\tilde{Y}_1 \cos \pi/8 - \tilde{X}_1 \sin \pi/8)\,]$$

$$J_{2n} = \text{sgn}[\, Y_{0n}(\tilde{Y}_1 \cos \pi/8 - \tilde{X}_1 \sin \pi/8) + X_{0n}(\tilde{Y}_1 \sin \pi/8 + \tilde{X}_1 \cos \pi/8)\,]$$

$$J_{3n} = \text{sgn}[\, Y_{0n}(\tilde{Y}_1 \cos 3\pi/8 - \tilde{X}_1 \sin 3\pi/8) + X_{0n}(\tilde{Y}_1 \sin 3\pi/8 + \tilde{X}_1 \cos 3\pi/8)\,]$$
$$\times \text{sgn}[\, Y_{0n}(\tilde{Y}_1 \sin 3\pi/8 + \tilde{X}_1 \cos 3\pi/8) - X_{0n}(\tilde{Y}_1 \cos 3\pi/8 - \tilde{X}_1 \sin 3\pi/8)\,]$$

(3.30)

Remember that $\tilde{X}_1$ and $\tilde{Y}_1$ are the estimates of projections of the first signal variant with the phase φ_1 onto the reference oscillations with an arbitrary phase, which are initially obtained by averaging the projections of the received synchronization signal, and then by averaging the reduced projections of the information signal.

The current values X_{1n} and Y_{1n} of the reduced projections are defined according to (3.13) by the current values of the projections X_{0n} and Y_{0n}. Table 3.3 shows the values for X_{1n} and Y_{1n} calculated by (3.13) for all eight possible signal phases. The last three columns specify the binary decisions made on the nth chip in three subchannels of the demodulator. These three-digit binary decisions define the values of the reduced projections on the nth chip by the projections of the received signal onto the reference oscillations with an arbitrary phase φ_0. After that the reduced projections are averaged within the interval of N information chips [see (3.24)], and the estimates $\tilde{X}_1$ and $\tilde{Y}_1$, obtained as a result of this averaging, are substituted into (3.30).

Thus, relations (3.30) together with Table 3.3 and formula (3.24) form the complete algorithm of coherent reception of the eight-phase signals. In this case we abstain from presenting the graphic representation of the scheme for the appropriate demodulator because it is comparatively bulky and adds little to the indicated algorithm. Its general structure corresponds to that of Figure 3.5.

The indicated algorithms and schemes for coherent demodulators can be used both as independent units in systems with PM and as components of algorithms

Table 3.3
Reduced Projections in the Eight-Phase System

1	2	1	2	3	4	5
i	$\Delta\tilde{\varphi}_n$	X_{1n}	Y_{1n}	J_{1n}	J_{2n}	J_{3n}
1	0	X_{0n}	Y_{0n}	+1	+1	+1
2	$\pi/4$	$\frac{1}{\sqrt{2}}(X_{0n}+Y_{0n})$	$\frac{1}{\sqrt{2}}(Y_{0n}-X_{0n})$	+1	+1	-1
3	$\pi/2$	Y_{0n}	$-X_{0n}$	+1	-1	-1
4	$3\pi/4$	$-\frac{1}{\sqrt{2}}(X_{0n}-Y_{0n})$	$-\frac{1}{\sqrt{2}}(X_{0n}+Y_{0n})$	+1	-1	+1
5	π	$-X_{0n}$	$-Y_{0n}$	-1	-1	+1
6	$5\pi/4$	$-\frac{1}{\sqrt{2}}(Y_{0n}+X_{0n})$	$-\frac{1}{\sqrt{2}}(Y_{0n}-X_{0n})$	-1	-1	-1
7	$3\pi/2$	$-Y_{0n}$	X_{0n}	-1	+1	-1
8	$7\pi/4$	$\frac{1}{\sqrt{2}}(X_{0n}-Y_{0n})$	$\frac{1}{\sqrt{2}}(X_{0n}+Y_{0n})$	-1	+1	+1

and schemes of coherent demodulators in systems with PDM, which are considered in the next section.

3.3 COHERENT DEMODULATORS OF FIRST-ORDER PDM SIGNALS

One of the features of synthesizing the algorithms of PDM-1 signal reception is that the interval of processing, necessary for decision making, is equal to the duration of two signal elements, although the decision itself is made for every element. Hence, according to the theory for optimum element-by-element reception, the algorithm of processing should cover two signal chips. However, as a rule, the optimum algorithm can be presented by the results of signal processing on every chip separately.

During the coherent reception of signals with PDM-1 it is possible to make a stronger statement: The two-chip optimum decision can be presented by a combination of two separate one-chip optimum decisions.

Let us consider this statement more thoroughly and prove it for binary PDM-1 signals with phase differences $\Delta\varphi_1 = 0$ and $\Delta\varphi_2 = \pi$.

In this case the signal within the interval of every chip can take one of two values $S_1(t)$ and $S_2(t)$, which differs by the sign $S_1(t) = -S_2(t)$, and within the interval of two chips can take one of four values, representing the combination of signals S_1 and S_2 (Fig. 3.16).

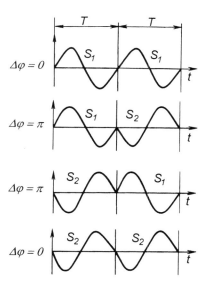

Figure 3.16 Signal variants at binary PDM-1.

Assume that signals $x_{n-1}(t)$ and $x_n(t)$ are received within intervals of the $(n-1)$th and nth chips. Then the optimum algorithm of signal processing includes the calculation of four convolutions within the interval of two chips:

$$
\begin{aligned}
V_{1n} &= X_{n-1,1} + X_{n,1} \\
V_{2n} &= X_{n-1,1} + X_{n,2} = X_{n-1,1} - X_{n,1} \\
V_{3n} &= X_{n-1,2} + X_{n,1} = -X_{n-1,1} + X_{n,1} \\
V_{4n} &= X_{n-1,2} + X_{n,2} = -X_{n-1,1} - X_{n,1}
\end{aligned}
\tag{3.31}
$$

where

$$
X_{n-1,i} = \int_{(n-1)T}^{nT} x(t)\, S_i(t)\ dt
\tag{3.32}
$$

$$
X_{n,i} = \int_{nT}^{(n+1)T} x(t)\, S_i(t),\ dt \qquad \text{for } i = 1,\ 2
$$

If one of the values V_{1n} or V_{4n} is maximum, the received phase difference $\Delta\varphi_n = 0$; if V_{2n} or V_{3n} appears to be maximum, then $\Delta\varphi_n = \pi$.

The values V_{1n} and V_{4n}, as well as V_{2n} and V_{3n}, differ only by a sign. That is why the comparison V_{1n} or V_{4n} with V_{2n} and V_{3n}, which determines in which of these couples there is a maximum value, can be replaced by the comparison of the absolute values V_{1n} and V_{3n}. Hence, the indicated algorithm can be formulated as follows: The received phase difference between the $(n-1)$th and nth chips is equal to $\Delta\varphi_n = 0$ if

$$|V_{1n}| > |V_{3n}| \tag{3.33}$$

Inequality (3.33) will not change if modules are replaced by squares, therefore from (3.33), and taking into account (3.31) and (3.32), we obtain

$$\left[\int_{(n-1)T}^{nT} x(t)S_1(t)\,dt + \int_{nT}^{(n+1)T} x(t)S_1(t)\,dt \right]^2$$

$$> \left[\int_{nT}^{(n+1)T} x(t)S_1(t)\,dt - \int_{(n-1)T}^{nT} x(t)S_1(t)\,dt \right]^2$$

or after squaring and reductions

$$\int_{(n-1)T}^{nT} x(t)S_1(t)\,dt \int_{nT}^{(n+1)T} x(t)S_1(t)\,dt > -\int_{(n-1)T}^{nT} x(t)S_1(t)\,dt \int_{nT}^{(n+1)T} x(t)S_1(t)\,dt \tag{3.34}$$

If we identify the phase difference $\Delta\varphi_n = 0$ with the binary symbol $J_n = +1$, and $\Delta\varphi = \pi$ with $J_n = -1$, the obtained optimum algorithm of processing the binary PDM signals can be represented as follows:

$$J_n = \text{sgn}\left[\int_{(n-1)T}^{nT} x(t)S_1(t)\,dt \int_{nT}^{(n+1)T} x(t)S_1(t)\,dt \right]$$

$$= \text{sgn}\int_{(n-1)T}^{nT} x(t)S_1(t)\,dt \times \text{sgn}\int_{nT}^{(n+1)T} x(t)S_1(t)\,dt \tag{3.35}$$

We can get the same result if we compare the values of (3.31) among themselves and consider the situations corresponding to the decisions made for the benefit of one or another phase difference.

If phase difference $\Delta\varphi_n = 0$ is received, the maximum convolution is equal to V_{1n} or V_{4n}. In the first case $X_{n,1} > -X_{n,1}$ comes from $V_{1n} > V_{2n}$, that is, $X_{n,1} > 0$; and $X_{n-1,1} > -X_{n-1,1}$ comes from $V_{1n} > V_{3n}$, that is, $X_{n-1,1} > 0$.

In the second case $-X_{n-1,1} > X_{n-1,1}$ comes from $V_{4n} > V_{2n}$, that is, $X_{n-1,1} < 0$; and $-X_{n,1} > X_{n,1}$ comes from $V_{4n} > V_{3n}$, that is, $X_{n,1} < 0$. Thus, the adjacent chip projections of the received signal onto the reference oscillations have identical signs if the phase difference between two chips $\Delta\varphi_n = 0$.

If the phase difference $\Delta\varphi_n = \pi$ is received, the maximum convolution is equal to V_{2n} or V_{3n}. In the first case $X_{n-1,1} > -X_{n-1,1}$ comes from $V_{2n} > V_{4n}$, that is, $X_{n-1,1} > 0$; and $-X_{n,1} > X_{n,1}$ comes from $V_{2n} > V_{1n}$, that is, $X_{n,1} < 0$. In the second case $-X_{n-1,1} > X_{n-1,1}$ comes from $V_{3n} > V_{1n}$, that is, $X_{n-1,1} < 0$; and $X_{n,1} > -X_{n,1}$ comes from $V_{3n} > V_{4n}$, that is, $X_{n,1} > 0$.

Thus, if the phase difference $\Delta\varphi_n = \pi$ is received as the result of optimum processing of two signal chips, the adjacent chip projections onto the reference oscillation have different signs. An important conclusion follows from this: To determine the transmitted binary symbol in the PDM-1 system with the minimum error probability, it is sufficient to multiply the signs of the projections of the adjacent signal chips onto the reference oscillation, which is equal to one of the signal variants [5]. The corresponding algorithm is

$$J_n = \mathrm{sgn}\ X_{n-1} \times \mathrm{sgn}\ X_n = \mathrm{sgn}\left[\int_{(n-1)T}^{nT} x(t)\ \sin(\omega t + \varphi_{1,2})\ dt \right]$$

$$\times \mathrm{sgn}\left[\int_{nT}^{(n+1)T} x(t)\ \sin(\omega t + \varphi_{1,2})\ dt \right] \qquad (3.36)$$

Here by the designation $\varphi_{1,2}$ we can emphasize that the initial phase of the reference oscillation can coincide with the initial phase of any of two signal variants—this is an important property of PDM.

The appropriate (3.36) scheme of the demodulator is shown in Figure 3.17. Externally it differs from the elementary scheme of a coherent demodulator of binary PM signals (see Fig. 3.9) only in the differential decoder, which consists of a delay unit and binary (sign) symbols multiplier. The essential difference is that in the scheme of Figure 3.17 the reference oscillation $S(t) = a \sin(\omega t + \varphi_{1,2})$,

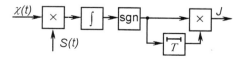

Figure 3.17 Coherent demodulator of binary PDM signals.

coming to the second input of the correlator, can coincide with any of two signal variants—this is the consequence of difference modulation application.

If when discussing the schemes in the previous section it was asserted that the reference oscillation should coincide with one signal variant chosen beforehand and with no other one, we emphasize here that any signal variant not determined beforehand can act as a reference oscillation.

Therefore to maintain the operation of the scheme shown in Figure 3.17 there is no need to have a special synchronization signal, by which we can establish a unique conformity between reference oscillations and transmitted signal variants, and the reference oscillation selector can operate by means of an information signal. This remark concerns all demodulators considered in the given section.

According to the algorithm obtained earlier, the common scheme of a coherent demodulator of binary PDM signals consists of the coherent demodulator of binary PM signals and differential decoder (Fig. 3.18). The coherent demodulator shown in Figure 3.7, which is conveniently realized in a digital form, can be used in this scheme.

For analog realization, the elementary scheme of Fig. 3.19(a), based on an open passive scheme for the ROS, is frequently used [6]. The selector includes a signal module calculator (i.e., rectifier) that performs the operation of removing the keying, a narrowband filter for the double frequency of a carrier 2ω, a frequency divider by two, and phase shifter for compensating the phase shift in the inertial circuit. This demodulator permits us to approximate a potential noise immunity if we manage to provide reasonably narrowband filtering in the ROS. The passband of this unit should be at least an order less than the information signal bandwidth.

Other approaches to implementation of the coherent reception of the PDM signals are also used. In the scheme of the demodulator shown in Figure 3.19(b), the ROS is based on a typical closed active scheme with a controlled oscillator and the signal remodulation occurs by means of multiplying the quadrature signal components [7]. Here the reference oscillation is generated by the controlled oscillator (CO), which with two correlators, a multiplier and a filter (F) form an automatic frequency control (AFC) unit. The oscillator is controlled through the filter of the AFC unit by a signal equal to the product of the received signal projections onto the orthogonal reference oscillations. This control signal is proportional to a sine of a double phase difference between the received signal and the reference oscillation.

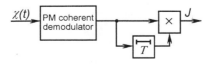

Figure 3.18 General structure of the coherent demodulator of binary PDM signals.

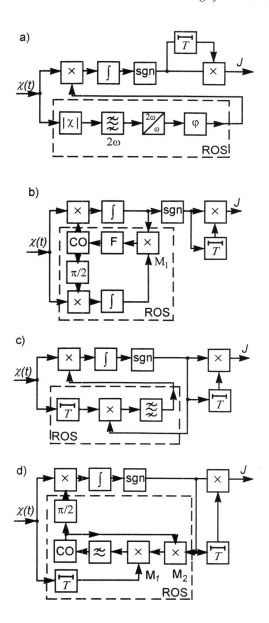

Figure 3.19 Coherent demodulators of binary PDM signals with various ROSs.

In the scheme shown in Figure 3.19(c), the ROS is a closed passive scheme with direct remodulation by means of reduction of the signal phase using the binary decision. The received signal is delayed and then multiplied by the result of its demodulation. At the correct demodulation it results in the removal of keying and the restoration of the carrier. The same requirements used in the Figure 3.19(a) scheme are presented here to a bandpass filter; however, the filtering is executed at a frequency that is two times below. A certain complication is associated with the introduction of a received signal delay line, however it can be used in parallel in a clock synchronization unit.

In the scheme shown in Figure 3.19(d), the ROS uses PM signal regeneration on the basis of the current symbol decision, but the ROS is an active closed AFC scheme with a controlled oscillator. In multiplier M_2 the keying of the CO output signal by the binary decision is performed, and in M_1 the received phase-modulated oscillation is compared with the received signal delayed by one chip. Then the signal from the AFC output controls the CO frequency, so the last one forms the oscillation, shifted by $\pi/2$ relative to the received signal; this shift is eliminated by a phase shifter. The defect in the scheme is a spurious frequency capture effect—a problem inherent to some degree in all ROSs with AFC.

The variety of methods and schemes for selecting the reference oscillation from the information signal (the main approaches to their synthesis are considered in Section 3.6) permits us to choose suitable variant depending on the system situation. Only in the case of the open passive scheme [see Fig. 3.19(a)] can ROS be fully selected as the independent unit. In the other cases some particular elements or connections of the main demodulation path are used in ROS.

Now let us consider coherent demodulators of signals with the first-order four-position PDM. Four-position PDM is used rather often in communications systems because it permits either a doubling of the transmission rate in comparison with two-position PDM in the same frequency band at relatively small loss of noise immunity, or it provides the same rate and the same noise immunity, as two-position PDM in a channel with a two times smaller frequency band. In multibeam channels the four-position PDM has additional advantages because the chip duration is twice as long.

Like the PDM demodulator of two-phase signals, the coherent demodulator of signals with four-phase PDM consists of a coherent demodulator of signals with four-phase PM and a decoder (Fig. 3.20).

To the decoder input from the PM signal coherent demodulator output and from the delay element output, we obtain discrete sign functions of the received signal projections onto the orthogonal reference oscillations by two adjacent chips: X_{n-1}, X_n, Y_{n-1}, and Y_n. These four binary numbers uniquely define one of four phase differences transmitted by a pair of chips. The decoder has to determine this difference and make it conform to a pair of binary symbols defined by the keying code. The algorithm of the decoder operation is illustrated later for the Gray code.

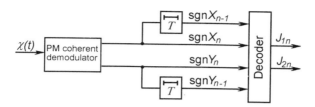

Figure 3.20 General scheme of the PDM coherent demodulator of four-phase signals.

Here and later we assume that phase differences 0, $\pi/2$, π, and $3\pi/2$ are applied in the four-phase PDM systems. Thus the signal within every chip has four initial phases: two orthogonal signals and two counterphase to them. In other cases, when a phase difference other than the one indicated is used, the number of signal variants within a chip can exceed the number of permitted phase differences. For example, when using phase differences $\pi/4$, $3\pi/4$, $5\pi/4$, and $7\pi/4$, the number of signal variants within a chip is equal to eight, because the initial signal phases can take (relative to the reference oscillation) the following values: 0, $\pi/4$, $\pi/2$, $3\pi/4$, π, $5\pi/4$, $3\pi/2$, and $7\pi/4$. Certainly, in this case also the demodulator can calculate only the received signal projections onto two mutually orthogonal reference oscillations; however, the remainder of the scheme becomes more complicated. In connection with its use in incoherent processing it is preferable to use signals with phase differences of 0, $\pi/2$, π, and $3\pi/2$.

The PM signal demodulator in Figure 3.20 can be implemented, for example, by the scheme indicated in Figure 3.11, in which the coherent processing of a PM signal is based on calculation of the signal quadrature projections on the reference oscillations with an arbitrary initial phase.

It is also possible to realize the coherent PM demodulator in the scheme of Figure 3.20 using ROS units. Figures 3.21 through 3.24 show some of the most popular variants for implementing the appropriate schemes.

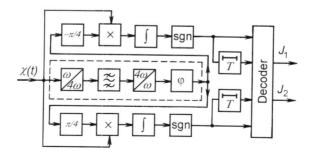

Figure 3.21 Coherent demodulator of four-level PDM signals with frequency multiplication.

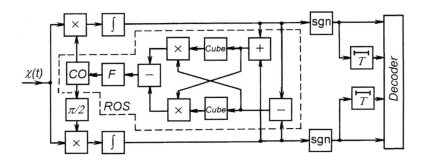

Figure 3.22 Coherent demodulator of four-level PDM signals with phase multiplication.

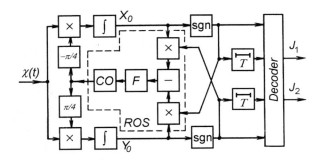

Figure 3.23 Coherent demodulator of four-level PDM signals with phase reduction.

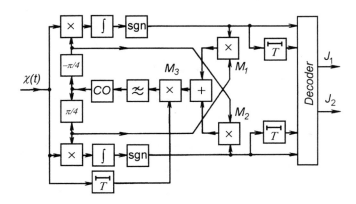

Figure 3.24 Coherent demodulator of four-level PDM signals with signal regeneration.

In the scheme of Figure 3.21, the ROS is constructed according to the open passive scheme with remodulation by multiplying the frequency of the received signal by four. The nonlinear element, containing the component of the fourth degree, or two quadrators connected in series can be used as a frequency multiplier by four. The difficulties arising in this ROS scheme during filtering and dividing of the four multiplied carrier frequency can be removed in some cases by downconverting the signal spectrum or by using the AFC system after the frequency multiplier.

The remodulation of a four-phase signal by multiplying its frequency and phase can be realized not only directly by the received signal, but also by its quadrature components (Fig. 3.22). This scheme uses a closed ROS with an oscillator, which is controlled by the circuit multiplying the phase by four on quadrature components of the received signal. The corresponding common algorithms are indicated in Section 3.6.

As can be seen from the scheme in this ROS the sum and difference of quadrature projections of a signal are raised to the third power (cube); then the first and third powers of these values are multiplied. The product difference, coming after filtering to the CO controlling input, is proportional to the sine of the quadruple phase difference between the received signal and reference oscillation. In the scheme considered, an oscillation is generated that is shifted by $\pi/4$ relative to the variants of the received signal without an additional phase shifter. The scheme in Figure 3.22 (as well as the one in Figure 3.21 at the correct parameter choice) provides a noise immunity that is rather close to the maximum possible noise immunity.

In the scheme of a demodulator of four-phase PDM signals, as shown in Figure 3.23, the ROS is implemented in the same way as in Figure 3.22, under a closed AFC circuit with a controlled oscillator; however, the remodulation of a signal is carried out here according to the phase reduction method.

The phase reduction is reached by multiplication of the quadrature projections by the binary symbols, coming to the decoder input. The difference of the obtained products is the controlling signal for CO. The appropriate algorithm, indicated in Section 3.6, is obtained for the case, when the received signal is projected onto the reference oscillations, shifted by $\pi/4$ relative to its possible variants, that entails the inclusion of two phase shifters. This scheme is more simple than the one indicated in Figure 3.22, but concedes to it a little on the noise immunity.

In the following demodulator scheme (Fig. 3.24), the ROS is implemented according to the closed circuit with a controlled oscillator and regeneration of the input PM signal.

The received signal $x(t)$ after being delayed by T is compared by the phase with the regenerated PM signal in the phase detector, consisting of a multiplier M_3 and LPF; the LPF output signal controls the frequency of CO. The regenerated PM signal is formed in that part of the scheme, which consists of multipliers M_1 and M_2 and an adder, by modulation of two orthogonal reference oscillations, received from the phase shifters outputs, by binary signals from the outputs of the PM signal demodulator. The scheme has practically the same noise immunity as the scheme of Figure 3.23, and both are close to optimum receivers with good signal-to-noise ratios.

A choice of one of the considered modified schemes for the quasicoherent PDM demodulator depends exclusively on realization reasons, because their noise immunity performances are close to the best achievable.

Let us now return to the common scheme of a coherent demodulator (see Fig. 3.20) and deduce algorithms for a decoder. As has been already noted, the decoder receives the signs of the signal projections onto the orthogonal reference oscillations, shifted relative to the signal variants by $\pi/4$. At all described methods of selecting the coherent reference oscillation there exists an initial phase uncertainty of the fourth order; that is, a pair of orthogonal reference oscillations f_1 and f_2 can take one of four positions relative to the set of four transmitted signal variants S_1, S_2, S_3, and S_4 (Fig. 3.25), and neither a priori nor a posteriori is it possible to define just which one, without additional information.

Therefore, unlike the reception of signals with the absolute PM at the presence of a synchronization signal, every one of the transmitted signals can correspond to four sign combinations of the projections X and Y, however at PDM it does not lead to the uncertainty.

It is not difficult to create the table of conformity between the sign combinations at the decoder input and the transmitted phase differences (Table 3.4): If the signs of both projections within two chips have not been changed, it is transmitted $\Delta\varphi = 0$; if both projections have had their signs changed, then $\Delta\varphi = \pi$; if the sign of one of two projections has been changed, the $\Delta\varphi = \pi/2$ or $\Delta\varphi = 3\pi/2$.

The transmitted phase differences are indicated in column 1 of Table 3.4. All 16 projection sign combinations at the decoder input are given in columns 2 through 5, and binary symbols at the decoder output corresponding to the Gray code are given in columns 6 and 7. Thus, the decoder in the demodulators schemes (see Figs. 3.20 through 3.24) can be realized as ROM, in which Table 3.4 is written.

If it is preferable to realize the decoder by discrete logic elements, then according to Table 3.4 we can make up an algorithm of a logic circuit with four inputs and two outputs. Using the rules of logic addition $\vee$, multiplication $\wedge$,

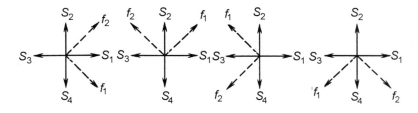

Figure 3.25 Probable positions of reference oscillations in four-phase systems.

Table 3.4
Decoding the Four-Phase Signal

1	2	3	4	5	6	7
$\Delta\varphi$	a sgn X_{n-1}	b sgn X_n	c sgn Y_{n-1}	d sgn Y_n	J_{1n}	J_{2n}
0	+	+	+	+	+	+
0	−	−	+	+	+	+
0	−	−	−	−	+	+
0	+	+	−	−	+	+
$\pi/2$	−	−	+	−	+	−
$\pi/2$	+	−	+	+	+	−
$\pi/2$	−	+	−	−	+	−
$\pi/2$	+	+	−	+	+	−
π	+	−	+	−	−	−
π	−	+	+	−	−	−
π	−	+	−	+	−	−
π	+	−	−	+	−	−
$-\pi/2$	+	+	+	−	−	+
$-\pi/2$	−	+	+	+	−	+
$-\pi/2$	−	−	−	+	−	+
$-\pi/2$	+	−	−	−	−	+

and inverting (designated by a line above the appropriate symbol), we obtain the following logic expressions:

$$J_{1n} = (\overline{a} \wedge \overline{b} \wedge \overline{c} \wedge \overline{d}) \vee (\overline{a} \wedge \overline{b} \wedge c \wedge \overline{d}) \vee (\overline{a} \wedge \overline{b} \wedge c \wedge d)$$
$$\vee (\overline{a} \wedge b \wedge \overline{c} \wedge \overline{d}) \vee (a \wedge \overline{b} \wedge c \wedge d) \vee (a \wedge b \wedge \overline{c} \wedge \overline{d})$$
$$\vee (a \wedge b \wedge \overline{c} \wedge d) \vee (a \wedge b \wedge c \wedge d)$$
$$J_{2n} = (\overline{a} \wedge \overline{b} \wedge \overline{c} \wedge \overline{d}) \vee (\overline{a} \wedge \overline{b} \wedge \overline{c} \wedge d) \vee (\overline{a} \wedge \overline{b} \wedge c \wedge \overline{d})$$
$$\vee (\overline{a} \wedge b \wedge c \wedge d) \vee (a \wedge \overline{b} \wedge \overline{c} \wedge \overline{d}) \vee (a \wedge b \wedge \overline{c} \wedge \overline{d})$$
$$\vee (a \wedge b \wedge c \wedge \overline{d}) \vee (a \wedge b \wedge c \wedge d)$$

Having applied the rules of logic algebra, we can simplify these expressions:

$$J_{1n} = (\overline{a} \wedge \overline{b} \wedge \overline{d}) \wedge (\overline{b} \wedge c \wedge d) \vee (b \wedge \overline{c} \wedge d) \vee (a \wedge b \wedge d) \quad (3.37\text{a})$$

$$J_{2n} = (\overline{a} \wedge \overline{b} \wedge \overline{c}) \vee (\overline{a} \wedge c \wedge d) \vee (a \wedge \overline{c} \wedge \overline{d}) \vee (a \wedge b \wedge c) \quad (3.37\text{b})$$

The decoder corresponding to algorithm (3.37) and consisting of logic elements AND, OR, NO is shown in Figure 3.26.

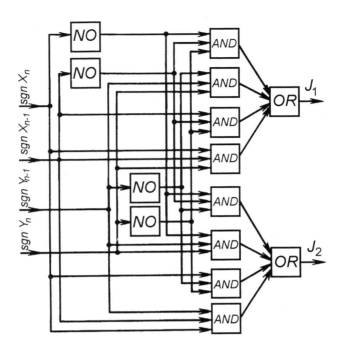

Figure 3.26 Decoder of signals with four-level PDM.

It is possible to form a decoder scheme similar to the point, but a little different by the form, using a general algorithm of decoding the signals with PDM and the keying Gray code [see (2.78)]. For phase differences 0, $\pi/2$, π, and $3\pi/2$ this algorithm is

$$J_{1n} = \operatorname{sgn}(\cos \Delta_n\varphi_\xi + \sin \Delta_n\varphi_\xi)$$
$$J_{2n} = \operatorname{sgn}(\cos \Delta_n\varphi_\xi - \sin \Delta_n\varphi_\xi)$$

(3.38)

where $\Delta_n\varphi_\xi$ is the phase difference of the $(n-1)$th and nth received signal chips. Let us express the cosine and sine, included in (3.38), by the projections of the received signal $x(t)$ onto the orthogonal reference oscillations f_1 and f_2 according to the formulas

$$\cos \Delta_n\varphi_\xi = A(X_nX_{n-1} + Y_nY_{n-1})$$
$$\sin \Delta_n\varphi_\xi = A(Y_nX_{n-1} - X_nY_{n-1})$$

where A is the proportionality factor

$$X_k = \int\limits_{kT}^{(k+1)T} x(t) f_1(t) \, dt$$

$$Y_k = \int\limits_{kT}^{(k+1)T} x(t) f_2(t) \, dt$$

(3.39)

As a result we get

$$J_{1n} = \operatorname{sgn}(X_n X_{n-1} + Y_n Y_{n-1} + Y_n X_{n-1} - X_n Y_{n-1})$$
$$J_{2n} = \operatorname{sgn}(X_n X_{n-1} + Y_n Y_{n-1} - Y_n X_{n-1} + X_n Y_{n-1})$$

If the reference oscillations f_1 and f_2 are chosen with a shift by $\pi/4$ relative to the signal variants (Fig. 3.25), the signs of projections (3.39) coincide with the binary decisions at the PM coherent demodulator output, which is why, having changed the projection values by their signs, we obtain

$$J_{1n} = \operatorname{sgn}[\operatorname{sgn} X_n \operatorname{sgn} X_{n-1} + \operatorname{sgn} Y_n \operatorname{sgn} Y_{n-1} + \operatorname{sgn} Y_n \operatorname{sgn} X_{n-1} - \operatorname{sgn} X_n \operatorname{sgn} Y_{n-1}]$$
$$J_{2n} = \operatorname{sgn}[\operatorname{sgn} X_n \operatorname{sgn} X_{n-1} + \operatorname{sgn} Y_n \operatorname{sgn} Y_{n-1} - \operatorname{sgn} Y_n \operatorname{sgn} X_{n-1} + \operatorname{sgn} X_n \operatorname{sgn} Y_{n-1}]$$

(3.40)

Expressions (3.40) in combination with (3.39) represent the algorithm of four-phase PDM signal coherent reception. It is easy to determine that the table of conformity, Table 3.4, follows from (3.40), and the algorithm of coherent reception (3.37) is equivalent to (3.40). The decoder, realizing (3.40), is shown in Figure 3.27.

Now let us consider the coherent reception of signals with eight-position PDM. In this case in the systems with coherent reception, as a rule, signals with phase differences of 0, $\pi/2$, $3\pi/4$, π, $5\pi/4$, $3\pi/2$, and $7\pi/4$ are used. Thus the variants of an initial phase within every chip also accept eight values relative to the reference oscillation.

The general algorithm for processing eight-position signals with PDM consists of the following. At first in the demodulator, just as in a coherent demodulator for eight-position signals with PM, we calculate the received signal projections onto four reference oscillations, shifted relative to each other by $\pi/4$ and shifted relative to the corresponding signal variants by $\pi/8$ [see Fig. 3.12(b)].

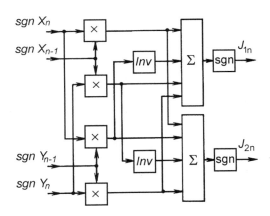

Figure 3.27 Modification of the decoder of four-phase PDM signals.

In this situation, there are two possible approaches. In the first one, the demodulator contains four reference oscillations, and the projections mentioned earlier are calculated with the help of four correlators. In the second approach, the demodulator contains two orthogonal reference oscillations (one of two pairs of the four original oscillations shifted by $\pi/2$ can form these reference oscillations), and with the help of two correlators, as in demodulators of the four-phase PDM, two quadrature projections are determined. Then the remaining two projections are calculated using linear conversion.

After that the demodulator determines the signs of all four projections, which is equivalent to the absolute phase decision. The signs of the given chip projections and the signs of the previous chip projections (totaling eight binary numbers) come to the decoder input, where we can determine the transmitted phase difference or, according to the used keying code, the transmitted three-digit binary combination.

The selection of coherent reference oscillations is executed using the same methods used for demodulators of the two-phase and four-phase signals; however, the appropriate carrier recovery units are considerably complicated.

First let us consider in more detail the algorithm for receiving the eight-phase PDM signals, based on averaging the converted received signal projections onto reference oscillations with an arbitrary initial phase, similar to algorithm (3.30) for receiving the eight-phase PM signals.

Thus, let X_0 and Y_0 be the received signal projections onto orthogonal reference oscillations with an arbitrary initial phase φ_0, determined by (3.3). We express by the X_0 and Y_0 the required projections X_1', X_2', X_3', and X_4' of the received signal $x(t)$ onto four reference oscillations f_1, f_2, f_3, and f_4 [see Fig. 3.12(b)] with the phases, correspondingly equal to $\varphi_1 - \pi/8$, $\varphi_1 + \pi/8$, $\varphi_1 + 3\pi/8$, and $\varphi_1 + 5\pi/8$, where φ_1 is the phase of a signal, conventionally taken as the first (starting) variant. We obtain

$$X_1' = \int_0^T x(t) \, \sin(\omega t + \varphi_1 - \pi/8) \, dt$$

$$= Y_0 \sin(\varphi_1 - \varphi_0 - \pi/8) + X_0 \cos(\varphi_1 - \varphi_0 - \pi/8)$$

$$X_2' = \int_0^T x(t) \, \sin(\omega t + \varphi_1 + \pi/8) \, dt$$

$$= Y_0 \sin(\varphi_1 - \varphi_0 + \pi/8) + X_0 \cos(\varphi_1 - \varphi_0 + \pi/8)$$

$$X_3' = \int_0^T x(t) \, \sin(\omega t + \varphi_1 + 3\pi/8) \, dt$$

$$= Y_0 \cos(\varphi_1 - \varphi_0 - \pi/8) - X_0 \sin(\varphi_1 - \varphi_0 - \pi/8)$$

$$X_4' = \int_0^T x(t) \, \sin(\omega t + \varphi_1 + 5\pi/8) \, dt$$

$$= Y_0 \cos(\varphi_1 - \varphi_0 + \pi/8) - X_0 \sin(\varphi_1 - \varphi_0 + \pi/8) \tag{3.41}$$

The sines and cosines included in (3.41) can be expressed by the estimates $\tilde{X}_1$ and $\tilde{Y}_1$ of the projections of the signal with the initial phase φ_1 onto the reference oscillation with the initial phase φ_0 under the following formulas:

$$\cos(\varphi_1 - \varphi_0 - \pi/8) = \frac{1}{A}(\tilde{Y}_1 \sin \pi/8 + \tilde{X}_1 \cos \pi/8)$$

$$\sin(\varphi_1 - \varphi_0 - \pi/8) = \frac{1}{A}(\tilde{Y}_1 \cos \pi/8 - \tilde{X}_1 \sin \pi/8)$$

$$\cos(\varphi_1 - \varphi_0 + \pi/8) = \frac{1}{A}(\tilde{X}_1 \cos \pi/8 - \tilde{Y}_1 \sin \pi/8)$$

$$\sin(\varphi_1 - \varphi_0 + \pi/8) = \frac{1}{A}(\tilde{X}_1 \sin \pi/8 + \tilde{Y}_1 \cos \pi/8) \tag{3.42}$$

where $A = \sqrt{\tilde{X}_1^2 + \tilde{Y}_1^2}$.

Thus, we obtain the following algorithms for calculating the received signal projections onto the reference oscillations, shifted relative to the signal by the angle $\pi/8$:

$$X'_1 = Y_0(\tilde{Y}_1 \cos \pi/8 - \tilde{X}_1 \sin \pi/8) + X_0(\tilde{Y}_1 \sin \pi/8 + \tilde{X}_1 \cos \pi/8)$$

$$X'_2 = Y_0(\tilde{X}_1 \sin \pi/8 + \tilde{Y}_1 \cos \pi/8) + X_0(\tilde{X}_1 \cos \pi/8 - \tilde{Y}_1 \sin \pi/8)$$

$$X'_3 = Y_0(\tilde{Y}_1 \sin \pi/8 + \tilde{X}_1 \cos \pi/8) - X_0(\tilde{Y}_1 \cos \pi/8 - \tilde{X}_1 \sin \pi/8)$$

$$X'_4 = Y_0(\tilde{X}_1 \cos \pi/8 - \tilde{Y}_1 \sin \pi/8) - X_0(\tilde{X}_1 \sin \pi/8 + \tilde{Y}_1 \cos \pi/8) \quad (3.43)$$

The estimates $\tilde{X}_1$ and $\tilde{Y}_1$ are obtained by averaging the reduced projections X_{1n} and Y_{1n} of the received information signal within N chips:

$$\tilde{X}_1 = \frac{1}{N}\sum_{n=1}^{N} X_{1n}$$
$$\tilde{Y}_1 = \frac{1}{N}\sum_{n=1}^{N} Y_{1n} \quad (3.44)$$

In its turn, the reduced projections X_{1n} and Y_{1n}, included in (3.44), are determined by the projections X_{0n} and Y_{0n} of the received signal onto the reference oscillation with the arbitrary initial phase φ_0 [see (3.13)] taking into account the decisions $\Delta\varphi_n$ made within the given chip. These last are completely defined by the signs of projections X'_1, X'_2, X'_3, and X'_4, which are calculated according to (3.43). Table 3.5 is the table of conformity between these signs and the values X_{1n} and Y_{1n}.

Table 3.5
Reduced Signal Projections in the Eight-Phase System

1	2	3	4	5	6	7
$\Delta\varphi_n$	sgn X'_1	sgn X'_2	sgn X'_3	sgn X'_4	X_{1n}	Y_{1n}
0	+	+	+	−	X_{0n}	Y_{0n}
$\pi/4$	+	+	+	+	$\frac{1}{\sqrt{2}}(X_{0n}+Y_{0n})$	$\frac{1}{\sqrt{2}}(Y_{0n}-X_{0n})$
$\pi/2$	−	+	+	+	Y_{0n}	$-X_{0n}$
$3\pi/4$	−	−	+	+	$\frac{1}{\sqrt{2}}(Y_{0n}-X_{0n})$	$-\frac{1}{\sqrt{2}}(Y_{0n}+X_{0n})$
π	−	−	−	+	$-X_{0n}$	$-Y_{0n}$
$5\pi/4$	−	−	−	−	$-\frac{1}{\sqrt{2}}(Y_{0n}+X_{0n})$	$\frac{1}{\sqrt{2}}(X_{0n}-Y_{0n})$
$3\pi/2$	+	−	−	−	$-Y_{0n}$	X_{0n}
$7\pi/4$	+	+	−	−	$\frac{1}{\sqrt{2}}(X_{0n}-Y_{0n})$	$\frac{1}{\sqrt{2}}(X_{0n}+Y_{0n})$

Thus, expressions (3.43) and (3.44) together with Table 3.5 determine the main part of the algorithm of a coherent demodulator of signals with eight-phase PDM.

To find the transmitted binary symbols, all that remains is to decode the sign of the values (3.43) obtained within two chips. The scheme of the demodulator corresponding to this algorithm is illustrated in Figure 3.28. The scheme contains two adders–accumulators Σ_N, averaging the reduced projections by the formulas (3.44) in the mode with the so-called "floating window": The interval of averaging is shifted by one element for each chip. A scheme for such a unit is indicated in Figure 3.8.

The scheme considered here for a coherent demodulator illustrates the opportunity to realize coherent processing of the signal by its quadrature components, determined relative to the reference oscillations with an arbitrary initial phase. This scheme is oriented to a digital implementation.

Another approach to realizing the coherent reception of signals with the eight-phase PDM is illustrated in the scheme shown in Figure 3.29. In this demodulator the correlators determine projections of the received signal onto four reference oscillations, generated by phase shifters and the selector of reference oscillations. The latter is implemented as the open passive system with signal remodulation accomplished by multiplying the carrier frequency eight times with the help of the appropriate nonlinear unit. Here the complexity of realization is connected with the necessity of narrowband filtering and dividing of the oscillations with the eight times multiplied frequency.

This complexity is overcome when using signal remodulation by multiplying its phase with the help of quadrature projections (Fig. 3.30). In this case the controlled oscillator generates two orthogonal oscillations, being in-phase with any one pair of eight possible signals. The oscillator is controlled by the signal u, which is proportional to the sine of the phase that has been multiplied eight times, which

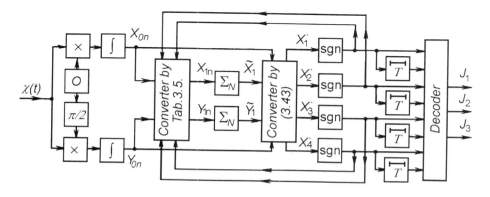

Figure 3.28 Coherent demodulator of eight-phase PDM signals.

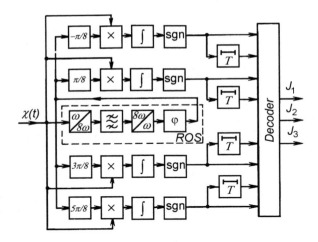

Figure 3.29 Coherent demodulator of eight-level PDM signals with frequency multiplication.

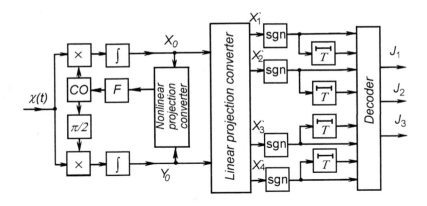

Figure 3.30 Coherent demodulator for eight-level PDM signals with phase multiplication.

is generated in a nonlinear converter of projections X_0 and Y_0 according to the following algorithm (the process of how we formed this algorithm is considered in detail in Section 3.6):

$$u = Y_0 X_0^7 - 4Y_0^3 X_0^5 + 4Y_0^5 X_0^3 - Y_0^7 X_0 \qquad (3.45)$$

The initial projections X_0 and Y_0 are converted in a linear converter into the received signal projections on four reference oscillations, shifted relative to the transmitted signal variants by $\pi/8$:

$$X'_1 = X_0 \sin \pi/8 - Y_0 \cos \pi/8$$

$$X'_2 = X_0 \sin \pi/8 + Y_0 \cos \pi/8$$

$$X'_3 = X_0 \cos \pi/8 + Y_0 \sin \pi/8$$

$$X'_4 = X_0 \cos \pi/8 - Y_0 \sin \pi/8 \qquad (3.46)$$

Finally, the sign functions of the projections (3.46) on two adjacent chips come to the decoder input. This part of the scheme is similar to the appropriate parts of the schemes of Figures 3.28 and 3.29.

As one more example of realizing the coherent demodulator of signals with eight-phase PDM, Figure 3.31 shows the demodulator scheme with modulated signal restoration (regeneration), which is similar in concept to the scheme of the four-phase signal demodulator (see Fig. 3.24).

In the scheme of Figure 3.31 as well as in the previous one, a closed-circuit AFC with a controlled oscillator is used. The control signal is generated in a phase detector, consisting of a multiplier and LPF, by convoluting the received signal $x(t)$, delayed by one chip, and the output signal of a phase modulator. The phase modulator generates an information signal by phase modulation of two orthogonal reference oscillations, generated in CO, and the rule of modulation is determined by the signs of the received signal projection onto four reference oscillations according to Table 3.5. The received signal projections, in their turn, are calculated in the linear converter, operating under the algorithm of (3.46).

Other schemes for eight-phase PDM demodulators are possible. They are represented as various combinations of ROS units and units for calculating the received signal projections. For example, it is possible to form a scheme similar to

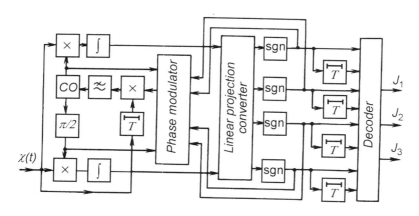

Figure 3.31 Coherent demodulator for eight-level PDM signals with regeneration.

the one indicated in Figure 3.31, with direct remodulation of the received signal by the same phase modulator.

Now let us consider the algorithms of a decoder in coherent demodulators of signals with the eight-position PDM. At the decoder input there are eight-digit binary numbers as the signs of signal projections onto four reference oscillations within two chips, and the output should be three-digit binary combinations according to the keying code used.

For this case it is not difficult to make up a table of conformity between eight inputs and three outputs of the decoder, similar to that of Table 3.4 for four-phase PDM. This table for eight-phase PDM will contain 11 columns and 256 lines. The appropriate decoder can be realized either as ROM, in which the table is written, or as a logic circuit, made up according to the table of conformity. In both cases the scheme is rather complex. Because of this, during decoding of the signals with multiposition PDM it is useful to determine in the decoder the index of the received phase difference (it is assumed that phase differences are numbered with increasing numbers from 1 up to 2^N, where N is the multiplicity of modulation). Then this index is converted into an N-digit binary combination according to the used keying code.

To determine the index of the received phase difference, we take advantage of the circumstance that the signs of the received signal projections onto the reference oscillations, shifted relative to the signal by half of the minimum phase angle, form the cyclic subspace of 2^N-digit binary numbers. As a result of this, the index of the received phase difference is equal to $A + 1$, where A is a number of cyclic rearrangements, transforming a combination of the given signal chip into a combination that corresponds to the previous chip.

At the eight-phase PDM, the binary combination

$$\mathbf{a}_n = (a_{n1}, a_{n2}, \ldots, a_{n8})$$

where $a_{ni} = \pm 1$ is the sign of the projection of the nth signal chip onto the ith reference oscillation f_i, takes the following values at the transmission of signals S_j (Fig. 3.32):

$$S_1 \rightarrow \mathbf{a}_n = 1, 1, 1, -1, -1, -1, -1, 1$$
$$S_2 \rightarrow \mathbf{a}_n = 1, 1, 1, 1, -1, -1, -1, -1$$
$$S_3 \rightarrow \mathbf{a}_n = -1, 1, 1, 1, 1, -1, -1, -1$$
$$\vdots$$
$$S_8 \rightarrow \mathbf{a}_n = 1, 1, -1, -1, -1, -1, 1, 1$$

Similar values are taken by the vector $\mathbf{a}_{n-1}$, corresponding to the $(n-1)$th signal chip. If there are transmitted phase differences

$$\Delta\varphi_1 = 0, \ \Delta\varphi_2 = \pi/4, \ \Delta\varphi_3 = \pi/2, \ \Delta\varphi_4 = 3\pi/4, \ \Delta\varphi_5 = \pi,$$
$$\Delta\varphi_6 = 5\pi/4, \ \Delta\varphi_7 = 3\pi/2, \ \Delta\varphi_8 = 7\pi/4$$

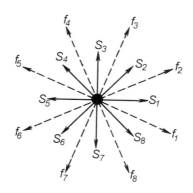

Figure 3.32 Signals and reference oscillations in the eight-phase system with PDM.

the index of the transmitted difference is one more than the number of rearrangements of the last digit into the first one, transforming $\mathbf{a}_n$ into $\mathbf{a}_{n-1}$. The scheme of the unit, realizing this operation, is shown in Figure 3.33.

In the ring shift register at the beginning of operation the eight-digit binary number is recorded. This number contains four 1's and four 0's following in succession along the ring. Four binary symbols a_{n1}, a_{n2}, a_{n3}, and a_{n4}, corresponding to the signs of projections onto reference oscillations f_1, f_2, f_3, and f_4 (see Fig. 3.32)

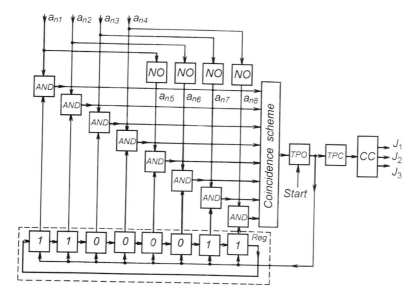

Figure 3.33 Decoder of signals with eight-phase PDM.

come from the coherent demodulator to the decoder input. The binary symbols a_{n5}, a_{n6}, a_{n7}, and a_{n8}, corresponding to the signs of projections onto the reference oscillations f_5, f_6, f_7, and f_8, are obtained by inverting (the NO circuit) the symbols a_{n1}, a_{n2}, a_{n3}, and a_{n4}.

The binary combination $\mathbf{a}_n$, corresponding to the nth chip, is compared digit by digit with the combination, recorded in register (Reg) by means of eight AND circuits. The transferring pulse oscillator (TPO) generates the pulses with a frequency that is at least seven times more than the keying frequency. It is started by an external locking pulse at the beginning of the decoding cycle, and is stopped by the pulse from the coincidence circuit output, which occurs when the combinations in the register and at the decoder input coincide.

Thus, the number of TPO pulses is equal to the number of steps of the combination transferring in the ring shift register before its coincidence with the received combination. This number is fixed by a transferring pulse counter (TPC); then it is converted into the received three-digit binary combination in a code converter (CC). If, for example, the Gray code is used, CC converts a combination of the natural binary code into a combination of the Gray code. In this decoder the ring shift register operates as a memory of the previous chip phase, therefore there is no need to have delay units placed before the decoder (see Figs. 3.28 through 3.31).

Using the specified approaches, it is possible to develop coherent demodulators of signals with multilevel PDM (Fig. 3.34). For any multiphase signal the input part of a demodulator is a quadrature splitter, which consists of two correlators with the orthogonal reference oscillations coinciding by phase with any two signal variants. Then, using the signal received projections on these reference oscillations, the linear projection converter calculates the received signal projections onto all reference oscillations (their number is equal, as a rule, to half of a number of admissible signal variants), shifted relative to the signal by half of the minimum phase angle. Then signs of the calculated projections are determined, and according to those in the decoder the transmitted information is determined.

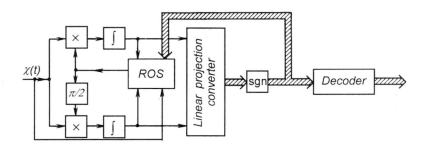

Figure 3.34 Coherent demodulator of signals with multilevel PDM.

The ROS can use different methods: multiplying the phase by quadrature projections, phase reduction from the results of calculating the projection signs, direct signal remodulation, or restoration (regeneration) of the information signal. These opportunities are illustrated in Figure 3.34.

The considered method of coherent reception of signals with PDM, based on averaging the signal projections onto the reference oscillations with an arbitrary initial phase (e.g., the scheme for eight-phase signals shown in Fig. 3.28), can also be realized by means of the scheme shown in Figure 3.34. The difference is that at this method the reference oscillations of correlators are generated by an autonomous oscillator; ROS is not in use at all, and its function is fulfilled by a nonlinear converter of the original quadrature projections and units for averaging the reduced projections.

Note, however, that in communications systems pure PM or PDM signals with more than eight phases are not used—the distance between the nearest signals appears to be too small and the noise immunity degradation too large. When the number of signal levels is more than eight, it is preferable to use the signals with combined amplitude-phase or amplitude-phase-difference modulation (APM or APDM), known also as quadrature amplitude modulation (QAM). The coherent reception of such signals is considered in Section 3.5.

3.4 COHERENT RECEPTION OF HIGH-ORDER PDM SIGNALS

High-order phase-difference modulation is, on the one hand, a generalization of PM and PDM-1 and, on the other hand, a particular case of difference methods of modulation.

Let us consider the generalization of coherent reception algorithms for modulating an arbitrary parameter γ, assuming that the information is put into the kth finite differences $\Delta^k \gamma$ of this parameter [8]. In this case γ is called the *determining parameter*, and $\Delta^k \gamma$ is the *information parameter* of a signal.

When using strictly coherent reception the transmitted signal variants are exactly known in the demodulator, therefore the received signals within every elementary chip can be mapped into the finite discrete set of the determining parameter γ. Consequently, the coherent demodulator of signals with the kth-order difference modulation can be presented in the form of series connection of the γ parameter coherent demodulator, at the output of which one of the admissible discrete values of γ is fixed, and the $\Delta^k \gamma$ difference calculator, which determines the transmitted $\Delta^k \gamma$ by the known formulas for calculating the final differences [see expressions (2.28), (2.29), and (2.30)]. The obtained value of $\Delta^k \gamma$ determines the transmitted binary symbols in accordance with the used keying code.

The indicated common structure of a coherent receiver undergoes a number of changes and refinements during transit to the specified conditions. If, for example, as for PDM-k, the set $\{\gamma\}$ of discrete values of the determining parameter γ is the algebraic ring of the order R, this set can be identified with the isomorphic ring of integers from 0 to $(R-1)$ with the operation of summation (subtraction) modulo R. In this case the algorithm for calculating the finite differences $\Delta^k \gamma$ includes the operation of subtraction modulo R of indicated integers, representing the numbers of discrete values of the parameter γ and its finite differences of the order from 1 to kth. So the generalized scheme of a coherent demodulator of signals with the kth-order difference modulation [Fig. 3.35(a)] contains a coherent demodulator of the determining parameter γ, k difference calculators, and a decoder.

In this case the demodulator of the parameter γ gives the number (from 0 to $R-1$) of the received discrete value γ. The difference calculators consist of a memory element of the R-ary number with a chip delay and modulo-R subtraction units by Figure 3.35(b). The decoder functions were indicated earlier.

Certainly, it is possible to implement the scheme, similar to the indicated one, such that intermediate finite differences are not calculated and to obtain at once the kth-order difference. For this purpose, the algorithm for calculating the differences (2.30) should be used. The appropriate scheme will also contain a coherent demodulator of the parameter γ, a memory unit and units for modulo-R adding and subtracting.

The opportunity to represent a set of γ parameter values at the coherent demodulator output as the isomorphic ring of integers is based on the following property of the finite differences set. If the set of the kth finite differences is an algebraic ring of Rth order, the set of the $(k-1)$th differences is also an algebraic ring of the same order.

a)

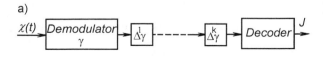

b)

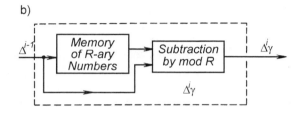

Figure 3.35 (a) Demodulator of signals with kth-order difference modulation and (b) a difference calculator.

The proof of this property follows from the common recurrent algorithm of signal modulation by finite differences. According to (2.23), we have

$$\Delta_n^{k-1}\gamma = \Delta_0^{k-1}\gamma + \sum_{i=1}^{n}\Delta_i^{k}\gamma \qquad (3.47)$$

The initial value of the $(k-1)$th difference $\Delta_0^{k-1}\gamma$ can be, without loss of generality, assumed equal to 0 or to 1 of the admissible values of the kth difference, as far as it is compulsorily established at the beginning of operation. The finite differences $\Delta_i^{k}\gamma$ represent the elements of the transmitted information and consequently are included in the sum (3.47) in all combinations. Thus, if the set $\{\Delta^k\gamma\}$ is the ring, the set $\{\Delta^{k-1}\gamma\}$ contains the same elements as the set $\{\Delta^k\gamma\}$ and, hence, is the ring of the same order.[1]

Expanding the proven property for the finite differences of all orders, less than k, we can assert that if the information kth difference of the parameter γ is the algebraic ring of the Rth order, then in a coherent demodulator all operations of calculating the kth differences are reduced to operations of modulo-R addition and subtraction of integers from 0 to $R-1$, with R being equal to a number of information parameter variants $m = 2^N$. Sometimes the set of γ parameter values at the coherent demodulator output forms the algebraic ring even if the kth order information differences are not the ring. However, as long as in this case all finite differences belong to the same set, as the values γ (by the property of the algebraic ring), their calculation is reduced to the operations with integers by the modulo, equal to the order of the γ parameter set. In the case considered, this order is more than the number of information parameter variants.

At PDM a set $\{\varphi\}$ of signal phases at the coherent demodulator output is the algebraic ring in all cases, when the information kth phase differences take the values

$$\Delta^k\varphi_i = \Delta^k\varphi_0 + 2\pi i/m \qquad (3.48)$$

where $i = 0, 1, \ldots, m-1$, and, besides, the initial phase difference $\Delta^k\varphi_0$ is either equal to 0 or is a 2π divisor. In fact, if $\Delta^k\varphi_0 = 0$, the set $\{\Delta^k\varphi\}$ is the ring of the mth order, and as a result of it the set $\{\varphi\}$ is also the ring of the mth order. If $\Delta^k\varphi_0 = 2\pi/l$, the $(k-1)$th phase differences will be of the form

$$\Delta^{k-1}\varphi_i = \frac{2\pi i}{r(l, m)}$$

---

1. The statement, reciprocal proven one, is not true. If the set of the $(k-1)$th differences is the ring, the set of the kth differences is not obligatory for the ring. It is caused by the fact that in the sum, forming the kth differences, not all combinations of the $(k-1)$th differences can be included, because the latter are formed in the transmitter depending on the chosen values of the kth differences.

where $i = 0, 1, \ldots, r(l, m)$ with $r(l, m)$ being the least common multiple of the numbers l and m. Therefore, the $(k - 1)$th phase differences and, consequently, the phases at the coherent demodulator output are in this case the ring of the $r(l, m)$th order.

Thus, during coherent reception of signals with PDM it is always practically possible to calculate finite phase differences by means of the operations with integers by this or that modulo. Let us consider, in a practical sense, the most important binary systems with high-order difference modulation, in which the determining parameter of a signal takes only two values: $\gamma^{(1)} = 0$ and $\gamma(2) = 1$, and all operations with them are executed by modulo 2. The system with binary PDM ($\varphi_1 = 0 \rightarrow \gamma^{(1)} = 0$; $\varphi_2 = \pi \rightarrow \gamma^{(2)} = 1$) or the system with FDM ($\omega_1 \rightarrow \gamma^{(1)} = 0$; $\omega_2 \rightarrow \gamma^{(2)} = 1$) refers to such systems. Let us replace the ring of numbers $\gamma = \{0,1\}$ with modulo-2 addition by the isomorphic ring of numbers $Z = \pm 1$ with multiplication operation using the sign function

$$Z = \text{sgn}(1/2 - \gamma) \tag{3.49}$$

Then the following statement is true: To determine the binary symbol transmitted on the nth chip at coherent reception of signals with binary difference modulation of the kth order, it is sufficient to calculate the product of the sign functions Z_{n-k+i} at the coherent demodulator output on the chips with such numbers $n - k + i$, where $i \in \{0, 1, 2, \ldots, k\}$, for which the binomial coefficients C_k^i are odd.

To prove this statement, let us use the common expression (2.30) for finite differences $\Delta^k \gamma$ of the parameter γ. Having replaced binary samples of the parameter γ_{n-k+i} by the samples of the sign function Z_{n-k+i} according to (3.49) and having similarly expressed the sign function of the binary difference $\Delta_n^k \gamma$, we obtain from (2.30)

$$\text{sgn}(1/2 - \Delta_n^k \gamma) = \prod_{i=0}^{k} [\text{sgn}(1/2 - \gamma_{n-k+i})]^{C_k^i} \tag{3.50}$$

If now we take into account that

$$(\text{sgn } \alpha)^\beta = \begin{cases} +1 & \text{for } \beta = 2l \ (l = \text{integer}) \\ \text{sgn } \alpha & \text{for } \beta = 2l - 1 \end{cases} \tag{3.51}$$

from (3.50) the statement mentioned earlier follows.

Remember that even (e) and odd (o) binomial coefficients C_k^i can be determined by means of the Pascal triangle. The latter can be represented for k from 1 to 16 in the following form:

$k = 1$	o o
2	o e o
3	o o o o
4	o e e e o
5	o o e e o o
6	o e o e o e o
7	o o o o o o o o
8	o e e e e e e o
9	o o e e e e e o o
10	o e o e e e e e o e o
11	o o o o e e e e o o o o
12	o e e e o e e e o e e e o
13	o o e e o o e e o o e e o o
14	o e o e o e o e o e o e o e o
15	o o o o o o o o o o o o o o o o
16	o e e e e e e e e e e e e e e o

Expression (3.50) has an important meaning in the theory of difference modulation. It is the most general record of algorithms for calculating the finite differences of a binary signal and allows us to determine the errors caused by interference in the systems with high-order difference modulation.

In particular, from (3.50) it follows that at all $k = 2^l$, where l is the integer, only two elements of a signal are used for calculating the kth-order difference:

$$\text{sgn}(1/2 - \Delta_n^{2^l}\gamma) = \text{sgn}(1/2 - \gamma_n)\,\text{sgn}(1/2 - \gamma_{n-2l}) \qquad \text{for } l = 0, 1, 2, \ldots$$

Based on this we can make an important conclusion, that binary systems with the difference modulation of the first-, second-, fourth-, and in general $(k = 2^l)$th-order at coherent reception in the channel with constant parameters and noncorrelated noises have identical noise immunities. Besides, it is obvious that the application of the $(k = 2^l)$ order difference modulation corresponds to the signal reception with the time diversity by 2^l chip intervals.

In a common case according to (3.50), the number of signal chips used for calculation of the kth differences is equal to a number of the odd binomial coefficients $(i = 0, 1, 2, \ldots, k)$. The number of such coefficients is determined by the number of odd numbers in the appropriate line of the Pascal triangle:

$$H_k = 2^{V(k)} \tag{3.52}$$

where $V(k)$ is the number of 1's in the binary representation of the number k, that is, the Hamming weight of k. For example, there are two 1's in the binary representation of the number 3 and one 1 in the binary representation of the number 4; hence, $V(3) = 2$, $H_3 = 4$, $V(4) = 1$, and $H_4 = 2$, that is, during coherent processing of the signals with third-order binary PDM, four signal chips are required for calculation of information phase differences, and during coherent processing of signals with fourth-order binary PDM two signal chips are sufficient.

For high-order binary PDM, the common algorithm (3.50) can be easily transformed to a quite constructive form. In this case the determining parameter is represented as the initial phase φ of the signal, and this phase and all of its finite differences have only two levels, 0 and π, which form the second-order algebraic ring with modulo-2 addition–subtraction operation:

$$0 \pm \pi = \pi \qquad \pi \pm \pi = 0 \qquad 0 \pm 0 = 0$$

Thus, from (2.30) we can obtain:

$$\Delta^k \varphi_n = \sum_{i=0}^{k} C_k^i \{\varphi_{n-k+i}\} \tag{3.53}$$

where $\{\varphi_{n-k+i}\}$ is the binary decision of the transmitted phase at the PM signal coherent demodulator output, C_k^i are the binomial coefficients, and addition is fulfilled by modulo 2. As in the case of (3.50), the addition of binary numbers φ_{n-k+i} is replaced by the multiplication of sign functions sgn cos φ_{n-k+i}, and the multiplication by raising to the power. Here we obtain the algorithm

$$J_n^k = \text{sgn cos } \Delta_n^k \varphi = \prod_{i=0}^{k} (\text{sgn cos } \varphi_{n-k+i})^{C_k^i} \tag{3.54}$$

where J_n^k is the binary symbol received on the nth chip. Because the sign of cos φ_{n-k+i} is equal to the sign of the projection X_{n-k+i} of the corresponding signal chip onto the coherent reference oscillation, from (3.54) we obtain the following algorithm of coherent reception of signals with kth-order binary PDM:

$$J_n^k = \text{sgn cos } \Delta_n^k \varphi = \prod_{i=0}^{k} (\text{sgn } X_{n-k+i})^{C_k^i} \tag{3.55}$$

where

$$X_r = \int_{(r-1)T}^{rT} x(t) \sin(\omega t + \varphi) \, dt \tag{3.56}$$

where φ is the phase of the coherent reference oscillation.

Further simplifying (3.55) is accomplished by using the relations (3.51). From (3.55) and considering (3.51), it follows that to determine the nth chip binary symbol in the system with PDM-k, it is sufficient to multiply the projection signs of signal chips on the intervals $n - k + i$ with such numbers i from the set $\{i\} = 0, 1, \ldots, k$, for which the binomial coefficients are odd numbers.

In particular, for PDM-1, PDM-2, PDM-3, and PDM-4 we obtain, respectively:

$$J_n^1 = \text{sgn } X_n \text{ sgn } X_{n-1}$$

$$J_n^2 = \text{sgn } X_n \text{ sgn } X_{n-2}$$

$$J_n^3 = \text{sgn } X_n \text{ sgn } X_{n-1} \text{ sgn } X_{n-2} \text{ sgn } X_{n-3}$$

$$J_n^4 = \text{sgn } X_n \text{ sgn } X_{n-4} \tag{3.57}$$

Figure 3.36 shows the block diagrams of the demodulators corresponding to (3.57).

To generate coherent reference oscillations in high-order PDM systems we can use the same methods and units for selecting the reference oscillations that have already been considered relative to the PM and PDM-1 systems. There are practically no peculiarities here because the elementary signal is still a part of a harmonic oscillation with definite frequency and with one of a number of admissible discrete levels of the initial phase.

As a whole, the coherent demodulator of signals with high-order PDM is not much more complex than the corresponding demodulator of signals with PM or PDM-1. For example, as can be seen from the comparison of the schemes shown in Figure 3.36(a,b), at the transition from PDM-1 to PDM-2 one element of delay of the binary symbol is added in the demodulator. At the same time as for the noise immunity concerning the AWGN these schemes are equivalent.

The last circumstance appears to be rather important when deciding the problem of using PDM-2 coherent demodulators in practice. The main property of PDM-2—invariance to the carrier frequency—is not naturally realized at coherent reception. At the same time in channels with the indefinite signal frequency, in which PDM-2 is used, there can be time intervals during which the frequency and the initial phase change relatively slowly, and then using the coherent reception is not only possible, but also permits noise immunity to increase up to the potential noise immunity of PDM. It makes expedient the use of a receiving aggregate, consisting of an autocorrelation demodulator of PDM-2, providing insensitivity (or small sensitivity) to frequency changes, and a coherent demodulator of PDM-2, providing high noise immunity at relatively slow changes of a signal phase.

3.5 COHERENT RECEPTION OF AMPLITUDE-PHASE-MODULATED SIGNALS

Signals with amplitude-phase modulation (also called quadrature amplitude modulation, QAM) are usually employed to achieve high specific rates of digital transmis-

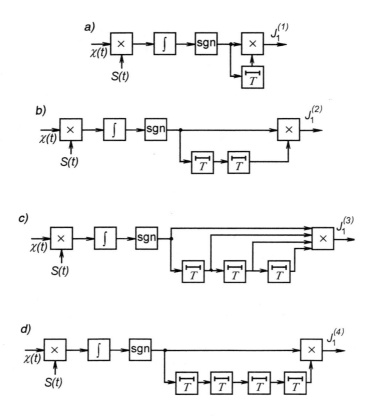

Figure 3.36 Coherent demodulators of signals with binary PDM of the (a) first order, (b) second order, (c) third order, and (d) fourth order.

sion (3 bps/Hz and more) in the channels with strict restriction of the frequency bandwidth and comparatively good energy budget. The 16-, 32-, and 64-position systems are the most popular and frequently used; the signals with 128, 256, and more amplitude-phase levels are also used.

Let us consider the most general case of digital transmission by an m-position signal with arbitrary amplitudes $a_1, a_2, \ldots, a_m$ and initial phases $\varphi_1, \varphi_2, \ldots, \varphi_m$, without regard for the differences between APM and APDM. In this case the ith variant of the transmitted signal can be represented as

$$S_i(t) = a_i \sin(\omega t + \varphi_i) \qquad \text{for } i = 1, 2, \ldots, m \qquad (3.58)$$

In the channel with AWGN the optimum algorithm for processing the signal (3.58) can be formulated as follows: The ith variant of a signal is fixed as the transmitted one, if at all $j \neq i$ the following inequality is true:

$$\int\limits_0^T [x(t) - S_i(t)]^2 \, dt < \int\limits_0^T [x(t) - S_j(t)]^2 \, dt \qquad (3.59a)$$

or, more concisely,

$$i = \arg\min_J \int\limits_0^T [x(t) - S_j(t)]^2 \, dt \qquad \text{for } j = 1, 2, \ldots, m \qquad (3.59b)$$

During digital processing it is convenient to pass from a high-frequency signal (3.58) to its representation in terms of coordinates in two-dimensional space, which in practice corresponds, for example, to the operations of spectrum converting or the division of orthogonal signals in a multichannel system. So, let the projections of the received signal $x(t)$ and signals (3.58) onto a reference oscillation with arbitrary amplitude a_0 and phase φ_0, calculated on the interval of one chip, be known

$$\left.\begin{aligned}
X_0 &= \int\limits_0^T x(t) \, a_0 \, \sin(\omega t + \varphi_0) \, dt \\
Y_0 &= \int\limits_0^T x(t) \, a_0 \, \cos(\omega t + \varphi_0) \, dt
\end{aligned}\right\} \qquad (3.60)$$

$$X_j = \int\limits_0^T S_j(t) \, a_0 \, \sin(\omega t + \varphi_0) \, dt \qquad (3.61)$$

$$Y_j = \int\limits_0^T S_j(t) \, a_0 \, \cos(\omega t + \varphi_0) \, dt \qquad \text{for } j = 1, 2, \ldots, m$$

Then the optimum algorithm (3.59b) can be presented as

$$i = \arg\min[(X_0 - X_j)^2 + (Y_0 - Y_j)^2] \qquad \text{for } j = 1, 2, \ldots, m \qquad (3.62)$$

In (3.62) the values X_0 and Y_0 are determined, as can be seen from (3.60), in the result of processing the current received signal chip, and $2m$ of the values X_j and Y_j should be known a priori or calculated (evaluated) during the process of receiving the previous chips of a signal. (If it is necessary to emphasize that the projections X_0 and Y_0 are calculated exactly on the nth signal chip, the symbol n will be added to their indexes: X_{0n} and Y_{0n}).

To estimate the projections X_j and Y_j let us use the method of reducing and averaging the projections of the received signal, which was discussed earlier concerning processing of PM and PDM signals. As our averaging values we choose the first signal variant projections, that is, X_1 and Y_1 [see (3.61)]; we shall also transform (reduce) the other variants of the received signal to the X_1 and Y_1 in the process of phase tuning by an information signal.

If the received signal contains within the interval of N chips the signal $S_1(t)$ in a mixture with AWGN (see Section 3.2), the maximum likely estimates of X_1 and Y_1 are equal to

$$\tilde{X}_1 = \frac{1}{N}\sum_{n=1}^{N} X_{0n} \qquad \tilde{Y}_1 = \frac{1}{N}\sum_{n=1}^{N} Y_{0n} \tag{3.63}$$

where X_{0n} and Y_{0n} are the projections (3.60) within the nth chip. The estimates of (3.63) are unbiased and effective. From them we can derive the unbiased and effective projection estimates of all the rest of the signal variants, included in the optimum algorithm (3.62). For this purpose, having introduced the designation $\varphi_j = \varphi_1 + \Delta\varphi_j$, let us transform the projection X_j in the following way:

$$X_j = \int_0^T a_j \sin(\omega t + \varphi_j) a_0 \sin(\omega t + \varphi_0)\, dt$$

$$= (a_j/a_1)\int_0^T a_1 \sin(\omega t) + \varphi_1 + \Delta\varphi_j) a_0 \sin(\omega t + \varphi_0)\, dt$$

$$= (a_j/a_1)\left[\cos \Delta\varphi_j \int_0^T a_1 \sin(\omega t + \varphi_1) a_0 \sin(\omega t + \varphi_0)\, dt\right.$$

$$\left. + \sin \Delta\varphi_j \int_0^T a_1 \cos(\omega t + \varphi_1) a_0 \sin(\omega t) + \varphi_0)\, dt\right]$$

$$= (a_j/a_1)\left[\cos \Delta\varphi_j \int_0^T a_1 \sin(\omega t + \varphi_1) a_0 \sin(\omega t + \varphi_0)\, dt\right.$$

$$\left. - \sin \Delta\varphi_j \int_0^T a_1 \sin(\omega t + \varphi_1) a_0 \cos(\omega t + \varphi_0)\, dt\right]$$

$$= (a_j/a_1)(X_1 \cos \Delta\varphi_j - Y_1 \sin \Delta\varphi_j) \tag{3.64}$$

Similarly, we obtain the projection Y_j. Now having replaced X_1 and Y_1 by their estimates we obtain

$$\tilde{X}_j = (a_j/a_1)(\tilde{X}_1 \cos \Delta\varphi_j - \tilde{Y}_1 \sin \Delta\varphi_j)$$
$$\tilde{Y}_j = (a_j/a_1)(\tilde{X}_1 \sin \Delta\varphi_j + \tilde{Y}_1 \cos \Delta\varphi_j)$$

(3.65)

where $\Delta\varphi_j$ is the known phase difference between the signals $S_j(t)$ and $S_1(t)$.

Note that when calculating the estimates by (3.65), there is no need to have the information about the amplitudes of the signal variants a_j and a_1, and it is enough to know the ratio of these amplitudes a_j/a_1.

The obtained algorithms solve the problem of coherent reception of multiposition APM signals, when we have a special synchronization (training) signal, which, for example, precedes information chip transmission:

1. The synchronization signal projections onto the reference oscillation with an arbitrary initial phase are calculated by (3.60).
2. The estimates of the first signal variant projections $\tilde{X}_1$ and $\tilde{Y}_1$ are determined by (3.63).
3. The estimates of the projections of all m variants of a signal are calculated by (3.65).
4. The obtained estimates $\tilde{X}_j$ and $\tilde{Y}_j$ of all signal variants are substituted instead of X_j and Y_j in the algorithm (3.62).

The appropriate structural scheme of a demodulator is shown in Figure 3.37. In this block diagram adders–accumulators, performing the operation for averaging according to (3.63), are denoted by Σ_N. These units, as well as the signal projection estimator by (3.65), operate only during synchronization signal transmission. The estimates come to the calculator of distances between the received signal and all m samples. This calculator makes the decision about the transmitted signal in accordance with (3.62). The decoder (Dec) has to transform this decision into a binary code combination according to the keying code used.

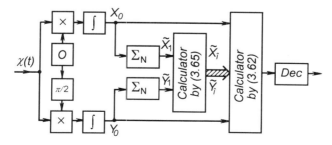

Figure 3.37 Coherent demodulator of multilevel APM signals.

The considered algorithm and scheme are oriented to receive the signal with absolute APM, because the presence of the training signal excludes the uncertainty of an initial phase.

Let us adapt this algorithm to the important case when there is no special synchronization signal and adjustment of the signal projections should be carried out just by the information chips. Here it is necessary to average not the projections of the received signal, but the reduced projections. In this case the reduction operation consists of converting the received projections to the projections, for example, of the first signal variant using the decision about the transmitted signal.

As before, let $\Delta\tilde\varphi_n$ be the phase difference between the signal variant, for the benefit of which a decision had been made on the nth chip, and the first signal variant; $\tilde a_n$ is the signal amplitude, for the benefit of which a decision has been made on the nth chip. Naturally, $\Delta\tilde\varphi_n$ belongs to the discrete set of the admissible phases, determined by (3.58). As to $\tilde a_n$, this value is equal to the actual amplitude of the nth chip, however hereinafter it is identified with the amplitude of that signal variant, for the benefit of which a decision has been made on the nth chip.

Then the reduced projections X_{1n} and Y_{1n} of the nth chip, similar to (3.13), are calculated by the received projections X_{0n} and Y_{0n} according to the formulas

$$X_{1n} = (a_1/\tilde a_n)(X_{0n}\cos\Delta\tilde\varphi_n + Y_{0n}\sin\Delta\tilde\varphi_n)$$
$$Y_{1n} = (a_1/\tilde a_n)(Y_{0n}\cos\Delta\tilde\varphi_n - X_{0n}\sin\Delta\tilde\varphi_n) \tag{3.66}$$

We emphasize that $\tilde a_n$ and $\Delta\tilde\varphi_n$ are determined by the decision about the transmitted signal variant on the nth chip, and this decision is made by processing of X_{0n} and Y_{0n}.

Then, as in the algorithm of reception by a training signal, the values (3.66) are averaged:

$$\tilde X_1 = \frac{1}{N}\sum_{n=1}^{N}\frac{a_1}{\tilde a_n}(X_{0n}\cos\Delta\tilde\varphi_n + Y_{0n}\sin\Delta\tilde\varphi_n)$$
$$\tilde Y_1 = \frac{1}{N}\sum_{n=1}^{N}\frac{a_1}{\tilde a_n}(Y_{0n}\cos\Delta\tilde\varphi_n - X_{0n}\sin\Delta\tilde\varphi_n) \tag{3.67}$$

The distinction of algorithm (3.67) from (3.63) is also seen in the fact that in (3.63) averaging is made on the training signal interval, and in (3.67) it is made on the N-chips interval, preceding the chip, which is processed at the given moment.

While calculating the estimates (3.67) there is no need for a priori information of the signal variant amplitudes, because only the amplitude ratios are included in this algorithm.

Thus, the algorithm of coherent reception of a multiposition APM signal is the following:

$$i = \arg \min_{j}[(X_{0n} - \tilde{X}_j)^2 + (Y_{0n} - \tilde{Y}_j)^2] \qquad \text{for } j = 1, 2, \ldots, m \qquad (3.68)$$

where $\tilde{X}_j$ and $\tilde{Y}_j$ are determined by (3.65) and (3.67). The scheme of the coherent demodulator corresponding to this algorithm is indicated in Figure 3.38.

The signals X_{0n} and Y_{0n} from the correlator outputs come to the input of the reduced projection calculator by (3.66), in which the phase $\tilde{\Delta}\varphi_n$ and ratio of the amplitudes $a_1/\bar{a}_n$ are also entered. After that the calculated reduced projections are averaged in two adder–accumulators, which operate in the "sliding window" mode and are realized, for example, by the scheme indicated in Figure 3.7. In the calculator, operating according to algorithm (3.68), a decision is made about the signal transmitted on the given chip. This decision then goes to the decoder and into the memory of phases and ratios of signal amplitudes. In memory, a selection is made of a pair of values $\tilde{\Delta}\varphi_n$ and $a_1/\bar{a}_n$, coming to the reduced projection calculator.

The scheme of the demodulator of Figure 3.38, which is different from that of Figure 3.37, is oriented to reception of the amplitude-phase-difference-modulated signal. The adjustment of signal projections in this demodulator is fulfilled by the information signal, not containing the information about the conformity between the input signal and the admissible variants of the signal. Consequently, the procedure of converting the received signal into the reduced projections is ambiguous: The received projections can be "reduced," with the same success, to several signal variants. The decision ambiguity arising here can be eliminated by difference encoding of the phase, that is, by the application of amplitude-phase-difference modulation, at which the result of demodulation on the given chip depends on the decision made on the previous chip. In the scheme of Figure 3.38 the operations of decision comparison on two chips and determination of the received code combination are executed by the decoder. The algorithm is not considered here, as far as it depends on the keying code used and signal structure.

Let us consider the peculiarities of using the synthesized common algorithms we have discussed on the example of the widely used 16-level QAM signals.

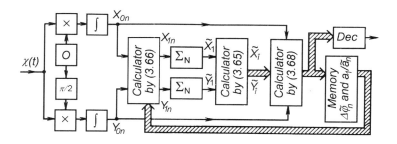

Figure 3.38 Coherent demodulator of QAM signals.

As the first example, we synthesize the coherent reception algorithm for the signal shown in Figure 3.39. APM signal variants are located in this case on two concentric circles. The relation between radii of the circles and between the amplitudes of signals with the odd and even numbers is equal approximately to 1.59 (see Section 1.5).

Let us choose the signal with number 1 as the sample, to which the rest of the variants are reduced. Then by (3.36) it is not difficult to calculate the reduced projections by the projections X_{0n} and Y_{0n} of the received signal onto the reference oscillation with an arbitrary initial phase. This calculation is illustrated in Table 3.6, which determines the algorithm of the reduced projection calculator of the synthesized coherent demodulator. The reduced projections are averaged according to (3.67), and projection estimates of all 16 signal samples are calculated by the obtained estimates $\tilde{X}_1$ and $\tilde{Y}_1$. The algorithm for calculating the estimates $\tilde{X}_j$ and $\tilde{Y}_j$ is shown in Table 3.7. After the calculation of all values $\tilde{X}_j$ and $\tilde{Y}_j$ ($j = 1, 2, \ldots, 16$), the decision is made by the general algorithm (3.68).

Thus, the algorithm of coherent reception of signals with 16-position APM (see Fig. 3.39) is determined by Tables 3.6 and 3.7 and formulas (3.67) and (3.68).

The general scheme of the appropriate demodulator remains the same, as in Figure 3.38. However, in this case there is only one distinction: The reduced projection calculator realizes Table 3.6, and the calculator of signal projection estimates realizes Table 3.7. Thus there is no memory of phase differences and amplitude ratios; it is replaced by the appropriate part of the calculation program of Tables 3.6 and 3.7. Apparently, at not very high transmission rates, the operation of all three calculators and two averaging unit–accumulators is possible and expedient to execute in one universal microprocessor or special microprocessor for digital signal processing.

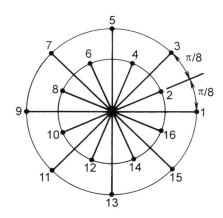

Figure 3.39 A 16-position signal on concentric circles.

Table 3.6
Reduced Projections for the 16-Level QAM Signal in Figure 3.39

Decision Made on nth Chip	Parameters of the Decision $\Delta\bar{\varphi}_n$	a_1/a_n	Reduced Projections X_{1n}	Y_{2n}
1	0	1	X_{0n}	Y_{0n}
2	$\pi/8$	1.59	$1.47X_{0n} + 0.61Y_{0n}$	$1.47\,Y_{0n} - 0.61\,X_{0n}$
3	$\pi/4$	1	$.707(X_{0n} + Y_{0n})$	$.707(Y_{0n} - X_{0n})$
4	$3\pi/8$	1.59	$.61Y_{0n} + 1.47X_{0n}$	$.61Y_{0n} + 1.47X_{0n}$
5	$\pi/2$	1	Y_{0n}	$-X_{0n}$
6	$5\pi/8$	1.59	$1.47Y_{0n} - 0.61X_{0n}$	$-0.61Y_{0n} - 1.47X_{0n}$
7	$3\pi/4$	1	$.707(Y_{0n} - X_{0n})$	$.707(Y_{0n} + X_{0n})$
8	$7\pi/8$	1.59	$.61Y_{0n} - 1.47X_{0n}$	$-1.47Y_{0n} - 0.61X_{0n}$
9	π	1	$-X_{0n}$	$-Y_{0n}$
10	$9\pi/8$	1.59	$-1.47X_{0n} - 0.61Y_{0n}$	$.61X_{0n} - 1.47Y_{0n}$
11	$5\pi/4$	1	$-0.707(X_{0n} + Y_{0n})$	$.707(X_{0n} - Y_{0n})$
12	$11\pi/8$	1.59	$-0.61X_{0n} - 1.47Y_{0n}$	$-0.61Y_{0n} - 1.47X_{0n}$
13	$3\pi/2$	1	$-Y_{0n}$	X_{0n}
14	$13\pi/8$	1.59	$.61X_{0n} - 1.47Y_{0n}$	$.61X_{0n} + 1.47Y_{0n}$
15	$7\pi/4$	1	$.707(X_{0n} - Y_{0n})$	$.707(X_{0n} + Y_{0n})$
16	$15\pi/8$	1.59	$1.47X_{0n} - 0.61Y_{0n}$	$1.47\,Y_{0n} + 0.61\,X_{0n}$

Table 3.7
Signal Projection Estimates for the 16-Level QAM Signal in Figure 3.39

Variant of a Signal j	Variant Parameters $\Delta\varphi_j$	a_j/a_1	Projection Estimates $\tilde{X}_j$	$\tilde{Y}_j$
1	0	1	$\tilde{X}_1$	$\tilde{Y}_1$
2	$\pi/8$	.63	$.58\,\tilde{X}_1 - 0.24\,\tilde{Y}_1$	$.24\,\tilde{X}_1 + 0.58\,\tilde{Y}_1$
3	$\pi/4$	1	$.707(\tilde{X}_1 - \tilde{Y}_1)$	$.707(\tilde{X}_1 + \tilde{Y}_1)$
4	$3\pi/8$	.63	$.24\,\tilde{X}_1 - 0.58\,\tilde{Y}_1$	$.58\,\tilde{X}_1 + 0.24\,\tilde{Y}_1$
5	$\pi/2$	1	$-\tilde{Y}_1$	$\tilde{X}_1$
6	$5\pi/8$	.63	$-0.24\,\tilde{X}_1 - 0.58\,\tilde{Y}_1$	$.58\,\tilde{X}_1 - 0.24\,\tilde{Y}_1$
7	$3\pi/4$	1	$-0.707(\tilde{X}_1 + \tilde{Y}_1)$	$.707\,(\tilde{X}_1 - \tilde{Y}_1)$
8	$7\pi/8$	.63	$-0.58\,\tilde{X}_1 - 0.24\,\tilde{Y}_1$	$.24\,\tilde{X}_1 - 0.58\,\tilde{Y}_1$
9	π	1	$-\tilde{X}_1$	$-\tilde{Y}_1$
10	$9\pi/8$	.63	$.24\,\tilde{Y}_1 - 0.58\,\tilde{X}_1$	$-0.24\,\tilde{X}_1 - 0.58\,\tilde{Y}_1$
11	$5\pi/4$	1	$.707\,(\tilde{Y}_1 - \tilde{X}_1)$	$-0.707(\tilde{X}_1 + \tilde{Y}_1)$
12	$11\pi/8$	.63	$.58\,\tilde{Y}_1 - 0.24\,\tilde{X}_1$	$-0.58\,\tilde{X}_1 - 0.24\,\tilde{Y}_1$
13	$3\pi/2$	1	$\tilde{Y}_1$	$-\tilde{X}_1$
14	$13\pi/8$	.63	$.24\,\tilde{X}_1 + 0.58\,\tilde{Y}_1$	$.24\,\tilde{Y}_1 - 0.58\,\tilde{X}_1$
15	$7\pi/4$	1	$.707(\tilde{X}_1 + \tilde{Y}_1)$	$.707(\tilde{Y}_1 - \tilde{X}_1)$
16	$15\pi/8$	.63	$.58\,\tilde{X}_1 + 0.24\,\tilde{Y}_1$	$.58\,\tilde{Y}_1 - 0.24\,\tilde{X}_1$

As the second example, we consider coherent processing of a 16-position signal with APM (Fig. 3.40). The APM signal variants are located in this case in the nodes of a rectangular array, and the length of vectors (amplitude of signals) has three values, characterized by the following relations:

$$a_3/a_1 = a_7/a_1 = a_{11}/a_1 = a_{15}/a_1 = 3$$
$$a_2/a_1 = a_4/a_1 = a_6/a_1 = a_8/a_1 = a_{10}/a_1 = a_{12}/a_1$$
$$= a_{14}/a_1 = a_{16}/a_1 = \sqrt{5}$$
$$a_5/a_1 = a_9/a_1 = a_{13}/a_1 = 1 \tag{3.69}$$

Phase difference variants $\Delta\varphi_j$, needed for calculation by (3.65) and (3.66), are easy to determine according to Figure 3.40. For some variants, it is convenient to calculate directly $\cos \Delta\varphi_j$ and $\sin \Delta\varphi_j$. For example, for the signal with number 4 we have

$$\sin \Delta\varphi_4 = \sin(\varphi_4 - \pi/4) = \frac{1}{\sqrt{2}}(\sin \varphi_4 - \cos \varphi_4)$$

$$= \frac{1}{\sqrt{2}}\left(\frac{y_4}{\sqrt{x_4^2 + y_4^2}} - \frac{x_4}{\sqrt{x_4^2 + y_4^2}}\right) = \frac{1}{\sqrt{5}} \tag{3.70}$$

$$\cos \Delta\varphi_4 = \cos(\varphi_4 - \pi/4) = (1/\sqrt{2})(\sin \varphi_4 + \cos \varphi_4) = 2/\sqrt{5}$$

Thus, the reduced projections according to (3.66), (3.69), and (3.70) are calculated by the formulas

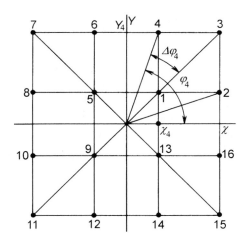

Figure 3.40 A 16-position signal in the nodes of a square array.

$$X_{1n} = \frac{1}{5}(2X_{0n} + Y_{0n}) \qquad Y_{1n} = \frac{1}{5}(2Y_{0n} - X_{0n}) \tag{3.71}$$

Similarly for the signal variant with number 2 we obtain $\sin \Delta\varphi_2 = -1/\sqrt{5}$, $\cos \Delta\varphi_2 = 2/\sqrt{5}$, and the reduced projections are equal to

$$X_{1n} = \frac{1}{5}(2X_{0n} - Y_{0n})$$
$$\tag{3.72}$$
$$Y_{1n} = \frac{1}{5}(2Y_{0n} + X_{0n})$$

The trigonometric functions of the other phase angles of the signal constellation of Figure 3.40 are obtained from the phase angles of signals 1 through 4. In fact, this signal constellation is formed by double phase modulation, so signals 5 through 8 are obtained by rotating signals 1 through 4 by 90 degrees; and signals 9 through 12 are obtained by rotating signals 1 through 4 by 180 degrees. As a result we have the algorithm for calculating the reduced projections, as shown in Table 3.8, which is similar to Table 3.6. These reduced projections are averaged, and therefore we get the estimates of the projections $\tilde{X}_1$ and $\tilde{Y}_1$ for the first signal variants. Then according to (3.65) the estimates of all projections $\tilde{X}_j$ and $\tilde{Y}_j$ are calculated. Results

Table 3.8
Reduced Projections for the 16-Level QAM in Figure 3.40

Decision Made on nth Chip	Parameters of the Decision $\Delta\tilde{\varphi}_n$	a_1/a_n	Reduced Projections X_{1n}	Y_{1n}
1	0	1	X_{0n}	Y_{0n}
2	$-\Delta\varphi_4$	$1/\sqrt{5}$	$1/5(2X_{0n} - Y_{0n})$	$1/5(2Y_{0n} + X_{0n})$
3	0	$1/3$	$1/3X_{0n}$	$1/3Y_{0n}$
4	$\Delta\varphi_4$	$1/\sqrt{5}$	$1/5(2X_{0n} + Y_{0n})$	$1/5(2Y_{0n} - X_{0n})$
5	$\pi/2$	1	Y_0	$-X_{0n}$
6	$\pi/2-\Delta\varphi_4$	$1/\sqrt{5}$	$1/5(X_{0n} + 2Y_{0n})$	$1/5(Y_{0n} - 2X_{0n})$
7	$\pi/2$	$1/3$	$1/3X_{0n}$	$-1/3X_{0n}$
8	$\pi/2 + \Delta\varphi_4$	$1/\sqrt{5}$	$1/5(2Y_{0n} - X_{0n})$	$-1/5(Y_{0n} + 2X_{0n})$
9	π	1	$-X_{0n}$	$-Y_{0n}$
10	$\pi - \Delta\varphi_4$	$1/\sqrt{5}$	$1/5(Y_{0n} - 2X_{0n})$	$-1/5(2Y_{0n} + X_{0n})$
11	π	$1/3$	$-1/3X_{0n}$	$-1/3 Y_{0n}$
12	$\pi + \Delta\varphi_4$	$1/\sqrt{5}$	$-1/5(2X_{0n} + Y_{0n})$	$1/5(2X_{0n} - Y_{0n})$
13	$3\pi/2$	1	$-Y_{0n}$	X_{0n}
14	$3\pi/2 + \Delta\varphi_4$	$1/\sqrt{5}$	$-1/5(X_{0n} + 2Y_{0n})$	$1/5(X_{0n} - 2Y_{0n})$
15	$3\pi/2$	$1/3$	$-1/3 Y_{0n}$	$1/3X_{0n}$
16	$3\pi/2 - \Delta\varphi_4$	$1/\sqrt{5}$	$1/5(X_{0n} - 2Y_{0n})$	$1/5(Y_{0n} + 2X_{0n})$

of these calculations are compiled in Table 3.9, which is similar to Table 3.7 (in Tables 3.8 and 3.9, $\Delta\varphi_4 = \arcsin 1/\sqrt{5}$).

Thus, Tables 3.8 and 3.9 together with expressions (3.67) and (3.68) form the algorithm for coherent reception of signals with 16-position QAM (Fig. 3.40).

The algorithms considered for coherent processing of the multiposition APM signals are especially convenient for multitone (multifrequency) modems with orthogonal channel signals because in these systems for dividing the orthogonal signals, we use the same procedure for calculating the received signal projections onto two orthogonal oscillations with an arbitrary initial phase [9]. The algorithms described for coherent processing do not require phase tuning of reference oscillations and permit quite simple realizations of both the orthogonal division of the channel signals and their coherent reception.

3.6 METHODS AND UNITS FOR SELECTING COHERENT REFERENCE OSCILLATIONS (CARRIER RECOVERY TECHNIQUES)

Classification of Methods and Units for Selecting Coherent Reference Oscillations

A reference oscillation coherent with the received signal must be generated for the operation of many coherent demodulator schemes [10–14]. The appropriate

Table 3.9
Signal Projection Estimates for the 16-Level QAM Signal in Figure 3.40

Variant of a Signal j	Variant Parameters $\Delta\varphi_j$	a_j/a_1	Projection Estimates $\tilde{X}_1$	$\tilde{Y}_j$
1	0	1	$\tilde{X}_1$	$\tilde{Y}_1$
2	$-\Delta\varphi_4$	$\sqrt{5}$	$2\tilde{X}_1 + \tilde{Y}_1$	$2\tilde{Y}_1 - \tilde{X}_1$
3	0	3	$3\tilde{X}_1$	$3\tilde{Y}_1$
4	$\Delta\varphi_4$	$\sqrt{5}$	$2\tilde{X}_1 - \tilde{Y}_1$	$\tilde{X}_1 + 2\tilde{Y}_1$
5	$\pi/2$	1	$-\tilde{Y}_1$	$\tilde{X}_1$
6	$\pi/2 - \Delta\varphi_4$	$\sqrt{5}$	$\tilde{X}_1 - 2\tilde{Y}_1$	$2\tilde{X}_1 + \tilde{Y}_1$
7	$\pi/2$	3	$-3\tilde{Y}_1$	$3\tilde{X}_1$
8	$\pi/2 - \Delta\varphi_4$	$\sqrt{5}$	$-\tilde{X}_1 - 2\tilde{Y}_1$	$2\tilde{X}_1 - \tilde{Y}_1$
9	π	1	$-\tilde{X}_1$	$-\tilde{Y}_1$
10	$\pi - \Delta\varphi_4$	$\sqrt{5}$	$-2\tilde{X}_1 - \tilde{Y}_1$	$\tilde{X}_1 - 2\tilde{Y}_1$
11	π	3	$-3\tilde{X}_1$	$-3\tilde{Y}_1$
12	$\pi + \Delta\varphi_4$	$\sqrt{5}$	$\tilde{Y}_1 - 2\tilde{X}_1$	$-\tilde{X}_1 - 2\tilde{Y}_1$
13	$3\pi/2$	1	$\tilde{Y}_1$	$-\tilde{X}_1$
14	$3\pi/2 - \Delta\varphi_4$	$\sqrt{5}$	$2\tilde{Y}_1 - \tilde{X}_1$	$-2\tilde{X}_1 - \tilde{Y}_1$
15	$3\pi/2$	3	$3\tilde{Y}_1$	$-3\tilde{X}_1$
16	$3\pi/2 + \Delta\varphi_4$	$\sqrt{5}$	$\tilde{X}_1 + 2\tilde{Y}_1$	$\tilde{Y}_1 - 2\tilde{X}_1$

units are known as units for selection of coherent reference oscillations or carrier recovery units. We will call them ROSs (reference oscillation selectors).

ROSs can operate by both a special synchronization signal and an information signal. If the synchronization signal is used, it contains the information about the selected reference oscillation phase, undistorted by a random information phase. Therefore, the ROS, operating by a special synchronization signal, contains a unit for measuring the current phase, a unit for accumulating measurement results, and a unit for properly generating a reference oscillation with a phase, determined by measurement and accumulation results.

There are two main types of ROSs: passive (or open) ROS and active (or closed) ROS. In elementary passive ROSs, the measurement and accumulation of a phase and the generation of a reference oscillation are implemented by a common unit, which is a narrowband filter or highly selective resonance unit, to which in some cases an adjustable phase shifter is added to compensate for a phase shift on the selected carrier frequency. Many methods are known to form such units: LC circuits, the circuits with distributed parameters, active RC circuits, including analog models based on operational amplifiers, quadrature splitters with low-pass filters, digital filters, and so on. The methods of forming and studying the passive ROS are based on the well-advanced theory of linear conversions [13].

Active ROSs are realized as automatic frequency control (AFC) units. The reference oscillation is formed in them by a controlled oscillator, the frequency of which changes in such a manner that it achieves the equality (with some accuracy) of this reference oscillation and synchronization signal phases. The oscillator is controlled by a phase discriminator (detector), which operates as a meter of phase difference between a reference oscillation and a signal. An AFC filter is frequently included between a discriminator and a controlled oscillator; it serves to average the results of the phase-difference measurements.

The problems of selecting coherent reference oscillations from a special synchronization signal do not have specific features at receiving PM and PDM signals.

For PM signals coherent processing methods for selecting reference oscillations by an information signal are of most interest. The passive and active circuits of ROSs are also used in this case; however, the formation of a signal, being a subject of the narrowband filtering, in passive ROSs and the control signal in active ROSs occurs by different means.

As long as the information phase of a signal is random, the methods of ROS design, based on direct filtering of the received signal or based on oscillator controlling by a signal, depending on phase difference between the received and reference signals, are unacceptable. To use these methods, first it is necessary to get rid of a random information phase or, as they say, remove phase modulation. The operation of modulation removal is called *signal remodulation*, and the methods of selecting reference oscillations out of an information signal are distinguished and classified by a remodulation way.

Five main classes of ROS units, distinguished by the means of their signal remodulation, are considered below:

1. Frequency multiplication;
2. Phase multiplication;
3. Phase reduction;
4. Direct remodulation of the received signal;
5. PM signal regeneration.

As a rule, at all methods of remodulation the ROS can be implemented according to passive and active circuits. For the first two remodulation methods, information about the decisions made in demodulator are not necessary. In the last three methods, such information is a main requirement—the appropriate ROSs are called the decision feedback units.

The selection of coherent reference oscillations can be considered to be a problem of optimum estimation of the signal initial phase at the noise influence. The maximum likelihood phase estimation is obtained as a result of the decision of the appropriate differential equation [10–14]. This estimation provides the minimum deviation of the reference oscillation phase. The schemes of the optimum ROS (in the sense of the maximum likelihood rule) are quite complex and are not widely used in practice. At the same time, many comparatively simple ROSs, based on the above-mentioned remodulation methods, do not practically concede by estimation accuracy and, consequently, by noise immunity to the optimum ones.

Selecting the Coherent Reference Oscillations by Signal Frequency Multiplication

The earliest developed method of remodulating the PM signal was based on its frequency multiplication. The essence of this method is as follows: To increase m times the argument of a trigonometric function, describing the PM signal with $2\pi/m$ phase hops, the phase modulation is removed, and the phase of the received signal does not depend on information phase hops. This phase repeats the initial phase of the signal carrier with an accuracy up to the value, multiple to $2\pi/m$.

The specified operation of multiplication of the argument of the function, describing the signal, is implemented by units, called *frequency multipliers*, hence, the name of the method. The frequency multiplier by m can be realized by means of a nonlinear element described by power series with a nonzero member of m power. At two-phase modulation ($m = 2$) the nonlinear element can be a quadrator or rectifier; at four-phase modulation ($m = 4$) it can be two quadrators connected in a series or one nonlinear element of the fourth order, and so on.

The scheme of a passive (open) ROS [Fig. 3.41(a)] contains a frequency multiplier by m, narrowband filter for frequency $m\omega$, frequency divider by m, and

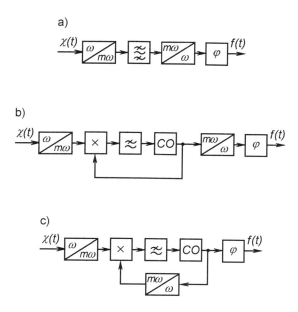

Figure 3.41 ROS versions with frequency multiplication.

adjustable phase shifter. The last one is necessary to compensate for a phase shift in the filter or other inertial elements of the scheme.

The schemes of active ROSs also contain a frequency multiplier that is connected to the AFC unit with a phase discriminator (multiplier and LPF) and a controlled oscillator (CO). The last one can generate the oscillations with the frequency of either $m\omega$ [Fig. 3.41(b)] or ω [Fig. 3.41(c)]. In the first case after CO it is necessary to connect a frequency divider by m; this divider can be also used for obtaining several reference oscillations shifted by a phase. In the second case, a frequency multiplier by m should be included in the AFC unit.

To obtain a very precise estimation of the initial phase in the considered schemes, it is necessary to provide high selectivity in accounting for the appropriate choice of filter parameters and integrating elements of the AFC. The passband of the ROS selective components should be at least an order less than the information signal passband. This is the main defect of ROSs with frequency multiplication: the necessity for signal narrowband filtering on the multiplied frequency. The other defect of these ROSs is connected with possible difficulties in the realization of the dividers, which need quite high frequencies. In some cases the specified defects can be overcome by the preliminary transformation of a signal to a lower intermediate frequency.

In the literature the passive scheme ROSs with frequency multiplication is called the *Pystolkors scheme* [6], and the active scheme of ROSs with frequency multiplication is called the *Siforov scheme* [15]. The Pystolkors scheme, which is widely used in practice, is comparatively simple, is steady in operation, and is not exposed to false pickups, as are the closed schemes with the use of controlled oscillators. The examples of coherent demodulators of signals with PDM, using the Pystolkors scheme, are illustrated in Section 3.3 [see Figs. 3.19(a), 3.21, and 3.29].

The ROS units with frequency multiplication permit us to create quasicoherent demodulators with rather insignificant energy losses when compared to the perfect coherent demodulator. At the same time, they require especially precise adjustment of the output oscillation phase at equipment tuning, and sometimes during operation.

Selecting the Coherent Reference Oscillations by Signal Phase Multiplication

ROSs with phase multiplication are subject to the same considerations discussed for ROSs with frequency multiplication: the remodulation principle—multiplying the argument of the trigonometric function, describing PM signal. However, the corresponding algorithms and schemes are different.

Briefly, the essence of operation of the ROS with phase multiplication can be formulated as follows: The coherent reference oscillation is generated by the oscillator, which is controlled by the sine of the m times increased phase difference between the output signal of the oscillator and the received signal; and the calculation of the specified phase difference is realized by quadrature components of the received signal.

Let us explain it in detail. Let, as before, the signal information phase take one of m values, that is, $\varphi_i = r\,2\pi/m$ where $r = 0, 1, \ldots, m-1$. Then the phase difference between the received signal and reference oscillation is equal to

$$\Delta\varphi = \Delta\varphi_0 + r\,2\pi/m$$

where $\Delta\varphi_0$ is an initial phase difference, independent of the transmitted information. If we make $\Delta\varphi$ phase m times more, the received m-valued difference $m\Delta\varphi$ depends only on the initial phase difference $\Delta\varphi_0$:

$$m\Delta\varphi = m(\Delta\varphi_0 + r\,2\pi/m) = m\Delta\varphi_0$$

and the sine of this m-valued difference can be used as the control signal in the AFC loop:

$$U = A\,\sin\,m\Delta\varphi \tag{3.73}$$

where A is an arbitrary coefficient.

If $m = 2^N$, (3.73) can be represented in the following form:

$$U = A \sin(2^N \Delta\varphi) = 2A \sin(2^{N-1}\Delta\varphi) \cos(2^{N-1}\Delta\varphi)$$
$$= 4A \sin(2^{N-2}\Delta\varphi) \cos(2^{N-2}\Delta\varphi) [\cos^2(2^{N-2}\Delta\varphi) - \sin^2(2^{N-2}\Delta\varphi)]$$
$$= \dots \tag{3.74}$$

Decreasing the argument of trigonometric functions (3.74) by a further power of 2, U can be expressed by $\cos\Delta\varphi$ and $\sin\Delta\varphi$, which, in their turn, are expressed by the projections X_0 and Y_0 of the received signal onto the orthogonal reference oscillations [see (3.3)] according to formulas

$$\cos\Delta\varphi = \frac{X_0}{\sqrt{X_0^2 + Y_0^2}} \tag{3.75}$$

$$\sin\Delta\varphi = \frac{Y_0}{\sqrt{X_0^2 + Y_0^2}}$$

Using (3.74) and (3.75), it is not difficult to obtain the following expressions for the control signals U_1, U_2, and U_3 for systems with two-, four-, and eight-phase PM:

$$U_1 = X_0 Y_0 \tag{3.76}$$

$$U_2 = X_0 Y_0^3 - Y_0 X_0^3 \tag{3.77}$$

$$U_3 = Y_0 X_0^7 - 4Y_0^3 X_0^5 + 4Y_0^5 X_0^3 - X_0 Y_0^7 \tag{3.78}$$

A general block diagram of ROS with phase multiplication is shown in Figure 3.42. Two correlators calculate projections X_0 and Y_0 of the received signal onto reference oscillations $f(t)$ and $f^*(t)$, generated by the controlled oscillator (CO) and phase shifter by $\pi/2$. The signals X_0 and Y_0 come to the input of a nonlinear converter (NC). The output signal of the last one controls the frequency of the

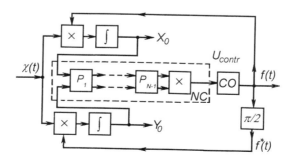

Figure 3.42 Reference oscillation selector with phase multiplication.

CO. The nonlinear converter, performing the operation of phase multiplying by quadrature projections, can operate in accordance with the general algorithm of (3.74) and (3.75), which takes the form of (3.76), (3.77), or (3.78) for various numbers of phases. In the scheme, the nonlinear converter contains $N - 1$ phase doublers (P) and a multiplier connected in series. The scheme of the phase doubler is shown in Figure 3.43.

As scheme input we have the projections of a signal with an initial phase; at the output we have the projections of a signal with doubled phase. At two-phase PDM the signal from the correlator comes directly to the multiplier; at four-phase PDM between the correlators and a multiplier, one phase doubler is included; at eight-phase PDM, two are included; and so on.

Here we consider the common scheme of ROS with phase multiplication, which generates the reference oscillation, coinciding by a phase with one of the received signal variants. In some cases in coherent demodulators reference oscillations are used that are shifted relative to signal variants by half of the minimum angle between them. In the ROS schemes with phase multiplication it can be quite simple to obtain such reference oscillations. For this it is enough to replace the projections X_0 and Y_0 with their linear combinations. For example, at four-phase PDM to make an additional phase shift $\pi/4$, in (3.77) or at the multiplier input (see Fig. 3.43), X_0 should be replaced by $(X_0 + Y_0)$, and Y_0 by $(X_0 - Y_0)$. This has been done in the scheme shown in Figure 3.22.

The coherent demodulators of signals with two-, four- and eight-phase PDM, using ROS with phase multiplication, were considered in Section 3.3 [see Figs. 3.19(b), 3.22, and 3.30]. Algorithms (3.76), (3.77), and (3.78) are realized in them.

The ROS with phase multiplication for the coherent demodulators of binary PM or PDM, shown in Figure 3.19(b) and operating according to algorithm (3.76), is called the *Costas scheme* [7]. Sometimes all the ROS units with phase multiplication as well as some other externally similar schemes are called Costas schemes. The preceding generalization of the carrier recovery algorithms by multiplying phase on the basis of quadrature projections of signals for multiphase systems is carried out by Ginzburg.

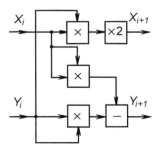

Figure 3.43 Phase doubler.

The ROSs with phase multiplication are widely used at the coherent processing of PM and PDM signals. They provide an error rate performance that is rather close to the potential one (that is why they are sometimes called *quasi-optimum ROSs*) and have some realization advantages. In particular, these ROSs do not require narrowband filtering of high-frequency oscillations, as do the ROSs with frequency multiplication. In addition, phase adjustment of an output oscillation is not required.

Defects of these units are that they have complex, nonlinear converters carrying out phase multiplication, and difficulties arise concerning the realization of steady operation of the active scheme with a controlled oscillator.

Selecting the Coherent Reference Oscillations by Phase Reduction

In the units considered here, as in the ROS with phase multiplication, a coherent reference oscillation is generated by a controlled oscillator operating in the close AFC scheme. The oscillator is also controlled by a signal, proportional to the sine of the phase difference between the reference oscillation and the remodulated received signal. However, the remodulation of a signal is realized not by phase multiplication, but by phase reduction using the decision about the transmitted signal.

In trigonometry the reduction operation is the replacement of trigonometric functions of an arbitrary angle by the trigonometric functions of some acute angle. This reduction operation consists of subtract some angle, divisible by 90 degrees, from an arbitrary angle. This operation is the basis for forming ROS algorithms with phase reduction.

Let, as before, the phase difference between the received signal and reference oscillation in the m-position system with PM or PDM be

$$\Delta\varphi = \Delta\varphi_0 + \varphi_{\text{inf}} = \Delta\varphi_0 + r2\pi/m \tag{3.79}$$

where $r = 0, 1, 2, \ldots, m - 1$. The first addend in (3.79) is an arbitrary initial phase difference, which has to be removed during the adjustment. The second addend is an information phase, varied from chip to chip by a value that is divisible by $2\pi/m$. Therefore, to remodulate the signal by phase reduction, this second angle (or some angle, distinguished from it by constant angle that is divisible by $2\pi/m$) should be subtracted from (3.79) on every chip. This condition of phase reduction will be executed if we subtract from (3.79) the phase $r\,2\pi/m$, for the benefit of which a decision has been made on the given chip and which either is equal to the information phase or differs from it by a constant value divisible by $2\pi/m$. We assume here, that the decision is correct. The mistaken decision results in deterioration of accuracy of signal phase estimation.

Thus, the signal, which controls the oscillator frequency in the AFC system, in this case is equal to

$$U = A \sin \Delta\varphi = A \sin(\Delta\varphi_0 + \varphi_{inf} - \tilde{r}2\pi/m)$$
$$= A[\sin(\Delta\varphi_0 + \varphi_{inf}) \cos(\tilde{r}2\pi/m) - \cos(\Delta\varphi_0 + \varphi_{inf}) \sin(\tilde{r}2\pi/m)] \quad (3.80)$$

Included in (3.80) $\sin(\Delta\varphi_0 + \varphi_{inf})$ and $\cos(\Delta\varphi_0 + \varphi_{inf})$ are expressed by projections X_0 and Y_0 of the received signal $x(t)$ with the information phase φ_{inf} onto the reference oscillation with the initial phase $\Delta\varphi_0$:

$$\begin{aligned} \sin(\Delta\varphi_0 + \varphi_{inf}) &= Y_0/A \\ \cos(\Delta\varphi_0 + \varphi_{inf}) &= X_0/A \end{aligned} \quad (3.81)$$

Using (3.81), the control signal can be presented in the following form:

$$U = Y_0 \cos(\tilde{r}2\pi/m) - X_0 \sin(\tilde{r}2\pi/m) \quad (3.82)$$

The general block diagram of the ROS with phase reduction is shown in Figure 3.44. Comparing this scheme with the scheme of Figure 3.42, note that in both schemes the oscillator, generating a reference oscillation, is controlled by a signal that is received from transformed projections of the received signal. The basic difference is that in the scheme with phase reduction the specified transformation uses the decision about the transmitted signal.

The advantage of this scheme is a more simple realization in comparison with ROS with phase multiplication. For example, at binary PM or PDM, when $m = 2$, $\sin(\tilde{r}2\pi/m) = 0$, and $\cos(\tilde{r}2\pi/m) = \text{sgn } X_0$, from (3.82) we obtain the following extremely simple algorithm for calculating the control signal:

$$U_1 = Y_0 \text{ sgn } X_0 \quad (3.83)$$

Algorithm (3.83) is realized by the demodulator, shown in Figure 3.45. This scheme is more simple than the scheme of the similar demodulator with phase multiplica-

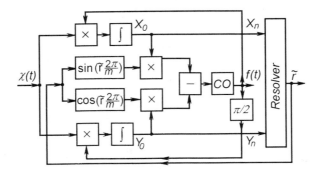

Figure 3.44 Reference oscillation selector with phase reduction.

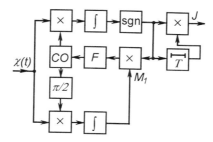

Figure 3.45 ROS with phase reduction for the binary PDM signals.

tion that is shown in Figure 3.19(b) because in this case the multiplier M_1 is the sign switch. The scheme in Figure 3.45 as well as in Figure 3.19(b) is sometimes called the Costas scheme. These schemes are very similar; however, we should not forget that various algorithms are realized in them, and that their performances are not identical.

At the 4-ary PM or PDM, when $m = 4$, the control signal according to (3.82) takes one of the values $Y_0, -X_0, -Y_0, X_0$ as $\bar{r}$ is equal to 0, 1, 2, and 3. The appropriate control signal driver can be easily realized as a switch of projections and their inverted values. Taking into account the direct relation between the decision $\bar{r}$ and signs of projections X_0 and Y_0, it is also possible to present the control signal in the following analytical form:

$$U_2 = (Y_0 + X_0)\ \text{sgn}(Y_0 - X_0) - (Y_0 - X_0)\ \text{sgn}(Y_0 + X_0) \tag{3.84}$$

At the control signal (3.84) the reference oscillation, generated by the oscillator, coincides with one of four possible signal variants. If in the demodulator reference oscillations are used, shifted by $\pi/4$ relative to the signal variants, the control signal is expressed by projections X_0 and Y_0 onto the shifted reference oscillations by the following formula:

$$U_2 = X_0\ \text{sgn}\ Y_0 - Y_0\ \text{sgn}\ X_0 \tag{3.85}$$

This algorithm is realized in a coherent demodulator of the 4-ary PDM signals, shown in Figure 3.23. Certainly, the scheme is appreciably easier than that of the similar demodulator with phase multiplication (see Fig. 3.22).

As can be seen, the realization advantages of the ROS with phase reduction in comparison with the ROS with phase multiplication are connected by replacing nonlinear conversion of the received signal projections with multiplying the last ones and sign functions, which, in their turn, are determined by demodulator decisions. These realization advantages, however, result in some degradation of accuracy in the estimation of the reference oscillation phase and the corresponding noise immunity losses.

Selecting the Coherent Reference Oscillations by Direct Signal Remodulation

The method of selection on the basis of the direct signal remodulation, conceptually contiguous to the method of phase reduction considered earlier, lies in that the received signal is modulated inversely to the modulation in the transmitter, and as a result, the effect of modulation removal is achieved.

Let us consider the received signal as a narrowband process, which can be presented within the chip interval as the following harmonic oscillation:

$$x(t) = a \sin(\omega t + \Delta\varphi_0 + \varphi_{\text{inf}}) \tag{3.86}$$

where, as before, $\Delta\varphi_0$ is the initial phase difference between $x(t)$ and the reference oscillation of a local oscillator, and φ_{inf} is the information phase. For modulation removal and converting (3.86) into a unmodulated signal it is possible, as noted earlier, to subtract the angle $\tilde{r}2\pi/m$ from the argument of this function, that is, the angle that corresponds to the decision $\tilde{r}$ received on the given chip. As a result of such operation, we receive the remodulated signal:

$$x_{\text{rem}}(t) = a \sin(\omega t + \Delta\varphi_0 + \varphi_{\text{inf}} - \tilde{r}2\pi/m) \tag{3.87}$$

The signal of (3.87) can be presented by the initial signal $x(t)$, determined by (3.86), and its Hilbert conversion $x^*(t)$. It should be taken into account that the decision $\tilde{r}$ is accepted at the end of a chip and, hence, it refers to a signal piece, delayed by one chip interval T. As a result we obtain

$$x_{\text{rem}}(t) = x(t - T) \cos(\tilde{r}2\pi/m) - x^*(t - T) \sin(\tilde{r}2\pi/m) \tag{3.88}$$

Now for obtaining the coherent reference oscillation it is possible to subject signal (3.88) to narrowband filtering or to transmit it to the AFC discriminator input.

Figure 3.46 illustrates the passive open circuit of ROS with direct signal remodulation. The remodulation is executed according to algorithm (3.88) by a delay

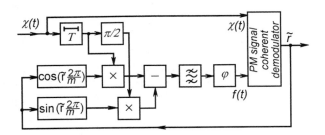

Figure 3.46 Passive ROS with direct signal remodulation.

line, the Hilbert converter (phase shifter by $\pi/2$), two multipliers, and a subtractor. The modulating signal is generated by converters $\sin(\tilde{r}2\pi/m)$ and $\cos(\tilde{r}2\pi/m)$, at the inputs of which there comes the decision $\tilde{r}$ from the demodulator output. Then the remodulated signal passes the narrowband filter and phase shifter; the last one serves to adjust for the initial phase of the reference oscillation $f(t)$, coming to the coherent demodulator.

Figure 3.47 illustrates the active closed scheme of ROS. The direct signal remodulation is fulfilled in it just the same as in the scheme of Figure 3.46, and then the remodulated signal comes to the AFC scheme, consisting of a multiplier, low-pass filter, and controlled oscillator.

The schemes of concrete ROSs with direct remodulation can be simplified in comparison with the schemes shown in Figures 3.46 and 3.47. For example, at binary PM or PDM, when $m = 2$, the second addend in (3.88) is equal to zero (hence, the phase shifter of the received signal is not needed), and the cosine in the first addend is equal to the sign of the signal X_0 at the correlator output of the coherent demodulator; that is,

$$x_{\text{rem}}(t) = x(t - T)\ \text{sgn}\ X_0 \tag{3.89}$$

According to this algorithm ROS operates in the coherent demodulator of the two-phase PDM [see Fig. 3.19(c)].

For signals of higher multiplicity it is also possible to synthesize the algorithm of direct remodulation using the projection signs of the received signal onto the reference oscillations. We shall make such an algorithm for an m-ary system with PM signal variants, separated from each other by $2\pi/m$. As before, let the received signal be presented within a chip interval as a narrowband oscillation

$$x(t) = a \sin(\omega t + \Delta\varphi_0 - r2\pi/m) \tag{3.90}$$

where $r = 0, 1, \ldots, m - 1$, and the reference oscillations are presented as follows:

$$f_j(t) = \sin(\omega t + \Delta\varphi_0 - \pi/2 - \pi/m + j2\pi/m) \tag{3.91}$$

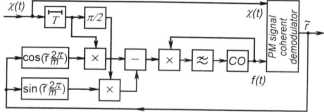

Figure 3.47 Active ROS with direct signal remodulation.

where $j = 1, 2, \ldots, m/2$. According to (3.90) and (3.91), for convenience in subsequent recordings the signal variants are numbered clockwise and the reference oscillations counterclockwise. The appropriate illustration for the case of the 8-ary system is depicted in Figure 3.48.

Now let us designate by X'_j the received signal projection (3.90) onto the reference oscillation (3.91), that is,

$$
X'_j = \int_0^T x(t)\, \sin\left(\omega t + \Delta\varphi_0 - \frac{\pi}{2} - \frac{\pi}{m} + j\frac{2\pi}{m} \right) dt \tag{3.92}
$$

Taking into account the designations (3.90), (3.91), and (3.92), it is possible to show [16] that the remodulated signal is

$$
x_{\text{rem}}(t) = \sum_{j=1}^{m/2} a \sin\left(\omega(t - T) + \Delta\varphi_0 - r\frac{2\pi}{m} + \left(\frac{m}{2} - j\right)\Delta\varphi \right)
$$
$$
\times \operatorname{sgn} X'_j = a\left(\cos\frac{\pi}{m}\right)^{-1} \times \sin\left(\omega t + \Delta\varphi_0 + \frac{\pi}{2} - \frac{\pi}{m} \right) \tag{3.93}
$$

The delay T is entered in (3.93) to coincide with the received signal chip and a result of the decision about projection sign related to this chip.

The remodulated signal is formed by summing $m/2$ samples of the received signal, shifted relative to each other by π/m with the signs of the received signal projections onto the corresponding reference oscillations. The amplitude of the remodulated signal is proportional to the cosecant of the angle π/m, and the phase coincides with the phase of one of the reference oscillations and does not depend on the information phase. For the case of an 8-ary PM signal, shown in Figure 3.48, the remodulated signal coincides by a phase with the reference oscillation f_4.

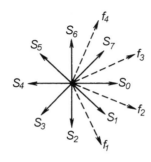

Figure 3.48 Reference oscillations in the eight-phase system.

The general block diagram of a passive ROS with direct signal remodulation is shown in Figure 3.49. The remodulator consists of $(m/2 - 1)$ phase shifters of the received narrowband signal, each of which changes its phase by $2\pi/m$, the same numbers of sign multipliers and an adder, at the output of which we have the remodulated signal $x_{\text{rem}}(t)$. After narrowband filtering the reference oscillation is formed from it. The remodulator and narrowband filter form the described ROS. The rest of the scheme includes phase shifters by $2\pi/m$ for formatting $m/2$ reference oscillations and $m/2$ correlators, calculating the received signal projections onto the reference oscillations. This part of the scheme directly concerns the coherent demodulator; its output signals come to a decoder.

In the scheme considered, the remodulated signal $x_{\text{rem}}(t)$ from the adder output can be transmitted to the AFC scheme to realize the active ROS with direct signal remodulation.

In terms of potential performances the ROSs with direct signal remodulation are equivalent to the ROSs with phase reduction. Their advantage is the comparatively easily selection of reference oscillations by both passive and active schemes with the controlled oscillator. The defect of the schemes with direct remodulation is the necessity of delay and the Hilbert conversion of the received signals.

Selecting the Coherent Reference Oscillations by PM Signal Regeneration

The essence of methods of selecting coherent reference oscillations by PM signal regeneration consists of the following: The control signal in the AFC unit is generated by comparing the received PM signal with the reference oscillation, modulated by phase in the same way as the received signal.

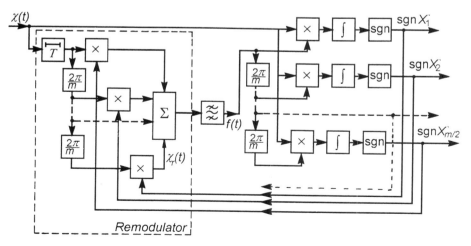

Figure 3.49 ROS with direct remodulation of the multiphase signal.

Hence, it is clear that for realizing the method it is necessary to determine the received signal modulation rule (to demodulate it) and then to modulate the reference oscillation according to this rule, that is, to regenerate the original PM signal.

The regenerated PM signal, coming to the discriminator input of the AFC unit, can be presented by analogy with (3.86) in the following form:

$$S_{reg}(t) = a \sin(\omega t + \Delta\varphi_0 + \tilde{r}2\pi/m) \tag{3.94}$$

The last addend in the argument of the sine is determined by the decision $\tilde{r}$ in the demodulator; it is either equal to the information phase φ_{inf} of the received signal (3.86) or differs from φ_{inf} by a constant value, a multiple of $2\pi/m$. As a result, the AFC unit, in the discriminator of which the one-chip delayed received signal (3.86) and the regenerated PM signal (3.94) are compared, will produce the reference oscillation, coinciding by the phase with one of the received signal variants. It is convenient to present (3.94) in the form

$$S_{reg}(t) = f(t) \cos(\tilde{r}2\pi/m) + f^*(t) \sin(\tilde{r}2\pi/m) \tag{3.95}$$

where $f(t) = a \sin(\omega t + \Delta\varphi_0)$ is the reference oscillation generated by the oscillator.

The general block diagram of the ROS with PM signal regeneration corresponding to the described algorithm is shown in Figure 3.50. In the phase modulator, in contrast to a similar modulator in the scheme of Figure 3.47, it is not the received signal that is modulated, but the reference oscillation, which is generated by the controlled oscillator. In all other items the schemes of Figures 3.50 and 3.47 are identical.

The modifications of the ROS with PM signal regeneration for the cases of two-, four-, and eight-phase PDM signals are considered in Section 3.3 [see the schemes in Figs. 3.19(d), 3.24, and 3.31].

The algorithms indicated in this section and the schemes of units for selecting reference oscillations can be realized by both analog and digital methods. However algorithmically they are oriented mainly to analog realization.

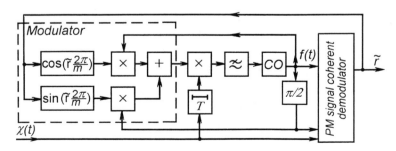

Figure 3.50 Reference oscillation selector with PM signal regeneration.

A more adequate approach for digital implementation consists of the algorithms considered in this chapter on coherent processing, based on the evaluation of signal parameters by means of the received signal projections onto the reference oscillations with an arbitrary phase. These algorithms do not contain the coherent reference oscillations and include only elementary arithmetic operations under two projections of the received signal. Consequently they are easily realized if the hardware and software of digital processing are used, particularly, special digital signal processors.

References

[1] Messel, A. F., and Yu. B. Okunev, "Effective Quasicoherent Algorithms of Phase Modulated Signal Reception," *Izvestiya VUZ, Radioelectronika, No. 5, Comm. Syst.,* Allerton Press Inc., 1991.

[2] Levin, B. R., "Theoretical Foundations of Statistical Radio Engineering," *Sov. Radio,* Vol. 1–3, 1976.

[3] Davenport, W. B., Jr., *Probability and Random Process,* New York: McGraw-Hill Book Company, 1970.

[4] Zayezdny, A. M., D. Tabak, and D. Wulich, *Engineering Application of Stochastic Processes,* New York: John Wiley & Sons, 1989.

[5] Petrovich, N. T., "New Methods of Phase Telegraphy," *Radiotechnika,* No. 10, 1957, pp. 47–54.

[6] Pystolkors, A. A., "Multivalued Telegraphy by Phase Changing," *News of Weak Current Electroindustry,* No. 3, 1935, pp. 51–58.

[7] Costas, J. P., "Synchronous Communications," *Proc. IRE,* Vol. 44, December 1956, pp. 1713–1718.

[8] Okunev, Yu. B., "Generalization of the Algorithms of the Operation of Coherent Demodulators of Binary Signals with Phase-Shift Keying," *Telecom. Radioeng.,* Vol. 33/34, No. 10, 1979.

[9] Berkman, L. N., and Yu. B. Okunev, "Quasicoherent Processing of QAM Signals in Multichannel Modems," *Telecom. Radioeng.,* No. 10, 1991.

[10] Viterbi, A. J., *Principles of Coherent Communication,* New York: McGraw-Hill Book Company, 1966.

[11] Stiffler, J. J., *Theory of Synchronous Communications,* Englewood Cliffs, NJ: Prentice-Hall, 1971.

[12] Lindsey, W. C., *Synchronization Systems in Communications,* Englewood Cliffs, NJ: Prentice-Hall, 1972.

[13] Ginzburg, V. V., and A. A. Kayatkas, *Demodulator Synchronization Theory,* Moscow: Svyaz, 1974.

[14] Tikhonov, V. I., and N. K. Kulman, "Nonlinear Filtering and Quasicoherent Reception of Signals," *Sov. Radio,* 1975.

[15] Siforov, V. I., *On Some New Synchronization Systems for Receiving Phase Telegraph Transmission,* Moscow: Svyazizdat, No. 5, 1937, pp. 66–71.

[16] Antonov, G. V., and I. V. Gurevich, Demodulator. USSR Patent 1317665.

CHAPTER 4

Optimum Incoherent Processing of PDM Signals

4.1 ESSENCE AND APPLICATION OF INCOHERENT SIGNAL PROCESSING

The common feature of incoherent signal processing, considered in this chapter, lies in the fact that the initial phase of the received signal is assumed to be unknown. Here we are concerned with two situations: (1) when the initial phase of the received element is random and cannot be readily determined or estimated beforehand and (2) when the initial phase can be basically determined with some degree of accuracy (as is done, for example, for quasicoherent reception), but for some reason such an opportunity is not used.

In the theory of optimum signal processing the two situations are united by the term *channel with random* (or *unknown, uncertain*) *phase*; and the corresponding methods of processing, which do not require knowledge of the initial signal phase, are referred to as *incoherent*.

Among the available incoherent methods, we can choose from bad ones and good ones from the point of view of noise immunity and other system parameters. If we enter the appropriate goal function and restrictions, it is possible to synthesize an optimum algorithm for incoherent processing.

Usually in the general theory of communications for optimum incoherent reception it is understood that the incoherent method (algorithm) of signal processing provides the minimum error probability in the AWGN channel with uncertain, random, and uniformly distributed initial phase of the received element.

All other noninformation parameters of a signal (except the initial phase) especially the carrier frequency, should be known precisely. As for the signal amplitude, when using the signals with equal energy it can be unknown.

From the definition of optimum incoherent reception, it follows that this method can be used only for phase-difference modulation or, more strictly speaking, only for determining the transmitted phases difference, instead of the absolute phase, which for this method of processing is essentially unknown.[1]

So, there exists an optimum (in the above-specified sense) incoherent method for PDM signal processing. There is a question: What is the relation of this method to the earlier considered coherent method of processing in terms of noise immunity? The answer to this question is comparatively simple only for two extreme cases, corresponding to the strict conditions at which coherent and incoherent techniques are optimum.

If the initial phase of the received signal elements is unknown and cannot be estimated by the previous history of the process, the coherent (or quasicoherent) demodulator is disabled, while the incoherent demodulator provides reasonably high noise immunity and is outside of competition.

If the initial phase of the receiving signal elements is known or can be estimated by the previous history of the process with high accuracy, then both the coherent and incoherent demodulators are able to operate; and the former, naturally, has higher noise immunity. The incoherent processing energy loss depends on a number of phase levels, a number of elements of the signal being processed, and other factors. These problems are considered in a separate chapter.

Here we note the qualitative feature of this problem. This feature means that the energy loss of optimum incoherent reception to the coherent one grows in accordance with the increase of the number of phase levels and dimension of the signal being processed. In particular, at four-level PDM the loss is more than at two-level PDM; when receiving the complex signal-code blocks as a whole the loss is more than that for element-by-element reception.

One of the important results of noise immunity theory is the following statement: During element-by-element reception of two-level PDM signals in a channel with known initial signal phase, the coherent reception energy gain relative to the optimum incoherent one is insignificant; it is not the basis on which we would end up preferring the coherent method; in this case the choice should be made by other parameters, for example, by complexity of realization.

Sometimes engineers try to synthesize rather complex quasicoherent demodulators of two-phase signals with PDM or PM, which does not make practical sense—the optimum incoherent demodulator actually solves the problem.

On the other hand, when receiving multilevel signals with PDM and QAM, the gain of coherent reception can appear to be significant, and here it makes sense to aim for its realization. At the reception of composite phase-modulated

---

1. It, by the way, shows that the PDM signal processing (in the common case) is not reduced, as is sometimes asserted, to differential decoding of the received PM signals. This occurs only for coherent reception. The PDM signals can be received by incoherent methods, which are principally impossible at PM.

signals with accumulation (for example, at diverse reception), the relationship between noise immunity of coherent and incoherent processing depends on the correlation between accumulated element phases. If element phases are random and uncorrelated, incoherent reception with incoherent element accumulation is used, which considerably concedes to the strictly coherent processing of a composite signal, if such processing were possible. If the initial phase is unknown but identical for all elements, incoherent reception with coherent accumulation is used, which only slightly concedes to the coherent reception.

Between the two specified cases—the exactly known or absolutely unknown initial phase—there is a large spectrum of intermediate situations, such as when the initial phase is unknown and changed on time, but can be estimated by the previous history of the process. The noise immunities of quasicoherent and optimum incoherent methods of processing at these intermediate situations depend on the dynamics of signal phase and signal frequency changing and a priori information about these changes. The real noise immunity also depends on element duration, communications session duration, the interval of phase correlation in the channel, and on the algorithmic and technical means for tracking the frequency and phase. As a result the scale is balanced between the coherent and optimum incoherent techniques. Some results of the analysis of similar situations are indicated in Chapter 6; however, every special case for choosing the reception method requires, as a rule, a particular system analysis.

We will specify the system situations for which incoherent reception is optimum. The unknown, uncertain initial phase of a signal takes place at the beginning of any communications session, during the entire reception interval for short-term sessions, during transmission of the information by separate pulses or batches, divided by passive pauses, and in all other cases when the previous history of a signal is either absent or is too short.

Similar situations can also take place in systems with continuous information transmission operating in the channels with variable parameters. For example, in channels with deep fading or signal interruptions, the signal has an uncertain initial phase (and sometimes also carrier frequency) if the interval of fading correlation (or interruption duration) surpasses or is compared with the interval of coherence (correlation) of the initial phase. At hop-like changes of a signal phase, there also appears a situation that corresponds to the conditions of incoherent reception application: At quasicoherent reception each hop results in error propagation because of the reorganizing (entry in synchronism) of the adaptive carrier recovery unit, whereas at incoherent reception there appears to be only one or few errors at the moment of a hop.

In general, in the channels with comparatively fast parameter changes, when the interval of coherence is too small to provide the quite exact phase estimate, incoherent reception can give less error rate than quasicoherent reception. The realization losses of quasicoherent reception because of the reference phase errors can be more than the theoretical loss of the optimum incoherent reception to the coherent one. Thus, for the situations mentioned, incoherent reception is preferable.

Sometimes incoherent reception is used for other reasons. For example, in the channel with constant parameters the coherent reception of binary PDM signals has greater noise immunity than the optimum incoherent one; however, the gain is so insignificant that, as they say, the game is not worth the candles. The incoherent demodulator does not require formation of the coherent reference signals and hence, it is more simple than the coherent one. In a number of cases, this last circumstance is the decisive factor when choosing between coherent and optimum incoherent methods of reception.

Besides the optimum incoherent algorithm there exist various suboptimum incoherent algorithms—comparable to it in terms of noise immunity—that can be realized more easily.

Finally, we should mention the following important circumstance, which was proven comparatively recently and is attracting additional attention to the incoherent signal processing technique. As is known, during coherent reception of signals without redundancy, the transition from element-by-element processing to multi-element processing of a signal as a whole does not result in an increase in noise immunity [1]. This appears not to be true for the optimum incoherent reception of PDM signals: The increase of the processing interval and the transition to the multisymbol optimum incoherent reception of the PDM signal as a whole leads to a decrease in the error probability, which approximates the error probability of the perfect coherent reception of the same signals.

Certainly, this conclusion has been made according to the assumption that the initial signal phase remains constant during the whole considered interval of processing. Some of these algorithms for multisymbol optimum incoherent reception of PDM signals as a whole are derived later.

Thus, the optimum incoherent reception of the PDM signals permits, by increasing the interval of processing, us to approximate as much as desired the potential noise immunity without estimating the signal carrier phase.

The optimum incoherent demodulators of PDM signals are widely applied in multitone (multichannel) systems with orthogonal channel signals, which are used for data transmission over HF radio channels and wire channels. The first system with PDM described in the literature was based on optimum incoherent signal processing [2–5].

4.2 OPTIMUM INCOHERENT RECEPTION OF FIRST-ORDER PDM SIGNALS

Synthesis of the General Optimum Algorithm

To obtain an algorithm for optimum incoherent reception of PDM signals we use the following result of the general theory for optimum signal processing [1,6].

Let one of m equiprobable and equally powerful signals with duration τ and a random, uniformly distributed initial phase φ

$$S_{j\varphi}(t) = S_j(t) \cos \varphi + S_j^*(t) \sin \varphi \qquad \text{for } j = 1, 2, \ldots, m$$

and the white Gaussian noise $n(t)$ come to the demodulator input:

$$x(t) = S_{j\varphi}(t) + n(t)$$

Then the error probability will be minimum, if the demodulator makes a decision in favor of such a signal S_i, for which the following inequality is correct for any $j \neq i$:

$$\left[\int_0^\tau x(t)\, S_i(t)\, dt \right]^2 + \left[\int_0^\tau x(t)\, S_i^*(t)\, dt \right]^2 > \left[\int_0^\tau x(t)\, S_j(t)\, dt \right]^2 + \left[\int_0^\tau x(t)\, S_j^*(t)\, dt \right]^2$$

(4.1)

where S_j^* is the Hilbert transformation of signal S_j.

For PDM-1 the value of τ is equal to the duration of two signal chips, that is, $\tau = 2T$. Therefore the algorithm of the optimum incoherent demodulator is as follows: Within the interval of the nth chip, a decision should be made in favor of the signal S_i with the phase difference $\Delta\varphi_i$

$$S_i(t) = \begin{cases} a \sin \omega t & \text{for } 0 < t \leq T \\ a \sin(\omega t + \Delta\varphi_i) & \text{for } T < t \leq 2T \end{cases}$$

(4.2)

if the following inequality takes place for any $j \neq i$:

$$\left[\int_{(n-1)T}^{nt} x(t) \sin \omega t\, dt + \int_{nt}^{(n+1)T} x(t) \sin(\omega t + \Delta\varphi_i)\, dt \right]^2$$

$$+ \left[\int_{(n-1)T}^{nt} x(t) \cos \omega t\, dt + \int_{nt}^{(n+1)T} x(t) \cos(\omega t + \Delta\varphi_i)\, dt \right]^2$$

$$> \left[\int_{(n-1)T}^{nt} x(t) \sin \omega t\, dt + \int_{nt}^{(n+1)T} x(t) \sin(\omega t + \Delta\varphi_j)\, dt \right]^2$$

$$+ \left[\int_{(n-1)T}^{nt} x(t) \cos \omega t\, dt + \int_{nt}^{(n+1)T} x(t) \cos(\omega t + \Delta\varphi_j)\, dt \right]^2$$

(4.3)

Having made in (4.3) the simple conversions and reductions and having used the designations

$$X_{n-1} = \int_{(n-1)T}^{nT} x(t) \sin \omega t \, dt \qquad X_n = \int_{nt}^{(n+1)T} x(t) \sin \omega t \, dt$$

$$Y_{n-1} = \int_{(n-1)T}^{nT} x(t) \cos \omega t \, dt, \qquad Y_n = \int_{nt}^{(n+1)T} x(t) \cos \omega t \, dt$$

(4.4)

we obtain the following algorithm:

$$\cos \Delta\varphi_i(X_{n-1}X_n + Y_{n-1}Y_n) + \sin \Delta\varphi_i(X_{n-1}Y_n - Y_{n-1}X_n)$$
$$> \cos \Delta\varphi_j(X_{n-1}X_n + Y_{n-1}Y_n) + \sin \Delta\varphi_j(X_{n-1}Y_n - Y_{n-1}X_n)$$
$$\text{for } j = 1, 2, \ldots, m \quad j \neq i$$

(4.5)

The obtained inequalities (4.5) represent the most general algorithm for optimum incoherent reception of PDM-1 signals and are suitable for any set of information phase differences $\Delta\varphi_1, \Delta\varphi_2, \ldots, \Delta\varphi_m$.

The essence of the obtained algorithm (4.5) is the following: In the demodulator the cosine functions of the angle differences are calculated between (1) the angle between the vectors of two adjacent received signal chips and (2) all permitted angles between the vectors of two adjacent transmitted signal chips. Thus, preference is given to that of admissible angles (alternatives), for which the specified cosine is maximum, that is, to that phase difference which is closest to the received one.

The peculiarity of the optimum incoherent reception is that the angles between the vectors of the adjacent signal chips are calculated by their projections onto two mutually orthogonal oscillations $\sin \omega t$ and $\cos \omega t$ (quadrature reception). In fact, the signal $x(t)$, received within a chip interval, can be considered to be a vector of two-dimensional Euclidean space. Then the values X and Y, calculated by (4.4), are the projections of the received signal chip onto the coordinate axes $\sin \omega t$ and $\cos \omega t$. Through these projections, using the formulas of vector algebra, it is possible to calculate trigonometric functions of the angle $\Delta\varphi_{n\xi}$ between the vectors of the nth and $(n-1)$th chips of the received signal:

$$\cos \Delta\varphi_{n\xi} = \frac{X_{n-1}X_n + Y_{n-1}Y_n}{\sqrt{X_{n-1}^2 + Y_{n-1}^2}\sqrt{X_n^2 + Y_n^2}}$$

$$\sin \Delta\varphi_{n\xi} = \frac{X_{n-1}Y_n - Y_{n-1}X_n}{\sqrt{X_{n-1}^2 + Y_{n-1}^2}\sqrt{X_n^2 + Y_n^2}}$$

(4.6)

Thus, the expressions within the brackets in (4.5) are proportional to the cosine and sine of the phase difference between the nth and $(n-1)$th chips of the received signal. Therefore (4.5) can be presented in the following equivalent form:

$$\cos \Delta\varphi_i \cos \Delta\varphi_{n\xi} + \sin \Delta\varphi_i \sin \Delta\varphi_{n\xi}$$
$$> \cos \Delta\varphi_j \cos \Delta\varphi_{n\xi} + \sin \Delta\varphi_j \sin \Delta\varphi_{n\xi}$$
$$\text{for } j = 1, 2, \ldots, m \quad j \neq i \tag{4.7}$$

or more simply

$$\cos (\Delta\varphi_{n\xi} - \Delta\varphi_i) > \cos (\Delta\varphi_{n\xi} - \Delta\varphi_j) \qquad \text{for } j = 1, 2, \ldots, m \quad j \neq i \tag{4.8}$$

Now we can derive from (4.5) the algorithms for special cases.

Binary PDM-1 Systems

At two-level PDM-1 with phase differences $\Delta\varphi_1 = 0$ and $\Delta\varphi_2 = \pi$ we can obtain the following algorithm from (4.5):

$$X_{n-1} X_n + Y_{n-1} Y_n > -(X_{n-1} X_n + Y_{n-1} Y_n) \tag{4.9}$$

that is, the transmitted binary symbol J_n is equal to

$$J_n = \text{sgn}(X_{n-1} X_n + Y_{n-1} Y_n) \tag{4.10}$$

The obtained algorithm of (4.10) differs also in its external elegance and plays an important role in the theory of PDM signals processing. This algorithm is realized by the demodulator shown in Figure 4.1. In this scheme, in contrast to the most usable schemes of quasicoherent demodulators, the reference oscillator

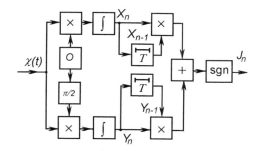

Figure 4.1 Optimum incoherent demodulator for binary PDM-1 signals.

(O) is uncontrolled, and it has random initial phase.[2] It essentially simplifies the realization of a demodulator. At the same time the frequency of the reference oscillator should coincide with the frequency of the PM signal carrier; slight frequency deviations are admissible, but they result in a noise immunity decrease. For operation of some demodulator units, such as an integrator with reset, memory element for chip duration *T*, sign determination unit "sgn," it is necessary to add a time recovery unit, which is not shown in the scheme.

Note the remarkable property of the obtained algorithm and the considered demodulator (see Chapter 6): The demodulator rather slightly (less than 1 dB) concedes in terms of noise immunity to the ideal coherent demodulator.

At the same time it is more simple than the coherent demodulators of two-phase signals, which is obvious from a comparison of the schemes of Figure 4.1 with Figures 3.7 and 3.19. In the scheme of Figure 4.1, the delay units and units for multiplying the analog values at the integrator output are of certain complexity at analog implementation.

It is basically possible to realize optimum incoherent reception without storing the results of previous chip processing, as in the Figure 4.1 demodulator. To carry this out, we should use the original algorithm (4.1), which, for the case of binary PDM-1 with phase differences $\Delta\varphi_1 = 0$ and $\Delta\varphi_2 = \pi$, can be transformed to the following form:

$$
J = \text{sgn}\left\{ \left[\int_0^{2T} x(t)\,\sin\,\omega t\,dt \right]^2 + \left[\int_0^{2T} x(t)\,\cos\,\omega t\,dt \right]^2 \right.
$$

$$
- \left[\int_0^{2T} x(t)\,\sin\,\omega t \left(\text{sgn}\,\sin\,\frac{\pi}{T}t \right) dt \right]^2
$$

$$
\left. - \left[\int_0^{2T} x(t)\,\cos\,\omega t \left(\text{sgn}\,\sin\,\frac{\pi}{T}t \right) dt \right]^2 \right\} \tag{4.11}
$$

The function sgn $\sin(\pi t/T)$ describes the change of the reference oscillation phase by 180 degrees at the boundary of two adjacent chips $(0,T)$ and $(T,2T)$. In the corresponding demodulator [Fig. 4.2(a)] this function is formed by a locking unit (LU), which generates a periodic oscillation sgn $\sin(\pi t/T)$. Reference oscillations

---

2. Quasicoherent reception, as was shown in Chapter 3, can also be realized by using an incoherent reference oscillator. The comparison of the obtained algorithm (3.18) with algorithm (4.10) is made in remarks to formula (3.18).

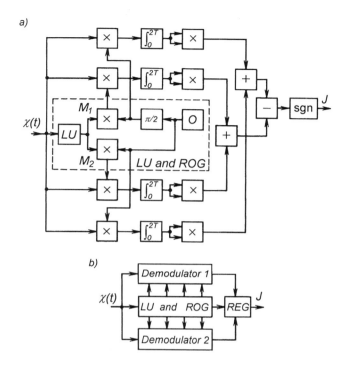

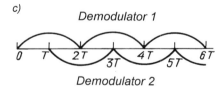

Figure 4.2 Optimum incoherent reception of binary PDM-1 signals by processing two chips as a whole.

$$f_1(t) = \sin \omega t$$
$$f_2(t) = \cos \omega t$$
$$f_3(t) = \sin \omega t \ \text{sgn} \ \sin(\pi t/T)$$
$$f_4(t) = \cos \omega t \ \text{sgn} \ \sin(\pi t/T)$$

arriving at the inputs of four correlators are formed by a local oscillator (O), $\pi/2$ phase shifter, and two multipliers (M_1 and M_2) of LU output oscillation and the direct and $\pi/2$-shifted oscillations from the local oscillator. This demodulator processes successive chip pairs: 1 and 2, 3 and 4, 5 and 6, and so on. For processing the adjacent chips of these pairs, that is, 2 and 3, 4 and 5, 6 and 7, and so on, one more demodulator is needed.

Thus, the receiving unit, realizing the optimum incoherent reception of PDM-1 signals on the basis of processing two chips as a whole, consists, as shown in Figure 4.2(b), of two identical demodulators operating with the one chip shift [Fig. 4.2(c)] and the locking and reference oscillation generation units (LU and ROG). Binary symbols, received from demodulator outputs through the chip, are combined by the output register (Reg).

Note that this demodulator looks more complex than the demodulator of Figure 4.1, even though it performs the same functions. However, its advantage lies in the absence of analog memory elements. For digital realization, algorithm (4.10) and the scheme of Figure 4.1 are doubtless more preferable than algorithm (4.11) and the scheme of Figure 4.2, because in this case the delay of correlator output signals and their multiplication are not complex.

The comparison of realization complexity of the considered algorithms is convenient to fulfill, if we write algorithm (4.11) in terms of the values of (4.4) in the following form:

$$J_n = \text{sgn}[(X_{n-1} + X_n)^2 + (Y_{n-1} + Y_n)^2 - (X_n - X_{n-1})^2 - (Y_n + Y_{n-1})^2] \tag{4.12}$$

It is not difficult to establish that algorithms (4.10) and (4.12) are equivalent.

Multiphase PDM-1 Systems

In four-phase PDM modems, basically two systems of signals are applied, as shown in Table 4.1. Both systems use the binary keying Gray code, which, as has been shown, is a unique optimum code for the four-phase system.

First let us find an algorithm for the optimum incoherent demodulator for the first system. Having substituted the appropriate phase differences

Table 4.1
Phase-Difference Options for Four-Level PDM

Phase Difference Designation	Phase Differences in the First System	Second System	Gray Code	
$\Delta\varphi_1$	$\pi/4$	0	+	+
$\Delta\varphi_2$	$3\pi/4$	$\pi/2$	+	−
$\Delta\varphi_3$	$5\pi/4$	π	−	−
$\Delta\varphi_4$	$7\pi/4$	$3\pi/2$	−	+

$$\Delta\varphi_i = (2i - 1)\,\pi/4 \qquad \text{for } i = 1, \ldots, 4$$

in general algorithm (4.7), we get the following four sums (a common multiplier $1/\sqrt{2}$ is omitted):

$$
\begin{aligned}
a &= \cos \Delta\varphi_{n\xi} + \sin \Delta\varphi_{n\xi} & &\to \Delta\varphi_n = \pi/4 \\
b &= -\cos \Delta\varphi_{n\xi} + \sin \Delta\varphi_{n\xi} & &\to \Delta\varphi_n = 3\pi/4 \\
c &= -\cos \Delta\varphi_{n\xi} - \sin \Delta\varphi_{n\xi} & &\to \Delta\varphi_n = 5\pi/4 \\
d &= \cos \Delta\varphi_{n\xi} - \sin \Delta\varphi_{n\xi} & &\to \Delta\varphi_n = 7\pi/4
\end{aligned}
\tag{4.13}
$$

Thus, the demodulator first calculates the values a, b, c, and d; then it determines the greatest of them and the corresponding transmitted difference $\Delta\varphi_n$; last, it makes decision about the transmitted binary combination according to Table 4.1. However, it is more convenient to express the transmitted binary symbols directly through the received signals. This can be done by comparing (4.13) with the data of Table 4.1. Let, for example, the greatest of four sums be a and, hence, the symbol "+" has been transmitted in both binary subchannels. On the basis of that we can conclude that $\cos \Delta\varphi_{n\xi} > 0$, and $\sin \Delta\varphi_{n\xi} > 0$. As can be seen, in this case the signs of the cosine and sine of the received phase difference coincide with the signs in binary subchannels. It is not difficult to determine that this is true in the other three situations as well.

Thus, the information symbols J_{1n} and J_{2n} in binary subchannels can be submitted in the following form:

$$
\begin{aligned}
J_{1n} &= \text{sgn} \sin \Delta\varphi_{n\xi} \\
J_{2n} &= \text{sgn} \cos \Delta\varphi_{n\xi}
\end{aligned}
\tag{4.14}
$$

or by the projections (4.4) of the received signal:

$$
\begin{aligned}
J_{1n} &= \text{sgn}(X_{n-1} Y_n - X_n Y_{n-1}) \\
J_{2n} &= \text{sgn}(X_{n-1} X_n + Y_n Y_{n-1})
\end{aligned}
\tag{4.15}
$$

The scheme of the corresponding demodulator is shown in Figure 4.3.

Algorithm (4.15) was realized for the first time in a multitone system MC with orthogonal signals and four-level PDM, which was intended for digital data transmission over HF radio channels [4,7]. The algorithm equivalent to (4.15) was also used in the Kineplex system [2,3].

We can find an algorithm that is similar to (4.15) for demodulating the second system of signals (see Table 4.1). Having substituted phase differences

$$\Delta\varphi_i = i\pi/2 \ (i = 0, 1, 2, 3)$$

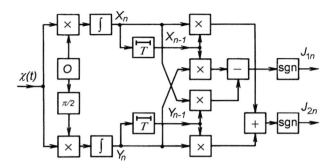

Figure 4.3 Optimum incoherent demodulator for four-level PDM-1 signals at phase differences $\pi/4$, $3\pi/4$, $5\pi/4$, and $7\pi/4$.

in the common algorithm (4.7), we should be able to calculate the following values in the demodulator:

$$
\begin{aligned}
a &= \cos \Delta\varphi_{n\xi} & &\rightarrow \Delta\varphi_n = 0 \\
b &= \sin \Delta\varphi_{n\xi} & &\rightarrow \Delta\varphi_n = \pi/2 \\
c &= -\cos \Delta\varphi_{n\xi} & &\rightarrow \Delta\varphi_n = \pi \\
d &= -\sin \Delta\varphi_{n\xi} & &\rightarrow \Delta\varphi_n = 3\pi/2
\end{aligned}
\tag{4.16}
$$

In (4.16) decisions are specified regarding what the demodulator should give if the corresponding value is greater than all others.

Let us compare (4.16) with data of Table 4.1. Let the greatest value be b, and, hence, the symbol "+" is transmitted by the first binary subchannel, and the symbol "–" by the second one. From $b > c$ and $b > a$ we get that $\sin \Delta\varphi_{n\xi} > -\cos \Delta\varphi_{n\xi}$ and $\sin \Delta\varphi_{n\xi} > \cos \Delta\varphi_{n\xi}$, that is,

$$
\begin{aligned}
\sin \Delta\varphi_{n\xi} + \cos \Delta\varphi_{n\xi} &> 0 \\
\cos \Delta\varphi_{n\xi} - \sin \Delta\varphi_{n\xi} &< 0
\end{aligned}
\tag{4.17}
$$

Thus, in this case the symbols, transmitted by binary subchannels, coincide with signs of values specified in (4.17). It is not difficult to determine that in the other three cases, when the maximum value is a, c, or d, the transmitted binary symbols are determined by the signs of expressions (4.17).

So, the required algorithm for optimum incoherent reception for four-level PDM-1 signals according to (4.17) and (4.6) is:

$$
\begin{aligned}
J_{1n} &= \operatorname{sgn}\left[\left(X_{n-1} Y_n - X_n Y_{n-1}\right) + \left(X_{n-1} X_n + Y_{n-1} Y_n\right)\right] \\
J_{2n} &= \operatorname{sgn}\left[\left(X_{n-1} X_n + Y_{n-1} Y_n\right) - \left(X_{n-1} Y_n - X_n Y_{n-1}\right)\right]
\end{aligned}
\tag{4.18}
$$

The scheme corresponding to the algorithm (4.18) is shown in Figure 4.4.

Optimum algorithm (4.18) can be presented in the following compact form, which was already used in Chapter 2:

$$J_{1n} = \text{sgn} \, \sin(\Delta\varphi_{n\xi} + \pi/4)$$
$$J_{2n} = \text{sgn} \, \cos(\Delta\varphi_{n\xi} + \pi/4)$$

(4.19)

As can be seen, the transmitted binary symbols are defined by the signs of the sine and cosine of the received phase difference $\Delta\varphi_{n\xi}$, shifted by the angle $\pi/4$. The transition from (4.19) to (4.18) can be carried out, having opened the sine and cosine of the angles sum in (4.19) and having substituted the values of $\cos \Delta\varphi_{n\xi}$ and $\sin \Delta\varphi_{n\xi}$ from (4.6) into the obtained expressions.

The general algorithms for PM signal decoding, when using the Gray code [see (2.78)], allow us to find algorithms for the optimum incoherent reception of PDM-1 signals with an arbitrary number of equidistant phase differences. Let, for example, phase differences be transmitted in a four-phase system:

$$\Delta\varphi_1 = \Delta\varphi_0 \qquad \Delta\varphi_2 = \Delta\varphi_0 + \pi/2$$
$$\Delta\varphi_3 = \Delta\varphi_0 + \pi \qquad \Delta\varphi_4 = \Delta\varphi_0 + 3\pi/2$$

(4.20)

where $\Delta\varphi_0$ is the arbitrary minimum phase difference. Then, having substituted in common algorithm (2.81) the expressions for $\cos \Delta\varphi_{n\xi}$ and $\sin \Delta\varphi_{n\xi}$ from (4.6), we obtain the following algorithm for the optimum incoherent reception of the four-phase signals (4.20) when using the Gray code:

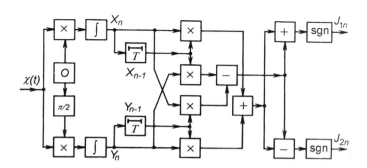

Figure 4.4 Optimum incoherent demodulator for four-level PDM-1 signals at phase differences 0, $\pi/2$, π, and $3\pi/2$.

$$J_{1n} = \text{sgn}[(X_{n-1}X_n + Y_{n-1}Y_n) \sin(\pi/4 - \Delta\varphi_0)$$
$$+ (X_{n-1}Y_n - Y_{n-1}X_n) \cos(\pi/4 - \Delta\varphi_0)]$$
$$J_{2n} = \text{sgn}[(X_{n-1}X_n + Y_{n-1}Y_n) \cos(\pi/4 - \Delta\varphi_0)$$
$$- (X_{n-1}Y_n - X_nY_{n-1}) \sin(\pi/4 - \Delta\varphi_0)] \tag{4.21}$$

Similarly, having substituted (4.6) into (2.84) it is possible to obtain the optimum incoherent algorithm of processing for an eight-phase system with PDM at the arbitrary minimum phase difference $\Delta\varphi_0$. Let us show the appropriate results for the two mostly usable eight-phase PM signals of Table 4.2.

For the first system of signals ($\Delta\varphi_0 = \pi/8$), we obtain

$$J_{1n} = \text{sgn}(X_{n-1}Y_n - X_nY_{n-1})$$
$$J_{2n} = \text{sgn}(X_{n-1}X_n + Y_nY_{n-1})$$
$$J_{3n} = \text{sgn}[(X_{n-1}X_n + Y_{n-1}Y_n) + (X_{n-1}Y_n - X_nY_{n-1})]$$
$$\times \text{sgn}[(X_{n-1}X_n + Y_{n-1}Y_n) - (X_{n-1}Y_n - X_nY_{n-1})] \tag{4.22}$$

The demodulator corresponding to (4.22) is indicated in Figure 4.5.

For the second signal system (see Table 4.2) we obtain:

$$J_{1n} = \text{sgn}[(X_{n-1}X_n + Y_{n-1}Y_n) \sin \pi/8 + (X_{n-1}Y_n - X_nY_{n-1}) \cos \pi/8]$$
$$J_{2n} = \text{sgn}[(X_{n-1}X_n + Y_{n-1}Y_n) \cos \pi/8 - (X_{n-1}Y_n - X_nY_{n-1}) \sin \pi/8]$$
$$J_{3n} = \text{sgn}[(X_{n-1}Y_n - X_nY_{n-1}) \cos 3\pi/8 + (X_{n-1}X_n + Y_{n-1}Y_n) \sin 3\pi/8]$$
$$\times \text{sgn}[(X_{n-1}X_n + Y_{n-1}Y_n) \cos 3\pi/8 - (X_{n-1}Y_n - X_nY_{n-1}) \sin 3\pi/8] \tag{4.23}$$

Similarly from formulas (4.22) and (4.23) we can obtain the algorithms for optimum incoherent reception of PDM-1 signals with an arbitrary eight-level system of phase differences

Table 4.2
Phase-Difference Options for Eight-Level PDM

Phase Difference Designation	Phase Differences in the		Gray Code		
	First System	Second System			
$\Delta\varphi_1 = \Delta\varphi_0$	$\pi/8$	0	+	+	+
$\Delta\varphi_2$	$3\pi/8$	$\pi/4$	+	+	−
$\Delta\varphi_3$	$5\pi/8$	$\pi/2$	+	−	−
$\Delta\varphi_4$	$7\pi/8$	$3\pi/4$	+	−	+
$\Delta\varphi_5$	$9\pi/8$	π	−	−	+
$\Delta\varphi_6$	$11\pi/8$	$5\pi/4$	−	−	−
$\Delta\varphi_7$	$13\pi/8$	$3\pi/2$	−	+	−
$\Delta\varphi_8$	$15\pi/8$	$7\pi/4$	−	+	+

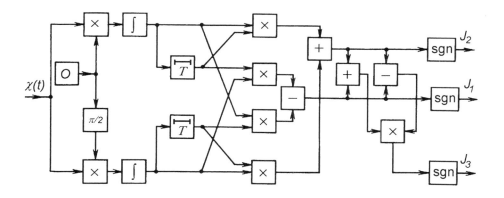

Figure 4.5 Optimum incoherent demodulator of eight-level PDM-1 signals.

$$\Delta\varphi_i = \Delta\varphi_0 + (i-1)\,\pi/4 \qquad \text{for } i = 1, 2, \ldots, 8$$

For this purpose formula (4.6) should be substituted into algorithm (2.84).

In general, to find an algorithm for optimum incoherent reception of the ith binary subchannel of the multiphase PDM system, using the Gray code, it is necessary to represent the common algorithm of decoding (2.78) through $\cos\Delta\varphi_\xi$ and $\sin\Delta\varphi_\xi$ and to replace them according to (4.6).

This is the common method of synthesizing algorithms for optimum incoherent reception of PDM-1 signals. It settles all possible situations for processing the signals within two elementary chips. Algorithms for multisymbol PDM signal processing (within more than two chip interval) are considered later.

The obtained algorithms can be realized in various ways. The arsenal of these ways was expanded during the last years in connection with the development of a means for digital signal processing. The indicated algorithms and block diagrams of the optimum incoherent demodulators are the basis for the corresponding hardware and software development. At the same time we should not forget that quickly developed analog microelectronics also provides new and considerable implementation opportunities. Modifications of the optimum incoherent demodulators, considered in the following section, provide us with additional material for choosing the best ways to implement PDM modems.

4.3 MODIFICATIONS OF THE OPTIMUM INCOHERENT DEMODULATORS OF FIRST-ORDER PDM SIGNALS

Demodulator schemes submitted in the previous section follow directly from synthesized algorithms for optimum incoherent reception of PDM-1 signals. These demodulators are sometimes called demodulators with an active filter, because their main unit is an active filter or, just the same, a correlator, consisting of a multiplier, an integrator, and a generator of reference oscillations.

The demodulators with active filters [4] are used rather often in modern digital transmission systems in light of the fact that "the frequency tuning" of an active filter is defined by an external drive oscillator of a reference oscillation, which is comparatively easily controlled. The phase of reference oscillation, naturally, can be an arbitrary one; but as to frequency, the required precision of its setup depends on modulation multiplicity and admissible energy losses: The higher the number of phase differences, the less admissible the deviation of the reference oscillation frequency from the carrier frequency. In any case this deviation should be only a small part of the value $1/T$, where T is chip duration, because at the $1/T$ deviation the correlator output signal is equal to zero due to the orthogonality of oscillations with such frequency shift. It is not less important to provide reasonably high precision of the $\pi/2$ phase shift between reference oscillations of correlators, as well as the equality of correlator losses.

At analog implementation of incoherent demodulators, the integrators with reset are based on integrated operational amplifiers, and multipliers are based on sign multiplication units, carrying out multiplication of the received analog signal by the periodic binary or step function, the first harmonic of which coincides with the signal carrier. If in the spectrum of the received signal there is not the third and higher carrier frequency harmonics, the replacement of an analog multiplier with the sign multiplier does not practically reduce the system noise immunity [4,8].

Further simplification of the correlator is connected with the replacement of an integrator with reset by a low-pass filter (LPF).

The units, consisting of the sign multiplier and LPF are often called phase detectors (PDs). Problems surrounding their use in incoherent and coherent demodulators are similar. The application of PD instead of a precise correlator undoubtedly reduces the noise immunity a little, however, this reduction can be insignificant and quite admissible in single-channel systems or multichannel systems with frequency division. In multitone systems with orthogonal signals the replacement of a correlator with a phase detector, as a rule, is inadmissible because in this case the correlators fulfill division of orthogonal signals. This division should be executed with very high precision in order to achieve the required dynamic range of demodulator operation.

Digital realization of the optimum incoherent demodulators with active filtering just described is becoming more common. In this case the correlators include ROM, storing reference oscillation samples in digital form, a multiplier of digital signal samples, and an adder–accumulator of multiplication results. When using sign functions similar to $\mathrm{sgn}(\sin \omega t)$ or $\mathrm{sgn}(\cos \omega t)$ as reference oscillations, a multiplier is not necessary, and the digital correlator operates as a signal sample adder. The signs of these samples are determined by ROM signals.

For digital realization it is considerably easier than for analog realization to store the results of the convolution of the received signal and the reference oscillation. It also simplifies multiplication and summation according to the algorithms described earlier. The optimum incoherent demodulators with active filters, realized on the basis of analog or digital technologies, are used in a number of practical systems of digital information transmission.

In addition to the demodulators with the active filters (correlators), demodulators with passive filters are used for optimum incoherent signal processing. Such demodulators are also called demodulators with switched matched filters (SMFs). The last ones, in contrast to the conventional linear matched filters (MFs), are parametric units. The pulse responses of MFs and SMFs coincide within the signal chip interval; at the end of this interval in SMFs a compulsory setup of the zero initial conditions (a mode with reset) is created. Historically, the optimum incoherent demodulators with passive filters were first realized on practice units of receiving signals with PDM [2,3].

Let us consider the optimum incoherent demodulators with passive filters. Figure 4.6 shows the schemes of such demodulators for two- and four-phase PDM-1. The algorithms of these demodulators, as will be shown later, are equivalent to algorithms (4.10) and (4.15), respectively.

The demodulators are based on the SMFs, which are resonance units with a high quality factor, tuned to the carrier frequency. They also contain PDs, including a multiplier and LPF, switches (K_1 and K_2), and an LU, controlling the operation of SMF, K_1 and K_2.

The $(n - 1)$th elementary input signal comes through K_1 to the input of SMF_1. This signal induces in the SMF the forced oscillation with linearly increasing amplitude. After the receiver input has been disconnected from SMF_1, in the last one free oscillation appears, coinciding by a phase with the $(n - 1)$th signal chip. We store the result of the functional conversion of a signal at the SMF_1. The next nth chip is processed at the SMF_2. After signal processing, free oscillations also appear in this filter, but with the phase of the nth signal chip. The free oscillations from the SMF_1 and SMF_2 outputs are transmitted to phase detectors PD_1 and PD_2. The voltage polarities at the PD_1 and PD_2 outputs define the binary symbols transmitted.

The voltage sign at the PD_2 output changes in a chip to the opposite by an inverter (Inv) and K_2. The necessity for such an inversion is explained as follows. From chip to chip the SMFs interchanged their functions, namely, if at the present moment in SMF_1 there is an oscillation with the phase of the previous chip φ_{prev}, and in SMF_2 an oscillation with the phase of the next chip φ_{next}, then in a chip the oscillation with the phase φ_{next} will be in SMF_1 and the oscillation with the phase φ_{prev} will be in SMF_2. As a result the PDs will calculate either phase difference $\Delta\varphi' = \varphi_{next} - \varphi_{prev}$ or phase difference $\Delta\varphi'' = \varphi_{prev} - \varphi_{next}$, with the distinction that PD_1 determines $\cos \Delta\varphi'$ or $\cos \Delta\varphi''$ and PD_2 determines $\sin \Delta\varphi'$ or $\sin \Delta\varphi''$. As $\cos \Delta\varphi' = \cos \Delta\varphi''$, the voltage sign at the PD_1 output defines the transmitted

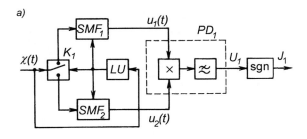

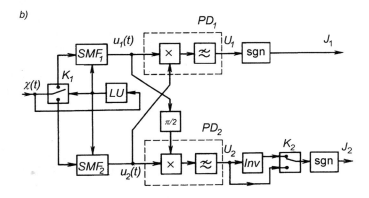

Figure 4.6 Optimum incoherent demodulators with the SMF for (a) two-phase and (b) four-phase PDM signals.

binary symbol irrespective of the chip number. As for the second binary subchannel, the voltage polarity at the PD_2 output has to be changed to the opposite in a chip because $\sin \Delta\varphi' = -\sin \Delta\varphi''$. This task is carried out by the inverter and switch K_2, operating synchronously with K_1.

We show that algorithms (4.10) and (4.15) are realized in demodulators on the base of the SMF. If we neglect the attenuation of a real SMF resonator and consider its pulse response to be equal to $h(t) = \cos \omega t$, then the SMF_1 and SMF_2 outputs are

$$u_1(t) = \int_0^t x_{n-1}(\tau) \cos[\omega(t - \tau)] \, dt \qquad (4.24)$$

$$u_2(t) = \int_0^t x_n(\tau) \cos[\omega(t - \tau)] \, dt \qquad (4.25)$$

where $x(\tau)$ is the received signal. Converting (4.24) we obtain

$$u_1(t) = \cos \omega t \int_0^t x_{n-1}(\tau) \cos \omega\tau \, d\tau + \sin \omega t \int_0^t x_{n-1}(\tau) \sin \omega\tau \, d\tau \qquad (4.26)$$

Because the signals come to the filter inputs only during an integration interval T, then at $t \geq T$ from (4.26) we obtain

$$u_1(t) = \cos \omega t \int_0^T x_{n-1}(t) \cos \omega t \, dt + \sin \omega t \int_0^T x_{n-1}(t) \sin \omega t \, dt \qquad (4.27)$$

or, taking into account the designations (4.4),

$$u_1(t) = Y_{n-1} \cos \omega t + X_{n-1} \sin \omega t \qquad (4.28)$$

Similarly from (4.25) we derive

$$u_2(t) = Y_n \cos \omega t + X_n \sin \omega t \qquad (4.29)$$

Expressions (4.28) and (4.29) describe free oscillations in the SMF after finishing the $(n-1)$th and nth chips correspondingly. It is not difficult to notice that at the absence of interference the signals of (4.28) and (4.29) coincide by the phases with the $(n-1)$th and nth chips of the transmitted signal.

Oscillations $u_1(t)$ and $u_2(t)$ are transmitted to the phase detectors. At the PD_1 output we obtain

$$\begin{aligned}
u_1(t)\,u_2(t) = 0.5(&Y_{n-1}\,Y_n + Y_{n-1}\,Y_n \cos 2\omega t \\
&+ X_{n-1}\,Y_n \sin 2\omega t + X_n\,Y_{n-1} \sin 2\omega t \\
&+ X_{n-1}\,X_n + X_{n-1}\,X_n \cos 2\omega t)
\end{aligned} \qquad (4.30)$$

At the PD_2 output, considering a phase shifter operation, we have

$$\begin{aligned}
u_1^*(t)\,u_2(t) = 0.5(&Y_{n-1}\,Y_n \sin 2\omega t + X_{n-1}\,Y_n \\
&+ X_{n-1}\,Y_n \sin 2\omega t + X_{n-1}\,Y_n \cos 2\omega t \\
&+ Y_{n-1}\,X_n \cos 2\omega t - Y_{n-1}\,X_n + X_{n-1}\,X_n \sin 2\omega t)
\end{aligned} \qquad (4.31)$$

Low-pass filters (see Fig. 4.6) pass only constant components of the signals (4.30) and (4.31), therefore

$$U_1 = 0.5(X_{n-1} Y_n - X_n Y_{n-1})$$
$$U_2 = 0.5(X_{n-1} X_n + Y_{n-1} Y_n)$$

Polarities of U_1 and U_2 at the moment of sampling determine the transmitted binary symbols, that is,

$$J_{1n} = \text{sgn}(X_{n-1} Y_n - X_n Y_{n-1})$$
$$J_{2n} = \text{sgn}(X_{n-1} X_n + Y_{n-1} Y_n)$$
(4.32)

which is the same as algorithm (4.15).

Here it is useful to pay attention to the fact that the demodulators, shown in Figure 4.6, are the optimum incoherent demodulators, but not autocorrelated, as is incorrectly asserted in a number of works. This misunderstanding occurs because of the apparent similarity of signal adjacent chip convolution in the autocorrelated and optimum incoherent demodulators. Actually the autocorrelated demodulators multiply input signals (see Chapter 5), but in the optimum incoherent demodulators with passive filters, output oscillations of the matched filters are multiplied.

We emphasize that this is not a terminology problem. The optimum incoherent and autocorrelated demodulators have different noise immunity, though under certain conditions the noise immunity of an autocorrelated demodulator can approximate (but is not equal to) the noise immunity of an optimum incoherent demodulator.

The modifications of demodulators with a passive filter are based on various highly selective resonance units. In connection with the fast development of analog integrated microcircuits it may be useful to use electronic models of resonant units in demodulators [9]. Figure 4.7 shows the scheme of a resonant unit as an electronic model, which functionally [Fig. 4.7(a)] consists of two integrators, a multiplier and an inverter, and practically [Fig. 4.7(b)] consists of three operational amplifiers (OA).

The frequency of self-induced oscillations of this system is equal to $\omega_0 = 1/R_1 C_1 R_2 C_2$, and the quality factors are determined by the OA gain. If signal $x(t)$ comes to the input of a such unit, the process in it will be described by the following differential equation:

$$d^2y/dt^2 + \omega_0^2 y = x(t)$$
(4.33)

Function (4.34) is the decision of this equation:

$$y(t) = U_{01} \sin \omega_0 t + U_{02} \cos \omega_0 t + \frac{1}{\omega_0} \int_0^t x(\tau) \cos \omega_0(t - \tau) \, d\tau$$
(4.34)

a)

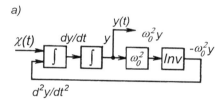

b)

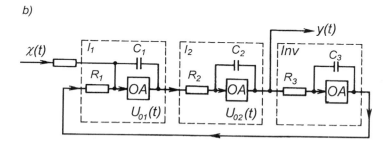

Figure 4.7 Highly selective resonance unit in the form of an electronic model: (a) functional block diagram and (b) implementation block diagram.

where U_{01} and U_{02} are the initial (at $t = 0$) voltages at integrators outputs. If the initial voltages are equal to zero, we have

$$y(t) = \frac{1}{\omega_0} \int_0^t x(\tau) \cos \omega_0(t - \tau) \, d\tau \qquad (4.35)$$

Thus, as seen from the comparison of (4.35) with (4.24) and (4.25), the output signal of the unit, shown in Figure 4.7, coincides with the output signal of the resonance unit of the switched matched filter.

The operation of the optimum incoherent demodulator of four-phase PDM signals on the basis of electronic models of an SMF (Fig. 4.8) does not differ from the operation of the demodulator, indicated in Figure 4.6, but it has some peculiarities. For example, a phase shifter is not required, because the oscillations shifted by $\pi/2$ can be directly received from the OA output and input.

Besides, in this demodulator we are able to realize not only the dynamic mode for storing the previous chip phase (as oscillations), but also the static mode for storing the previous chip phase (as numbers X_n and Y_n), as in demodulators with an active filter. As memory cells at static storage we can use integrating capacitors, which are in the feedback circuits of the OA. In this case, the loop, consisting of

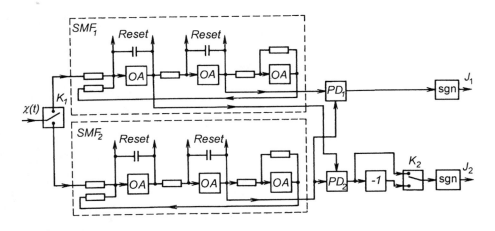

Figure 4.8 Optimum incoherent demodulator of four-phase PDM signals in the form of an electronic model.

integrators and inverters, is disconnected by the corresponding keys for the period of one chip. Then the loop is connected again, and at the integrator outputs there appears an oscillation with a phase determined by initial voltages at the integrating capacitors. The static storage permits us to reduce an error of the phase difference measurement, stipulated by nonidentity of the passive filter parameters.

The demodulators with the SMFs can be realized both in analog and digital forms. In the former case, the main element of such demodulators is the OA. Parameters of the existing OA chips permit us to increase the SMF frequency up to hundreds of megahertz. For digital implementation, the main element of incoherent demodulators is a universal microprocessor or the specialized processor of digital signal processing (DSP), on the basis of which a digital filter corresponding to SMF is made up [10,11].

The use of electronic modeling techniques for direct implementation of the PDM demodulators provides a number of practically useful modifications. These techniques permit us to combine the advantages of demodulators with active and passive filters.

Figure 4.9 illustrates a scheme for an optimum incoherent demodulator for four-level PDM-1 signals, in which the convolution of the received signal with reference oscillation is executed by correlators, as in the demodulator with an active filter, and phase storage and phase difference calculations are executed by SMF and PD, as in the demodulator with a passive filter. We shall explain the operation of this scheme.

At the $(n-1)$th chip let all keys be in position 1, and at the nth chip let all keys be in position 2. Then within the $(n-1)$th chip the multiplier outputs are connected to the inputs of the first and second integrators of OA_1 and OA_2 and

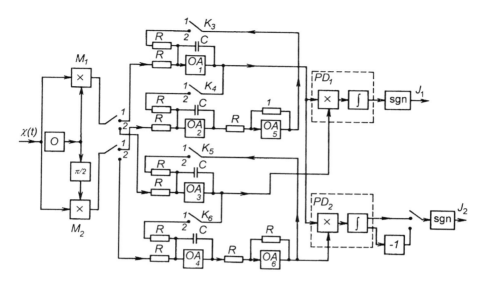

Figure 4.9 Optimum incoherent demodulator for four-level PDM signals based on two electronic models of SMF.

compose with them the correlators. At the moment T these correlators accumulate the voltages, proportional to the values X_{n-1} and Y_{n-1}, equal to the convolutions of the received signal and coordinate functions $\sin \omega t$ and $\cos \omega t$. At this moment the keys are switched to position 2. As a result, the integrators of OA_1 and OA_2 and the inverter of OA_5 form, as is obvious from the scheme, the electronic model of SMF, which generates (at the outputs of the first and second integrators) the following free oscillations:

$$u_1(t) = X_{n-1} \cos \omega t - Y_{n-1} \sin \omega t$$

$$u_2(t) = X_{n-1} \sin \omega t + Y_{n-1} \cos \omega t$$

The phase of one of these oscillations coincides with the phase of the $(n-1)$th signal chip, and the phase of the other one is additionally shifted by $\pi/2$. The free oscillations in the scheme, made up from OA_1, OA_2, and OA_5, last during the nth chip. For this period the active filter is composed of multipliers M_1 and M_2 with integrators of OA_3 and OA_4. By the end of the nth chip at the integrator outputs voltages X_n and Y_n are established. After closing K_5 and K_6 the third and fourth integrators also form the electronic model of SMF, in which the following free oscillations take place:

$$u_3(t) = X_n \cos \omega t - Y_n \sin \omega t$$

$$u_4(t) = X_n \sin \omega t + Y_n \cos \omega t$$

one of which coincides by the phase with the nth chip. During the guard time interval, the free oscillations from the integrator output are transmitted to the phase detectors (the oscillations u_1 and u_3 come to PD_1, and u_2 and u_4 to PD_2) in a manner similar to how it occurs in the usual demodulator with a passive filter. As a result, the PD_1 output is proportional to the cosine of the phase difference of the nth and $(n-1)$th signal chips, and the PD_2 output is proportional to the sine of the phase difference of the nth and $(n-1)$th chips. If the phase differences are $\pi/4$, $3\pi/4$, $5\pi/4$, and $7\pi/4$, the signs of these output signals determine the transmitted binary symbols.

Note that the demodulator of Figure 4.9 includes elements of demodulators with active and passive filters, and the integrator of an active filter is at the same time one of two integrators of a passive filter electronic model. As a consequence of this, the considered demodulator combines positive properties of demodulators with active and passive filters: the high precision of signal detection of the former one and the simplicity of the phase storage and phase difference calculations of the last one.

Figure 4.10 shows the scheme for an optimum incoherent demodulator of PDM-1 signals, in which, in contrast to the scheme of Figure 4.9, the convolution of a signal with reference oscillation, that is, the calculation of X_n and Y_n, is realized by the electronic model of SMF, and storage of X_n and Y_n and phase difference calculation occurs by a phase difference calculation unit (PDCU), as in the scheme of a demodulator with an active filter. The scheme of a PDCU is illustrated in Figure 4.11.

The unit shown in Figure 4.11 is part of all optimum incoherent demodulators with active filters (see Figs. 4.3, 4.4, and 4.5). The received signal projections X_n and Y_n come to its inputs, and the output signals are proportional to the cosine and sine of the phase difference of the nth and $(n-1)$th chips.

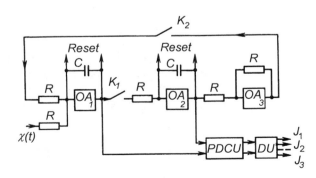

Figure 4.10 Optimum incoherent demodulator of PDM signals with PDM on the basis of one electronic model of SMF.

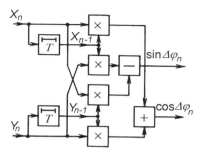

Figure 4.11 Phase difference calculation unit.

We now explain the operation of the demodulator in Figure 4.10. During the elementary processing interval the keys K_1 and K_2 are closed, and the scheme with OA_1, OA_2, and OA_3 forms the SMF model. At the end of the chip the keys are disconnected, and the constant voltages from the OA_1 and OA_2 outputs, proportional to X_n and Y_n, come to the PDCU input. Then the output signals of PDCU are decoded in a decoding unit (DU) depending on the keying code used. In contrast with the demodulators with a passive filter considered earlier, the last demodulator has only one SMF. This is explained by the fact that in this case the SMF is used only for calculation the values X_n and Y_n, and their storage is executed by memory cells of PDCU.

4.4 OPTIMUM INCOHERENT RECEPTION OF SECOND-ORDER PDM SIGNALS

To synthesize algorithms for optimum incoherent reception of the PDM-2 signals, we use the same general algorithm of optimum incoherent processing (4.1) that was used in Section 4.2 for the synthesis of algorithms of PDM-1 signal reception.

In this case the duration of a signal processing interval is equal to three chips ($\tau = 3T$), because the transmitted second phase difference, in common cases, is determined by not less than three chips of a signal. Thus, it is necessary to represent all possible PDM-2 signal variants within three chips and to substitute these variants in (4.1). The first of the three chips has an arbitrary, but identical, initial phase for all variants.

For binary PDM-2 with second phase differences of 0 or π, four signal variants, shown in Figure 4.12, take place. These variants can be represented in the following compact form:

$$
\begin{aligned}
S_1(t) &= a \sin \omega t & \Delta^2 \varphi = 0 \\
S_2(t) &= a \sin \omega t \, \text{sgn} \sin(\pi t / T) & \Delta^2 \varphi = 0 \\
S_3(t) &= a \sin \omega t \, \text{sgn} \sin(\pi t / 2T + \pi/2) & \Delta^2 \varphi = \pi \\
S_4(t) &= a \sin \omega t \, \text{sgn} \sin(\pi t / 2T) & \Delta^2 \varphi = \pi
\end{aligned}
\tag{4.36}
$$

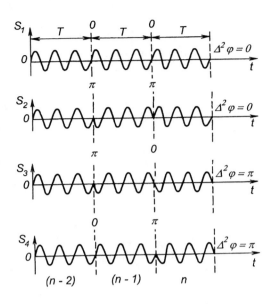

Figure 4.12 Variants of a binary PDM-2 signal.

The specified value of the second phase difference corresponds to each of four signals in (4.36). Having substituted (4.36) in (4.1) and having used the following designations:

$$X_n = \int_{nT}^{(n+1)T} x(t) \sin \omega t \, dt$$

$$Y_n = \int_{nT}^{(n+1)T} x(t) \cos \omega t \, dt \qquad (4.37)$$

we determine that in a required demodulator the following four values should be calculated:

$$V_1 = (X_{n-2} + X_{n-1} + X_n)^2 + (Y_{n-2} + Y_{n-1} + Y_n)^2$$
$$V_2 = (X_{n-2} - X_{n-1} + X_n)^2 + (Y_{n-2} - Y_{n-1} + Y_n)^2 \qquad (4.38)$$
$$V_3 = (X_{n-2} - X_{n-1} - X_n)^2 + (Y_{n-2} - Y_{n-1} - Y_n)^2$$
$$V_4 = (X_{n-2} + X_{n-1} - X_n)^2 + (Y_{n-2} + Y_{n-1} - Y_n)^2$$

The decision is made as follows: If V_1 or V_2 is the greatest, the second phase difference $\Delta^2\varphi_n = 0$ is considered to be transmitted; if V_3 or V_4 appears as the greatest, $\Delta^2\varphi_n = \pi$ is considered to be transmitted.

Algorithm (4.38) can be presented in another equivalent form if we calculate squares of sums and exclude from the obtained expressions the identical components—squares of values X and Y. Then we determine that in the demodulator these values should be compared:

$$
\begin{aligned}
V_1 &= (X_n X_{n-2} + Y_n Y_{n-2}) + (X_n X_{n-1} + Y_n Y_{n-1}) + (X_{n-1} X_{n-2} + Y_{n-1} Y_{n-2}) \\
V_2 &= (X_n X_{n-2} + Y_n Y_{n-2}) - (X_n X_{n-1} + Y_n Y_{n-1}) - (X_{n-1} X_{n-2} + Y_{n-1} Y_{n-2}) \qquad (4.39)\\
V_3 &= -(X_n X_{n-2} + Y_n Y_{n-2}) - (X_n X_{n-1} + Y_n Y_{n-1}) + (X_{n-1} X_{n-2} + Y_{n-1} Y_{n-2}) \\
V_4 &= -(X_n X_{n-2} + Y_n Y_{n-2}) + (X_n X_{n-1} + Y_n Y_{n-1}) - (X_{n-1} X_{n-2} + Y_{n-1} Y_{n-2})
\end{aligned}
$$

The demodulator scheme, corresponding to the synthesized algorithm (4.39) is illustrated in Figure 4.13. This scheme contains correlators, a reference oscillator with arbitrary phase, delay lines, and a calculator of the components V_1, V_2, V_3, and V_4, operating according to algorithm (4.38) or (4.39). A comparison scheme makes a decision about the transmitted binary symbol.

We should emphasize that the frequency of the reference oscillation should coincide with the carrier frequency of the received signal, that is, this demodulator, as well as the optimum incoherent demodulator of PDM-1 signals (see Fig. 4.1), is not invariant to the carrier frequency, and its noise immunity decreases at frequency shifts.

The remarkable property of algorithms (4.38) and (4.39) and the corresponding demodulator (Fig. 4.13) is as follows: The noise immunity of this demodulator of PDM-2 signals is higher than the noise immunity of the optimum incoherent demodulator of PDM-1 signals (Fig. 4.1) and is almost indistinctive from the noise immunity of a coherent demodulator of PDM-1. This statement is proved in Chapter 6. It allows us to come to the fundamental conclusion of increasing the

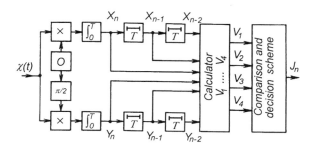

Figure 4.13 Optimum incoherent demodulator of binary PDM-2 signals.

noise immunity by means of increasing the interval of optimum incoherent processing of signals (see Section 4.5).

The received optimum demodulator scheme (see Fig. 4.13) is based on using active filters (correlators). Principally the implementation of algorithm (4.38) can be fulfilled by using passive matched filters. The appropriate scheme includes four filters, matched with signals, shown in Figure 4.12, and a comparison scheme. The samples of envelopes at the matched filter outputs at the end of the three-chip interval are proportional to the values of (4.38).

For four-level PDM-2 with the second phase differences 0, $\pi/2$, π, and $3\pi/2$ on three chips, of which the first one is arbitrary but identical for all variants the initial phase, we have 16 signal variants. The final algorithm, received as a result of substitution of all 16 variants in the general algorithm (4.1), is

$$
\begin{aligned}
V_1 &= (X_{n-2} + X_{n-1} + X_n)^2 + (Y_{n-2} + Y_{n-1} + Y_n)^2 \\
V_2 &= (X_{n-2} + Y_{n-1} - X_n)^2 + (Y_{n-2} - X_{n-1} - Y_n)^2 \\
V_3 &= (X_{n-2} - X_{n-1} + X_n)^2 + (Y_{n-2} - Y_{n-1} + Y_n)^2 \\
V_4 &= (X_{n-2} - Y_{n-1} - X_n)^2 + (Y_{n-2} + X_{n-1} - Y_n)^2
\end{aligned}
\qquad \Delta^2\varphi_n = 0 \qquad (4.40)
$$

$$
\begin{aligned}
V_5 &= (X_{n-2} + X_{n-1} + Y_n)^2 + (Y_{n-2} + Y_{n-1} - X_n)^2 \\
V_6 &= (X_{n-2} + Y_{n-1} - Y_n)^2 + (Y_{n-2} - X_{n-1} + X_n)^2 \\
V_7 &= (X_{n-2} - X_{n-1} + Y_n)^2 + (Y_{n-2} - Y_{n-1} - X_n)^2 \\
V_8 &= (X_{n-2} - Y_{n-1} - Y_n)^2 + (Y_{n-2} + X_{n-1} + X_n)^2
\end{aligned}
\qquad \Delta^2\varphi_n = \pi/2 \qquad (4.41)
$$

$$
\begin{aligned}
V_9 &= (X_{n-2} + X_{n-1} - X_n)^2 + (Y_{n-2} + Y_{n-1} + Y_n)^2 \\
V_{10} &= (X_{n-2} + Y_{n-1} + X_n)^2 + (Y_{n-2} - X_{n-1} - Y_n)^2 \\
V_{11} &= (X_{n-2} - X_{n-1} - X_n)^2 + (Y_{n-2} - Y_{n-1} + Y_n)^2 \\
V_{12} &= (X_{n-2} - Y_{n-1} + X_n)^2 + (Y_{n-2} + X_{n-1} - Y_n)^2
\end{aligned}
\qquad \Delta^2\varphi_n = \pi \qquad (4.42)
$$

$$
\begin{aligned}
V_{13} &= (X_{n-2} + X_{n-1} - Y_n)^2 + (Y_{n-2} + Y_{n-1} + X_n)^2 \\
V_{14} &= (X_{n-2} + Y_{n-1} + Y_n)^2 + (Y_{n-2} - X_{n-1} - X_n)^2 \\
V_{15} &= (X_{n-2} - X_{n-1} - Y_n)^2 + (Y_{n-2} - Y_{n-1} + X_n)^2 \\
V_{16} &= (X_{n-2} - Y_{n-1} + Y_n)^2 + (Y_{n-2} + X_{n-1} - X_n)^2
\end{aligned}
\qquad \Delta^2\varphi_n = 3\pi/2 \qquad (4.43)
$$

In these equations, for every four V_j there are corresponding second phase differences. If even one of the V_i values, belonging to the set of four values, is more than all other 15 values of V_j, the decision is made such that it benefits the corresponding second difference.

On construction of a similar algorithm for a system with second phase differences of $\Delta\varphi_n = \pi/4$, $3\pi/4$, $5\pi/4$, and $7\pi/4$, the number of values compared in the demodulator is considerably increased.

The scheme for an optimum incoherent demodulator for four-level PDM-2 signals with the second phase differences 0, $\pi/2$, π, and $3\pi/2$ coincides with the scheme shown in Figure 4.13; the only difference in this case is that the calculator and the comparison scheme form and compare 16 values [as in (4.40) through (4.43)] and determine one of four transmitted second phase differences.

As is obvious from the comparison of algorithms (4.38) with (4.40) through (4.43), the transition from two-level to four-level PDM-2 results in demodulator complications. However, for digital implementation on the basis of a high-speed processor and RAM, the values of (4.40) through (4.43) can be successively calculated during one signal chip, and the equipment complication will appear insignificant. The following general presentation of the values of (4.40) through (4.43) can be useful at digital implementation:

$$V = (X_{n-2} + acX_{n-1} + \overline{a}cY_{n-1} + bdX_n + \overline{b}dY_n)^2$$
$$+ (Y_{n-2} + acY_{n-1} - \overline{a}cX_{n-1} + bdY_n - \overline{b}dX_n)^2 \tag{4.44}$$

where a and b represent binary symbols 0 or 1; c and d represent binary symbols ± 1; and the line above a and b refers to inversion. The combinations of binary symbols a, b, c, and d gives 16 possible values [(4.40) through (4.43)].

Using the method considered, it is not difficult to synthesize the optimum incoherent algorithms for eight-phase and other multiphase PDM-2 signals. These algorithms are rather bulky; for example, at eight-phase PDM-2 the demodulator should calculate 64 values as (4.40) through (4.43) and choose the maximum value from them. The demodulator schemes also become more complex. In connection with this we are interested in suboptimum algorithms for incoherent reception of PDM-2 signals, some of which concede a little the optimum algorithms by noise immunity, but are more simple in realization.

One of the approaches to suboptimum incoherent algorithms synthesis is similar to the method proposed in the previous chapter for coherent demodulators of high-order PDM signals. It is based on the fact that decisions at the output of the optimum incoherent demodulator of PDM-1 signals, that is, the permitted set of the first phase differences, are an algebraic ring of R order, and it can be replaced by the isomorphic ring of integers from 0 to $R - 1$. Hence, the determination of the transmitted second phase difference can be carried out by modulo R calculating the difference between two R-ary numbers, in series, that appear at the output of the PDM-1 demodulator. The appropriate block diagram is illustrated in Figure 4.14.

Figure 4.14 includes the optimum incoherent demodulator for first-order PDM signals, a memory of one-chip duration, and a unit for modulo-R subtraction. The number R depends on modulation multiplicity N and the minimum value for

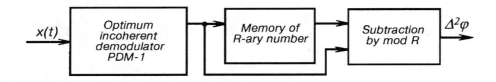

Figure 4.14 Incoherent processing of PDM-2 signals.

the admissible phase difference $\Delta^2\varphi_0$: if $\Delta^2\varphi_0 = 0$, then $R = 2^N$; and if $\Delta^2\varphi_0 \neq 0$, then $R = 2\pi/\Delta^2\varphi_0$. For example, in the four-phase ($N = 2$) system $R = 4$ when using information differences 0, $\pi/2$, π, and $3\pi/2$, and $R = 8$ when using phase differences $\pi/4$, $3\pi/4$, $5\pi/4$, and $7\pi/4$.

At binary PDM-2 with second phase differences of 0 and π, the demodulator algorithm is simplified, and the subtraction of binary numbers 0 and 1 can be replaced by multiplying the numbers ± 1; as a result, the transmitted binary symbol at the nth chip is equal to

$$J_n = \mathrm{sgn}(X_{n-1}X_n + Y_{n-1}Y_n)\,\mathrm{sgn}(X_{n-2}X_{n-1} + Y_{n-2}Y_{n-1}) \qquad (4.45)$$

The corresponding demodulator is shown in Figure 4.15.

At the binary PDM-2 with second phase differences of 0 and π, we can also use the optimum incoherent demodulator of PDM-1 signals, which processes signals in a chip duration. In this case, the second phase difference is equal to the phase difference in a chip duration. Let φ_{n-2}, φ_{n-1}, and φ_n be initial phases of three series signal chips. It is obvious that

$$\varphi_{n-1} = \varphi_{n-2} + \Delta^1\varphi_{n-1} + \omega T$$
$$\varphi_n = \varphi_{n-1} + \Delta^1\varphi_n + \omega T$$

where ω is carrier frequency and $\Delta^1\varphi_{n-1}$ and $\Delta^1\varphi_n$ are the transmitted first phase differences. Hence, the second phase difference is equal to

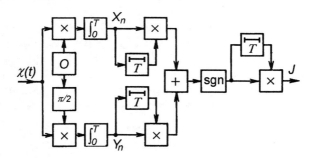

Figure 4.15 Suboptimum incoherent demodulator of binary PDM-2 signals.

$$\Delta^2 \varphi_n = \varphi_n - 2\varphi_{n-1} + \varphi_{n-2} = \Delta^1 \varphi_n - \Delta^1 \varphi_{n-1} \tag{4.46}$$

and the first phase difference in a chip duration is equal to

$$\varphi_n - \varphi_{n-2} = \Delta^1 \varphi_n + \Delta^1 \varphi_{n-1} + 2\omega T \tag{4.47}$$

From comparison of (4.47) and (4.46) it is obvious that the second phase difference is the same as the first difference in a chip duration, if the first differences are 0 or π and $2\omega T = 2k\pi$ (where k is an integer). Thus, on fulfillment of these conditions the PDM-2 signal can be demodulated according to the following algorithm:

$$J_n = \mathrm{sgn}(X_{n-2} X_n + Y_{n-2} Y_n) \tag{4.48}$$

The corresponding block diagram is shown in Figure 4.16. This suboptimum demodulator of PDM-2 signals obviously has the same noise immunity as the optimum incoherent demodulator of PDM-1 signals (Fig. 4.1). The suboptimum demodulator (Fig. 4.15) concedes a little to it. Besides, all suboptimum demodulators (i.e., those of Figs. 4.14, 4.15, and 4.16), lose to the scheme, indicated in Figure 4.13 (see Chapter 6).

Another way of constructing the algorithms for suboptimum incoherent demodulators of PDM-2 signals is based on direct calculation of trigonometric functions of the second phase differences by trigonometric functions of the first phase differences, which, in their turn, are determined by the projections of the received signal onto orthogonal reference oscillations with an arbitrary initial phase. If Gray codes are used (see Section 2.5), the binary symbols in subchannels are expressed by the cosine and sine of the second phase differences according to the following formulas: For two-phase PDM-2,

$$J_{1n} = \mathrm{sgn}(A_1 \sin \Delta_n^2 \varphi + B_1 \cos \Delta_n^2 \varphi) \tag{4.49}$$
$$A_1 = \cos(\pi/2 - \Delta^2 \varphi_0)$$
$$B_1 = \sin(\pi/2 - \Delta^2 \varphi_0)$$

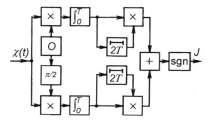

Figure 4.16 Modification of the suboptimum incoherent demodulator of binary PDM-2 signals.

for four-phase PDM-2

$$J_{1n} = \text{sgn}(C_2 \cos \Delta_n^2\varphi + D_2 \sin \Delta_n^2\varphi) \tag{4.50}$$
$$J_{2n} = \text{sgn}(D_2 \cos \Delta_n^2\varphi - C_2 \sin \Delta_n^2\varphi)$$

$$C_2 = \sin(\pi/4 - \Delta^2\varphi_0) \tag{4.51}$$
$$D_2 = \cos(\pi/4 - \Delta^2\varphi_0)$$

for eight-phase PDM-2

$$J_{1n} = \text{sgn}(C_3 \sin \Delta_n^2\varphi + D_3 \cos \Delta_n^2\varphi) \tag{4.52}$$
$$J_{2n} = \text{sgn}(C_3 \cos \Delta_n^2\varphi - D_3 \sin \Delta_n^2\varphi)$$
$$J_{3n} = \text{sgn}(E_3 \sin \Delta_n^2\varphi + F_3 \cos \Delta_n^2\varphi) \, \text{sgn}(E_3 \cos \Delta_n^2\varphi - F_3 \sin \Delta_n^2\varphi)$$

$$C_3 = \cos(\pi/8 - \Delta^2\varphi_0) \qquad D_3 = \sin(\pi/8 - \Delta^2\varphi_0) \tag{4.53}$$
$$E_3 = \cos(3\pi/8 - \Delta^2\varphi_0) \qquad F_3 = \sin(3\pi/8 - \Delta^2\varphi_0)$$

Here $\Delta^2\varphi_0$ is the minimum second phase difference, and $\sin \Delta_n^2\varphi_0$ and $\cos \Delta_n^2\varphi_0$ can be determined in the following way:

$$\cos \Delta_n^2\varphi = \cos \Delta_n^1\varphi \cos \Delta_{n-1}^1\varphi + \sin \Delta_n^1\varphi \sin \Delta_{n-1}^1\varphi$$
$$\sin \Delta_n^2\varphi = \sin \Delta_n^1\varphi \cos \Delta_{n-1}^1\varphi - \sin \Delta_{n-1}^1\varphi \cos \Delta_n^1\varphi \tag{4.54}$$

The suboptimum incoherent demodulator of PDM-2 signals, which realize algorithms (4.49) through (4.54), is depicted in Figure 4.17. It contains the quadrature scheme, consisting of a local oscillator with an arbitrary phase, two correlators, and a phase shifter by $\pi/2$. Two PDCUs connected in series operate according to

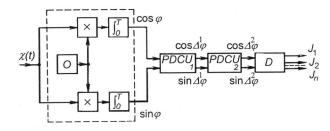

Figure 4.17 PDM-2 signal incoherent demodulator with successive calculation of finite phase differences.

algorithm (4.54), and a decoder (D) operates according to one of algorithms (4.49), (4.50), or (4.52).

PDCU can be implemented using the scheme indicated in Figure 4.11. Figure 4.18 shows, as examples, the demodulators of two-phase and four-phase PDM-2 signals.

The last of the submitted demodulators (Figs. 4.17 and 4.18) have larger energy losses than the suboptimum demodulators (Figs. 4.14, 4.15, and 4.16). The optimum demodulator of PDM-2 signals under the scheme of Figure 4.13 has higher noise immunity than the optimum incoherent demodulator of PDM-1 signals, and the demodulator under the scheme of Figure 4.18(a) has approximately the same noise immunity as the optimum incoherent demodulator of FM signals. The remaining suboptimum demodulators take an intermediate position between these two schemes.

4.5 OPTIMUM MULTISYMBOL INCOHERENT RECEPTION OF PDM SIGNALS

The algorithms and schemes for optimum incoherent reception of PDM-1 signals were obtained in Section 4.2 for the case of signal processing within two adjacent

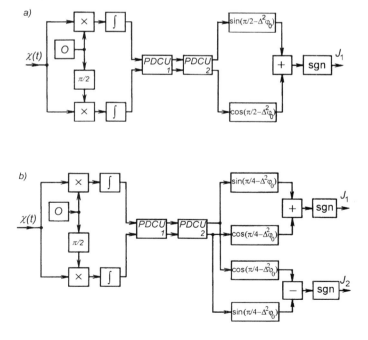

Figure 4.18 Incoherent demodulators of (a) two-phase and (b) four-phase PDM-2 signals.

chips. These algorithms are strictly optimum, that is, they provide a minimum of error probability under the assumption that the initial phase of two processed elements is a random value with uniform distribution and is constant within the two elements interval.

The specified assumption contains a contradiction: The second element of a processed pair is simultaneously the first element of the next pair; hence, if the initial phase is constant within the two elements interval, it should be also constant within the three elements interval, and so on. Physically this contradiction is removed due to the fact that the initial signal phase changes comparatively slowly, and therefore it may be considered to be almost constant within several elements. If this interval of constancy is much longer than two chips, there is an opportunity for optimum incoherent reception of PDM signals as a whole not within just two, but within a greater number of chips.

The phrase "reception as a whole" usually refers to the processing of encoded signals or composite signals with redundancy. The main point of such processing is the reception of the whole signal, including several chips, for example, the signal, corresponding to the code combination of a block code. As is known from communication theory, at proper encoding with redundancy, the optimum reception as a whole has higher noise immunity than the optimum element-by-element reception with the following optimum discrete decoding. At the same time, it is known that when there is no redundancy the optimum coherent reception as a whole has the same noise immunity as the optimum coherent element-by-element reception [1]. Therefore the coherent reception as a whole is usually used for signals corresponding to correcting code combinations, for composite signals, or for signal-code structures with internal redundancy.

During incoherent reception the point is different. The increase of the processing interval results in the improvement of initial phase estimation, if, certainly, this phase really remains constant within this interval. Therefore optimum incoherent reception as a whole can provide less error probability than the optimum incoherent element-by-element reception, even with signals without redundancy. Moreover, when the interval of optimum incoherent processing increases, the error probability decreases, approximating the error probability of ideal coherent reception.

Here we would like to emphasize an extremely important issue of the communication theory and communication engineering—the problem of achieving potential noise immunity without estimating the initial signal phase.

As is known, the potential noise immunity is quantitatively characterized by the minimum error probability that can be reached for a given signal set and interferences with known statistic characteristics. The theory and engineering of optimum signal processing in the AWGN channel suggest two ways to achieve the potential noise immunity.

The first way is based on the traditional algorithm of the ideal receiver: The receiver has exact copies (samples) of the transmitted signals, and that sample is considered to have been transmitted, which has the minimum distance (according to Euclid) from the received signal. This way is not practically realized in communication systems, because the signal in the channel is subjected to changes (distortions) that cannot be absolutely exactly predicted and, hence, it is impossible to have a priori knowledge of the exact signal samples in the receiver.

The second, quite real way to approximate the potential noise immunity consists of generating signal samples in the receiver on the basis of current estimates of the signal parameters. There are many methods and algorithms for forming signal samples both by a special synchronization signal, transmitted through a communication channel, and by appropriate information signal conversions. When using harmonic oscillations with this or that amplitude, frequency and phase as signal variants, the appropriate methods are termed *quasicoherent reception,* because in this case the reference oscillations are "almost coherent" with the received useful signal. The up-to-date techniques of the quasicoherent reception of signals with phase and phase-difference modulation permit us to approximate the potential noise immunity at a continuous transmission mode in the channel with slowly varying parameters. However these techniques, based on estimation of the signal phase and carrier frequency by their previous history, are not effective in the channels with quickly varying parameters and at the pulse (batch) information transmission. In these conditions we should apply incoherent methods of processing, which, as a rule, lose to the perfect coherent methods.

Here we would like to emphasize that there is a different way to achieve the potential noise immunity that does not need the estimation of signal parameters. This way is based on increasing the signal incoherent processing interval for the element-by-element decisions. The principles of this approach are considered in [12–16].

Let us formulate the basic statement, generalizing the above-mentioned phenomena, and deduce particular algorithms for the optimum incoherent and auto-correlated reception of PDM signals, solving the problem of practical achievement of the potential noise immunity without initial phase evaluation.

Formulation of the Basic Statement and General Algorithm

To approximate the potential noise immunity as close as desired, we do not conceptually need a priori information or current estimates of unknown signal parameters, and, in particular, we do not need coherent or quasicoherent processing. For the specified approximation, we can use the algorithms for receiving the signals with indefinite parameters, expanding these algorithms for several elements. In particular, we can use the optimum incoherent processing of several signal chips at element-by-element decision making within every chip.

Note that here we do not mean reception of signals with redundancy as a whole. The formulated statement concerns the noise immunity of element-by-element reception. We can say that the matter consists of the element-by-element decisions on the basis of the results of several elements incoherent processing as a whole (the multisymbol signal processing with the element-by-element decisions).

We explain the essence of the considered algorithms. Let us use a discrete set of m equiprobable and equipower signals S_i ($i = 1, 2, \ldots, m$) with the same duration T. Within the k element interval with duration kT there can be m^k signal variants, which, taking into account the indefinite initial phase φ, we designate $S_{j\varphi}(t)$, $j = 1, 2, \ldots, m^k$. Further let these signals be transmitted over the AWGN channel, and the channel properties be such that the initial phase φ is a random uniformly distributed value that is constant within the interval of k elements. Then the signal $S_{j\varphi}$ can be presented in the form of a linear combination of two signals $S_j(t)$ and $S_j^*(t)$, connected by the Hilbert conversion:

$$S_{j\varphi}(t) = S_j(t) \cos \varphi - S_j^*(t) \sin \varphi$$

At the specified conditions, the minimum error probability on receiving the signals $S_{j\varphi}$, according to the maximum likelihood rule, can be achieved by means of the optimum incoherent processing, which consists of choosing the maximum value for the following values:

$$V_{nj} = \left[\int_{(n-K)T}^{nT} x(t) S_j(t) \, dt \right]^2 + \left[\int_{(n-K)T}^{nT} x(t) S_j^*(t) \, dt \right]^2 \qquad (4.55)$$

where n is the current number of the receiving element and $x(t)$ is the received signal and noise mixture.

Let us modify algorithm (4.55) so that on its basis we could make element-by-element decisions concerning the signal, transmitted within the last, that is, the nth, element of the interval $[(n - k)T; nT]$. For this purpose we divide a set of values V_{nj} into m subsets. These subsets are determined by the reference signals within the nth element, namely, the ith subset has signal S_i within the nth element. If the maximum value of the V_{nj} values appears in the subset of the signal S_i, the element decision within the nth chip is made for the benefit of this signal S_i. To make the element decision within the next, $(n + 1)$th chip the interval of processing as a whole should be shifted by one element, that is, it becomes $[(n - k + 1)T; (n + 1)T]$. The set of values $V_{(n+1)j}$ is then calculated, and maximum of them determines the $(n + 1)$th transmitted symbol, and so on.

If we now return to the point of the statement made at the beginning of this subsection, we recall that the bit error probability of the above-described strictly incoherent processing tends to the bit error rate (BER) of the strictly coherent processing when increasing the number k elementary signals, used at every element-by-element decision making step.

Optimum Algorithms for PDM Signal Multisymbol Processing

Further consideration of the preceding general algorithm is oriented to PDM signal processing. First let us consider two-level PDM with phase differences of 0 and π. We use the following designations for convolutions of the received signal $x(t)$ and the reference signals with an arbitrary initial phase φ:

$$X_i = \int\limits_{(i-1)T}^{iT} x(t) \sin(\omega t + \varphi) \, dt$$

$$(4.56)$$

$$Y_i = \int\limits_{(i-1)T}^{iT} x(t) \cos(\omega t + \varphi) \, dt$$

In the case of conventional two-chip processing, the values of (4.55) can be easily presented by (4.56) in the following form:

$$V_{n1} = (X_{n-1} + X_n)^2 + (Y_{n-1} + Y_n)^2$$
$$V_{n2} = (X_{n-1} - X_n)^2 + (Y_{n-1} - Y_n)^2$$

$$(4.57)$$

Expression (5.57) is a modification of algorithm (4.10) for the optimum incoherent processing of binary PDM signals: The phase difference 0 is considered to be transmitted within the nth chip, if $V_{n1} > V_{n2}$, and the phase π in the opposite case.

Now let us consider binary PDM-1 signals within a three-chip interval. In this case there are four variants of signals with an arbitrary initial phase (see Fig. 4.12), and the values (4.55) become the following:

$$V_{n1} = (X_{n-2} + X_{n-1} + X_n)^2 + (Y_{n-2} + Y_{n-1} + Y_n)^2$$
$$V_{n2} = (X_{n-2} - X_{n-1} + X_n)^2 + (Y_{n-2} - Y_{n-1} + Y_n)^2$$
$$V_{n3} = (X_{n-2} - X_{n-1} - X_n)^2 + (Y_{n-2} - Y_{n-1} - Y_n)^2$$
$$V_{n4} = (X_{n-2} + X_{n-1} - X_n)^2 + (Y_{n-2} + Y_{n-1} - Y_n)^2$$

$$(4.58)$$

The required algorithm for decision making is formulated as follows: If V_{n1} or V_{n3} is the greatest of the values of (4.58), the phase difference 0 is fixed as having been transmitted, otherwise the phase difference is π.

Note that algorithm (4.58) almost coincides with algorithm (4.40) for optimum incoherent reception of binary PDM-2 signals. It differs only by the procedure of decision making. Therefore the demodulator shown in Figure 4.13 is the optimum incoherent demodulator for PDM-1 signal processing as a whole within three chips, if the comparison and decision-making scheme operates according to the preceding algorithm.

The error probability of the obtained algorithm is the same as that for the optimum incoherent reception of binary PDM-2 signals: It is less than the error probability of the optimum two-chip incoherent reception of binary PDM-1 signals. The appropriate quantitative relations are derived in Chapter 6. Obviously there is no sense in further increasing the processing interval in the case of binary PDM because the error probability of three-chip processing differs negligibly from the error probability of strictly coherent processing.

A four-level PDM-1 signal with first phase differences of 0, $\pi/2$, π, $3\pi/2$ has 16 variants within three chips, the first of which is arbitrary, but identical, for all variants of the initial phase. These variants coincide with those considered in Section 4.4 for signal variants of four-level PDM-2 signals. Therefore the optimum incoherent demodulator for four-level PDM-1 signals with processing as a whole within three chips should calculate the same values of (4.40) through (4.43) as the optimum incoherent demodulator of four-level PDM-2 signals. Based on (4.40) through (4.43), the decision-making algorithm has been represented in Table 4.3. This table specifies phase differences $\Delta_n^1\varphi$, for the benefit of which a decision is made within the nth chip depending on which of the values of (4.40) through (4.43) is the maximum value. Code combinations of the Gray code, indicated in last two table columns, correspond to four-phase-difference decisions. The demodulator block diagram is shown in Figure 4.19.

The algorithm for optimum incoherent reception of eight-level PDM-1 signals with processing within three chips is synthesized in a similar manner. The appropriate demodulator has the same input part as Figure 4.19, forming values X_n, X_{n-1}, X_{n-2}, Y_n, Y_{n-1}, Y_{n-2}, the calculator of values $V_1, \ldots, V_{64}$ corresponding to 64 variants of the eight-phase signal within the $(n-1)$th and nth chips, and the comparison and decision-making scheme. In the last one, depending on which of the values of $V_1, \ldots, V_{64}$ is the maximum one, the decision is made for the benefit of one of eight possible phase differences: 0, $\pi/4$, $\pi/2$, $3\pi/4$, π, $5\pi/4$, $3\pi/2$, or $7\pi/4$.

As is already known, at multiphase PDM-1 signals optimum incoherent reception loses considerably to the coherent one with regard to noise immunity. There-

Table 4.3
Decision Algorithm for Four-Phase PDM-1 Signals at Three-Symbol Processing

Maximum Value of (4.40)–(4.43)	$\Delta_n^1\varphi$	J_n^1	J_n^2
V_1, V_8, V_{11}, V_{14}	0	+	+
V_2, V_5, V_{12}, V_{15}	$\pi/2$	+	−
V_3, V_6, V_9, V_{16}	π	−	−
V_4, V_7, V_{10}, V_{13}	$3\pi/2$	−	+

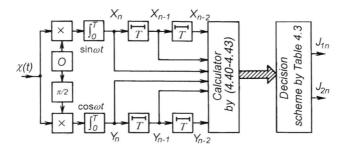

Figure 4.19 Optimum incoherent demodulator of four-level PDM-1 signals with three-chip interval processing.

fore the algorithms for incoherent processing of such signals within intervals lasting longer than three chips are of interest. These multisymbol processing algorithms can provide further approximations to the noise immunity corresponding to the perfect coherent reception. Using the preceding method, it is not difficult to synthesize the algorithms for 2^N phase signals at l-chip processing. With these conditions, a number of signal variants and values, being a subject to compare in the demodulator, are equal to $2^{N(l-1)}$. The demodulators, naturally, are becoming complicated. For example, at four-chip processing the number of values to be compared increases to 64 for four-level and up to 512 for eight-level PDM-1.

The opportunity to achieve potential noise immunity without unknown signal parameters evaluation is not limited to the algorithms we have considered. In other algorithms, synthesized for the conditions of greater a priori uncertainty, the specified result also takes place. For example, it is possible to approximate the potential noise immunity by increasing the interval of autocorrelated signal processing. The appropriate algorithms can be obtained from the rule of the generalized maximum likelihood [1] for signals with unknown form [8].

Omitting the intermediate calculations, we give the result for a binary system, using arbitrary antipodal signals encoded similarly to those of the usual two-phase PDM signals. At two-chip autocorrelated processing, values similar to (4.57) are as follows:

$$V_{n1} = \int_{(n-1)T}^{nT} [x(t) + x(t - T)]^2 \, dt$$

$$V_{n2} = \int_{(n-1)T}^{nT} [x(t) - x(t - T)]^2 \, dt$$

(4.59)

In this case the decision algorithm is the same as relations (4.57). At three-chip interval processing, the following values are calculated:

$$V_{n1} = \int_{(n-1)T}^{nT} [x(t) + x(t - T) + x(t - 2T)]^2 \, dt$$

$$V_{n2} = \int_{(n-1)T}^{nT} [x(t) - x(t - T) + x(t - 2T)]^2 \, dt$$

$$V_{n3} = \int_{(n-1)T}^{nT} [x(t) + x(t - T) - x(t - 2T)]^2 \, dt \quad (4.60)$$

$$V_{n4} = \int_{(n-1)T}^{nT} [x(t) - x(t - T) - x(t - 2T)]^2 \, dt$$

If V_{n1} or V_{n4} is maximum, then the zero phase difference is fixed at the nth chip; if V_{n2} or V_{n3} is maximum, then the signal sign change is fixed.

Application of algorithm (4.60) results in lowering the error probability in comparison with the application of algorithm (4.59), similar to the process of lowering the error probability at transition from algorithm (4.57) to algorithm (4.58). It is not difficult to make up similar autocorrelated algorithms for intervals of four and more chips. These multisymbol processing algorithms allow us to lower error probabilities further and to approximate the minimum possible error probability for the antipodal signals.

Looking once again at formulas (4.59) and (4.60), we should note that the preceding effect is reached by means of the algorithms in which no a priori information about signal parameters is used, and the unknown parameters are not estimated in any way.

Noise Immunity Estimates

To evaluate the different means of achieving the potential noise immunity, let us consider the case of two-level PDM. The minimum error probability achievable in an AWGN channel using the ideal coherent receiver is in this case equal to (see Chapter 6 for details):

$$P_{\min} = 2F(\sqrt{2}h)\,[1 - F(\sqrt{2}h)] \quad (4.61)$$

where $h^2 = P_c T / N_0$ is the ratio of chip energy to the spectral density of noise power,

$$F(\alpha) = \frac{1}{\sqrt{2\pi}} \int_{\alpha}^{\infty} \exp\left(-\frac{x^2}{2}\right) dx$$

The error probability when using algorithm (4.57)—the optimum incoherent reception with two-chip interval processing—is also known and is equal to

$$P_2 = \frac{1}{2} \exp(-h^2) \tag{4.62}$$

For the case of three-chip interval processing according to algorithm (4.58), we can use the upper bound for the system with the second-order PDM [12]:

$$P_3 < \exp\left(-\frac{3h^2}{2}\right) \left[I_0\left(-\frac{h^2}{2}\right) + 2 \sum_{n=1}^{\infty} (3 - \sqrt{8})^n I_n\left(\frac{h^2}{2}\right) \right] \tag{4.63}$$

where I_0 and I_n are modified Bessel functions of the corresponding order. The calculation results of formulas (4.61), (4.62), and (4.63) are represented in Table 4.4. As is obvious, the transition from two-chip incoherent processing to three-chip incoherent processing leads to a decrease in the error probability of element-by-element reception and also to a reduction in the gap between noise immunities of coherent and incoherent receptions.

Table 4.4
Noise Immunity of Various Algorithms for Processing of Binary PDM Signals

| | | *Optimum Incoherent Reception* | |
SNR $h^2 = E/N_0$	*Ideal Coherent Reception*	*Three Chips (Upper Bound)*	*Two Chips*
2.00	4.46×10^{-2}	7.31×10^{-2}	6.77×10^{-2}
2.56	2.7×10^{-2}	3.73×10^{-2}	3.86×10^{-2}
3.24	1.08×10^{-2}	1.69×10^{-2}	1.96×10^{-2}
4.00	4.67×10^{-3}	7.11×10^{-3}	9.16×10^{-3}
4.84	1.86×10^{-3}	2.79×10^{-3}	3.95×10^{-3}
5.76	6.88×10^{-4}	1.02×10^{-3}	1.58×10^{-3}
6.25	4.06×10^{-4}	6.01×10^{-4}	9.65×10^{-4}
8.00	6.35×10^{-5}	9.26×10^{-5}	1.68×10^{-4}
9.00	2.21×10^{-5}	3.22×10^{-5}	6.17×10^{-5}
11.00	2.73×10^{-6}	3.95×10^{-6}	8.35×10^{-6}
13.00	3.41×10^{-7}	4.92×10^{-7}	1.13×10^{-6}

The indicated results permit is to formulate an important conclusion: Algorithm (4.58) practically completely solves the problem of achieving the potential noise immunity without estimating the unknown initial signal phase!

The effect of lowering the error probability takes place during transition from algorithm (4.59) to algorithm (4.60). As has been shown, the power gain is about 1 dB at the error probability level of 10^{-3} [17].

In multiphase systems it is expedient to use multisymbol incoherent reception with four-chip and higher interval processing, because in these systems the energy loss of incoherent reception relative to coherent is more than that of two-phase systems. To determine the necessary processing interval, we should find the mentioned energy loss as a function of a number of elementary chips participating in processing. In the case of multiphase PDM as the upper bound for the dependence of the error probability on the processing interval, it is possible to use expressions for the error probability of quasicoherent reception with the synchronization signal [18]. For example, for four-phase PDM the error probability of the optimum incoherent reception with processing within a k-chip interval has the following estimate:

$$P_k < Q\left[\frac{h}{\sqrt{2}} \cdot \sqrt{k+1-\sqrt{2k}}; \quad \frac{h}{\sqrt{2}} \cdot \sqrt{k+1+\sqrt{2k}};\right]$$
$$-\frac{1}{2}\exp\left[-\frac{(k+1)h^2}{2}\right]I_0\left[-\frac{h^2}{2}\sqrt{k^2+1}\right] \tag{4.64}$$

where $Q(\alpha;\beta)$ is the Marcum function.

Having calculated by (4.64) the signal-to-noise ratio h_k^2, needed for providing some given error probability, and having compared it with the ratio h_{min}^2, needed for providing the same error probability at ideal coherent reception, we obtain the energy loss of the incoherent reception to the ideal coherent one:

$$\Delta h^2 = 10 \log\frac{h_k^2}{h_{min}^2} \tag{4.65}$$

The energy losses of (4.65) versus a number of elementary chips in the processing interval for various error probabilities are shown in Figure 4.20.

We can conclude that the energy losses are equal to 1 dB for three-chip interval processing, and they are reduced up to 0.5 dB for five-chip interval processing. These numbers are quite adequate for approximation of the potential noise immunity.

When using multisymbol incoherent processing, we should take into account the fact that noise immunity increases only at a constant initial signal phase during the whole processing interval. In real conditions the initial phase changes, and at a large processing interval this change can bring to nought the positive effect.

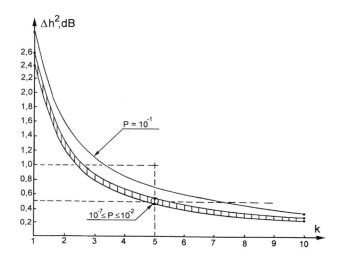

Figure 4.20 Energy losses of the multisymbol incoherent reception as a function of processing interval.

References

[1] Fink, L. M., *Discrete Message Transmission Theory*, 2nd ed., Moscow: Sov. Radio, 1970.

[2] Doelz, M. L., E. T. Heald, and D. L. Martin, "Binary Data Transmission Techniques for Linear Systems," *Proc. IRE*, Vol. 45, No. 5, 1957, pp. 656–661.

[3] Mosier, R. R., and R. G. Claubaugh, "Kineplex: A Band-Width Efficient Binary Transmission System," *Commun. Electron.*, No. 34, 1958, pp. 723–727.

[4] Ginzburg, V. V., et al., *Digital Information Transmission Equipment MC-5*, Moscow: Svyaz Publ., 1970.

[5] Ginzburg, V. V., V. S. Girshov, and Yu. B. Okunev, "Use of Multichannel Modems for High Speed Digital Data Transmission over Wire Channels," *Telecom. Radioeng.*, Vol. 39, No. 11, 1984, pp. 43–47.

[6] Proakis, J. G., *Digital Communications*, New York: McGraw-Hill Book Company, 1983.

[7] Rakhovich, L. M., "Method of Detecting Phase-Modulated Signals, Transmitted by Relative Phase Keying Technique," USSR Patent 177471.

[8] Okunev, Yu. B., *Theory of Phase-Difference Modulation*, Moscow: Svyaz, 1979.

[9] Kustov, O. V., V. Z. Lundin, and Yu. B. Okunev, "Electronic Modeling Applied to Radio Circuit Realization," *Radiotechnika*, No. 1, 1969, pp. 42–51.

[10] Goldenberg, L. M., B. D. Matyushkin, and M. N. Polyak, *Digital Signal Processing*, Moscow: Radio & Svyaz, 1990.

[11] Oppenheim, A. V., and R. W. Schafer, *Digital Signal Processing*, Englewood Cliffs, NJ: Prentice-Hall, 1979.

[12] Okunev, Yu. B., and L. M. Fink, "Noise Immunity of Various Receiving Methods for Binary System with Second Order Phase Difference Modulation," *Telecom. Radioeng.*, Vol. 39, No. 9, 1984.

[13] Kam, P. Y., "Maximum Likelihood Digital Data Sequence Estimation over the Gaussian Channel with Unknown Carrier Phase," *IEEE Trans. Commun.*, Vol. 35, July 1987.

[14] Divsalar, D., and M. K. Simon, "Multiple Symbol Differential Detection of MPSK," *IEEE Trans. Commun.*, Vol. 38, No. 3, March 1990, pp. 300–308.

[15] Okunev, Yu. B., *Digital Transmission of Information by Phase Modulated Signals*, Moscow: Radio & Svyaz, 1991.

[16] Simon, M. K., S. M. Hinedi, and W. C. Lindsey, *Digital Communication Techniques,* Englewood Cliffs, NJ: Prentice-Hall, 1995.

[17] Okunev, Yu. B., N. M. Sidorov, and L. M. Fink, "Noise Immunity of Incoherent Reception of Signals with Single Phase Difference Modulation," *Telecom. Radioeng.,* Vol. 40, No. 12, 1985.

[18] Messel, A. F., and Yu. B. Okunev, "Effective Quasicoherent Algorithms of Phase Modulated Signal Reception," *Radioelectronika,* No. 5, 1991.

CHAPTER 5

Autocorrelated Processing of PDM Signals

5.1 ESSENCE AND APPLICATION OF AUTOCORRELATED SIGNAL PROCESSING

A great number of optimum algorithms has resulted from the study of the general theory of signal reception. This circumstance, in which we have a variety of optimum algorithms but lack the single and best algorithm for any cases, is not a consequence of dealing with incomplete theory. On the contrary, it is a consequence of the fundamental statement of systems theory: The optimum choice is always relative, and the optimum algorithms should be different depending on the optimization criteria and restrictions, stipulated by channel and signal characteristics, and peculiarities of a communications system as a whole.

The most widely used classification for optimization conditions is based on what we know about the signal before we receive it, that is, on the degree of a priori indetermination. It is said that the signal about which everything is known beforehand should not be received, and the signal about which nothing is known is impossible to receive. All of the situations studied in theory are between these extreme cases and are arranged in decreasing order of a priori information about the received signal. The less such information the receiver has, the lower the potential noise immunity of the corresponding optimum processing algorithm.

The hierarchy of optimum algorithms, classified in this way, starts with a coherent processing algorithm, which is optimum when the transmitted signal variants are completely known. Next is the optimum incoherent algorithm, which is optimum when signal variants have a random and uniformly distributed initial phase. If we go further, reducing the a priori information of the receiving signal, we can synthesize other reception methods that are optimum under certain conditions. Thus the less a priori information of signal parameters we have (or use), the lower the noise immunity. This hierarchy of optimum methods ends with the processing algorithms used for unknown shape signals; the algorithms for PDM signal autocorrelated processing considered in this chapter refer to them too.

First we specify the concept of an unknown shape signal. Let us consider two variants of the transmitted signal $S_1(t)$ and $S_2(t)$ within the interval $(0, \tau)$. These signals can be presented as the expansions by orthonormalized functions $\{\varphi_i\}$ and $\{\psi_i\}$:

$$S_1(t) = \sum_{i=1}^{2B} \alpha_i \varphi_i(t)$$
$$S_2(t) = \sum_{i=1}^{2B} \beta_i \psi_i(t) \tag{5.1}$$

where $2B$ is the dimension of the expected signal depending on its bandwidth and duration:

$$\alpha_i = \int_0^\tau S_1(t)\, \varphi_i(t)\ dt$$
$$\beta_i = \int_0^\tau S_2(t)\, \psi_i(t)\ dt \tag{5.2}$$

The signals $S_1(t)$ and $S_2(t)$ are referred to as the signals with unknown shape, if there is not any a priori information about the factors α_i and β_i.

At the same time, however, basic functions φ_i and ψ_i are known: the system of functions $\{\varphi_i\}$ defines the subspace of the possible signals $S_1(t)$, and the system of functions $\{\psi_i\}$ defines the subspace of the possible signals $S_2(t)$. We can say that in the case considered, we know the subspaces of the expected signals, determined by the appropriate set of basic functions and the time interval of their existence.

If distributions of coefficients α_i and β_i are known, then to synthesize the optimum algorithm of signals $S_1(t)$ and $S_2(t)$ reception we can use the conventional rule of maximum likelihood. If the distributions of α_i and β_i are unknown, we should apply the generalized maximum likelihood rule, according to which of two possible hypotheses (S_1 or S_2) we choose the one that has the maximum likelihood function [conventional probability of this hypothesis at $x(t)$ signal reception] taken by all unknown parameters [1].

The synthesis of the algorithm of unknown shape signals reception according to the generalized maximum likelihood rule is not given here [2,3]. For the case of signals with identical energy, when

$$\sum_{i=1}^{2B} \alpha_i^2 = \sum_{i=1}^{2B} \beta_i^2 \tag{5.3}$$

the obtained algorithm is formulated as follows: The signal $S_1(t)$ should be considered as having been transmitted if

$$\sum_{i=1}^{2B} \left[\int_0^\tau x(t)\, \varphi_i(t)\ dt \right]^2 > \sum_{i=1}^{2B} \left[\int_0^\tau x(t)\, \psi_i(t)\ dt \right]^2 \tag{5.4}$$

and signal $S_2(t)$ is considered in the opposite case.

It is not difficult to notice that we have the expansion coefficients of the received signal (according to basic functions φ_i and ψ_i) in square brackets in (5.4). Therefore the sums of squares of these coefficients, that is, the left and right parts of (5.4), are equal to the energy of the signal received in subspaces $\{\varphi_i\}$ and $\{\psi_i\}$, respectively. As a result, the algorithm (5.4) can be presented in the form:

$$\int_0^\tau x_1^2(t)\ dt > \int_0^\tau x_2^2(t)\ dt \tag{5.5}$$

where $x_1(t)$ and $x_2(t)$ are the components of the received signal, which appeared in the subspaces $\{\varphi_i\}$ and $\{\varphi_i\}$ correspondingly.

Thus, the physical sense of the obtained algorithm is as follows: It is necessary to calculate the energy of signals, received in subspaces of the first and second signal variants, and to consider that variant as having been transmitted in the subspace for which the calculated energy is larger. According to this, algorithms (5.4) and (5.5) are called *energy* or *autocorrelated* algorithms, meaning in the latter case that the receiver calculates the convolution of the received signal with itself at this or that time shift [in (5.5) this shift is equal to zero]. These algorithms are true at various types of signal modulation, and to obtain the appropriate special algorithm the corresponding basic functions should be substituted into (5.4).

We can explain this by means of an example of signal processing with frequency modulation. Let the carrier frequencies of signals $S_1(t)$ and $S_2(t)$ be unknown, however the subspaces of these signals are known: $S_1(t)$ exists in the frequency band Δf_1, and $S_2(t)$ in the band Δf_2, and Δf_1 and Δf_2 are not overlapped. Such signals are a special case of the unknown shape signals. Their reception according to the indicated algorithm of energy (autocorrelated) processing is made up as follows: Within the chip interval T, the energies of the signals received in the frequency bands Δf_1 and Δf_2 are calculated. The signal S_1 is considered to have been transmitted if the energy appears to be larger in band Δf_1, and the signal S_2 is transmitted in the opposite case. In this case the integral in the left part of (5.5) is the energy of a signal received in the frequency band Δf_1, and the integral in the right part of (5.5) is the energy of a signal received in the frequency band Δf_2. Respectively, the basic functions φ_i in (5.4) are the harmonics of the frequency $2\pi/T$, belonging to the band $2\pi\Delta f_1$, and the functions ψ_i are the harmonics, belonging to the band $2\pi\Delta f_2$.

Note that the algorithms considered here operate not only at the unknown shape, but also at the changing shape of signals; however, strictly speaking, only such changes are admissible at which each of the transmitted signal variants does not exceed the limits of the subspace defined for the signal. As illustrated later, at PDM-1 this means that the signal shape should be repeated within two chips. From this we have to satisfy, in particular, rather severe requirements to obtain the necessary stability of a carrier frequency at autocorrelated reception of PDM-1 signals.

Autocorrelated demodulators of PDM-1 signals are attractive first of all because of their extreme simplicity. They do not need units for either selecting coherent reference oscillations, as in coherent demodulators, or correlators with quadrature reference oscillations (or switched matched filters), as in optimum incoherent demodulators.

On the other hand, we should certainly take into account that potentially autocorrelated demodulators have less noise immunity than coherent and optimum incoherent ones. One problem with them is that the noise immunity of autocorrelated (energy) demodulators depends not only on the ratio h^2 of the signal energy to the spectral density of noise power, but also on the system parameter, which is equal to the double product of bandwidth F of the receiver input filter and the signal duration T, that is, $2FT$. The larger this parameter $2FT$, the less noise immunity at the same h^2. In other words, the more complex the signal and the larger its dimension, the higher, under other equal conditions, the error probability—these are the peculiarities of autocorrelated reception. At the same time, when receiving comparatively narrowband signals with $2FT \approx 2$, autocorrelated demodulators are very close to the optimum incoherent ones in the sense of noise immunity.

On evaluating the comparative noise immunity of coherent and optimum incoherent demodulators, on the one hand, and autocorrelated demodulators, on the other hand, we should not forget that the former have advantages only under certain conditions, which have been discussed in detail in previous chapters. If these conditions are not carried out and we really have a priori unknown signal shape, the autocorrelated demodulator can provide less error probability. In a general case it is difficult to define under what conditions this situation will occur, because for a signal with unknown shape we cannot calculate the error probability at coherent or optimum incoherent reception. We can assert that the autocorrelated reception (as a method, following from the generalized maximum likelihood rule) provides a strict minimum of error probability at uniformly distributed unknown parameters (the decomposition coefficients of an unknown shape signal).

The algorithms of autocorrelated processing of PDM-1 signals (see Section 5.2) directly follow from the generalized maximum likelihood (GML) rule, and in this sense they are optimum for the unknown shape signals. The similar optimum (in the sense of GML) algorithm for processing the PDM-2 signals results in the scheme of an energy demodulator (see Section 5.3). As in the case for optimum incoherent reception, the energy demodulator of PDM-2 signals surpasses the auto-correlated demodulator of PDM-1 signals in its noise immunity. At the same time the noise immunity of both types of demodulators depends on the stability of a carrier frequency. To overcome this dependence, a PDM-2 autocorrelated demodulator is used.

If at PDM-1 the autocorrelated reception is attractive, mainly because of its simplicity, then at PDM-2 the autocorrelated reception has a unique property of invariance to a carrier frequency, a property that we cannot find either at coherent or at optimum incoherent methods of processing. Therefore the autocorrelated algorithms for receiving the signals with PDM-2 have a basic meaning in communications theory and communications engineering. It is expedient to use them in channels with uncertain carrier frequency.

The system situations, corresponding to the channel with an uncertain frequency, are similar to the situations with an uncertain initial phase, as described in Section 4.1. Because of instability and a nonstandard driving oscillator frequency in all communications system elements, the uncertain carrier frequency takes place at the beginning of any communications session, and during the whole reception interval for short-term sessions. The uncertain carrier frequency takes place also during pulse or batch transmission mode, especially if the batches are generated by different transmitters.

The signal frequency uncertainty may be a result of a Doppler effect, arising as a result of communication with high-speed moving objects or when retransmitting the signals through a moving repeater. Unpredictable changes in the carrier frequency can be the consequence of both the high-speed movement of the communications object and the changing movement direction. These changes are difficult to compensate by automatic frequency control. In some cases, automatic frequency control is also disabled in the systems with continuous data transmission, when the last ones operate in the channels with variable parameters. For example, when communicating with a high-speed moving object through a channel with deep fading, after every reduction of a signal amplitude below some threshold the carrier frequency differs from the frequency, which was before the amplitude reduction. The automatic frequency control unit is not able to help in this case.

Thus, at digital transmission via various communications channels there are a number of system situations in which the receiver may be required to process a signal with an unknown or not exactly known carrier frequency—the channels with uncertain frequency. The second-order PDM in combination with autocorrelated method of reception is adequate to such channels [2–6].

At present, the autocorrelated modems are applied in various communications systems and channels. The autocorrelated PDM-2 modems, having the property of relative or absolute invariance to carrier frequency changes, are used in aircraft HF and UHF communications systems, in satellite communications systems, and in other digital transmission systems when communicating with high-speed moving objects. A new application for the methods of autocorrelated reception of signals with PDM-1 and PDM-2 is digital communications in an optic frequency range and fiber optic communications [4].

The conditions of expediency and the particular areas for applying the auto-correlated algorithms of processing will now be discussed in further detail.

5.2 OPTIMUM AUTOCORRELATED (ENERGY) DEMODULATORS OF FIRST-ORDER PDM SIGNALS

Let us use the preceding general algorithm for unknown shape signals reception (5.4) to synthesize the appropriate algorithm for receiving the binary PDM-1 signals. In this case the signal variants within the interval of two chips can be submitted by the following formulas:

$$S_1(t) = \begin{cases} f(t) & 0 < t < T \\ f(t-T) & T < t < 2T \end{cases}$$

$$S_2(t) = \begin{cases} f(t) & 0 < t < T \\ -f(t-T) & T < t < 2T \end{cases} \tag{5.6}$$

where $f(t)$ is an arbitrary function. To represent the signals S_1 and S_2 we can choose the following basic functions:

$$\text{for } S_1(t) \quad \begin{aligned} \varphi_i(t) &= (1/T) \sin i\omega_0 t & 0 < t < 2T \\ \varphi_i^*(t) &= (1/T)\cos i\omega_0 t & 0 < t < 2T \end{aligned}$$

$$\text{for } S_2(t) \quad \psi_i(t) = \begin{cases} (1/T) \sin i\omega_0 t & 0 < t < T \\ (-1/T) \sin i\omega_0 t & T < t < 2T \end{cases}$$

$$\psi_i^*(t) = \begin{cases} (1/T) \cos i\omega_0 t & 0 < t < T \\ (-1/T) \cos i\omega_0 t & T < t < 2T \end{cases} \tag{5.7}$$

where $\omega_0 = 2\pi/T$. The functions (5.7) at $i = 1$ are shown in Figure 5.1. The basic functions (5.7) determine two noncrossing subspaces, corresponding to two transmitted signal variants, one of which has the 180-degree phase hop of all frequency components, and the second has no phase hop.

In addition to the specified division of transmitted PDM signals at PDM by two noncrossing subspaces, we can suggest another division at which these subspaces

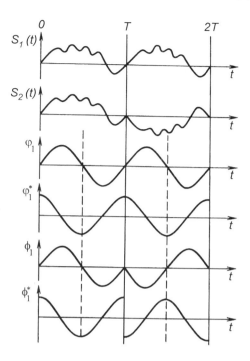

Figure 5.1 Variants of an unknown shape signal with PDM and their basic functions within a two-chip interval.

have various frequency components. As long as the signal $S_1(t)$ within the interval $(T, 2T)$ is the same, as within the interval $(0, T)$, then in the signal decomposition by harmonics of the frequency π/T there are no odd harmonics. In a similar decomposition of the signal $S_2(t)$, which, within the interval $(T, 2T)$, repeats the signal within the interval $(0, T)$ with the opposite sign, there are no even harmonics. Thus, at PDM the subspace basis of the signal $S_1(t)$ contains the even harmonics of the frequency π/T, and the subspace basis of the signal $S_2(t)$ contains odd ones:

$$
\begin{aligned}
S_1(t) &= \sum_{i=1}^{B} a_{2i} \sin 2i\frac{\pi}{T}t + b_{2i} \cos 2i\frac{\pi}{T}t \\
S_2(t) &= \sum_{i=1}^{B} a_{2i-1} \sin(2i-1)\frac{\pi}{T}t + b_{2i-1} \cos(2i-1)\frac{\pi}{T}t
\end{aligned}
\tag{5.8}
$$

The schematic view of the appropriate subspaces of signals S_1 and S_2 is shown in Figure 5.2, where $\omega_0 = \pi/T$.

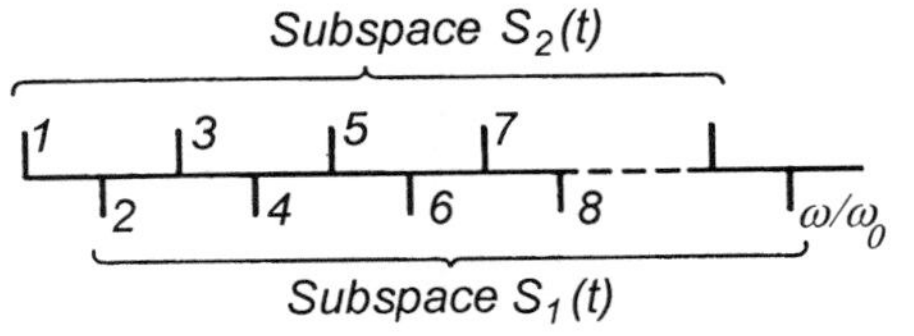

Subspace $S_2(t)$

Subspace $S_1(t)$

Figure 5.2 Subspaces of unknown shape signals with PDM.

Let us find the optimum algorithm for receiving the unknown shape signals with PDM. For this we shall rewrite general algorithm (5.4) in the following form:

$$\sum_{i=1}^{B} \left\{ \left[\int_0^{2T} x(t)\, \varphi_i(t)\, dt \right]^2 + \left[\int_0^{2T} x(t)\, \varphi_i^*(t)\, dt \right]^2 \right\}$$

$$> \sum_{i=1}^{B} \left\{ \left[\int_0^{2T} x(t)\, \psi_i(t)\, dt \right]^2 + \left[\int_0^{2T} x(t)\, \psi_i^*(t)\, dt \right]^2 \right\} \tag{5.9}$$

In comparison with (5.4) the integration interval $\tau = 2T$ is used in (5.9), because with PDM every information symbol is determined by two signal chips. Having substituted (5.7) into (5.9), we obtain

$$\sum_{i=1}^{B} \left\{ \left[\int_0^{T} x(t)\ \sin i\omega_0 t\ dt + \int_T^{2T} x(t)\ \sin i\omega_0 t\ dt \right]^2 \right.$$

$$+ \left[\int_0^{T} x(t)\ \cos \omega_0 t\ dt + \int_T^{2T} x(t)\ \cos i\omega_0 t\ dt \right]^2 \right\}$$

$$> \sum_{i=1}^{B} \left\{ \left[\int_0^{T} x(t)\ \sin i\omega_0 t\ dt - \int_T^{2T} x(t)\ \sin i\omega_0 t\ dt \right]^2 \right.$$

$$+ \left[\int_0^{T} x(t)\ \cos \omega_0 t\ dt - \int_T^{2T} x(t)\ \cos i\omega_0 t\ dt \right]^2 \right\} \tag{5.10}$$

Having squared and reduced the identical members in the left and right parts of (5.10), we obtain

$$J = \text{sgn} \sum_{i=1}^{B} \left[\int_{0}^{T} x(t) \, \sin \, i\omega_0 t \, dt \int_{T}^{2T} x(t) \, \sin \, i\omega_0 t \, dt \right.$$

$$\left. + \int_{0}^{T} x(t) \, \cos \, i\omega_0 t \, dt \int_{T}^{2T} x(t) \, \cos \, i\omega_0 t \, dt \right] \qquad (5.11)$$

The integrals in (5.11) within the limits $(0, T)$ are proportional to the decomposition coefficients of the signal received within the interval 0 to T, and the integrals within the limits $(T, 2T)$ are proportional to the decomposition coefficients of the signal received within the interval T to $2T$. Hence, the sum of products of these coefficients is proportional to the scalar product of two adjacent signal chips, that is, (5.11) can be written

$$J = \text{sgn} \int_{T}^{2T} x(t) x(t - T) \, dt \qquad (5.12)$$

which coincides with the algorithm of the autocorrelated reception of binary PDM signals [7], obtained on an algebraic basis in Section 1.3. Thus, the use of the optimum algorithm of unknown shape signal reception, realizing the generalized maximum likelihood rule, results in the autocorrelated scheme of PDM-1 signal processing as shown in Figure 5.3.

The input signal comes to a passband filter, which fulfills two functions in an autocorrelated demodulator. First, it carries out the usual function of frequency selection of a useful signal. Such selection is needed when a group signal of the transmission system with frequency division comes at the demodulator input, or when the spectrally concentrated interference from other stations is in a radio channel. For this function the similar input passband filter is also applied at the input of coherent and optimum incoherent demodulators. As a rule, the necessary selectivity of a system as a whole is provided by selective circuits of a radio receiver or by selective filters of multiplexing equipment. In this case we do not need the special filter at the demodulator input. Second, the input passband filter serves to restrict the white noise spectrum (and, hence, power as well) arriving at the

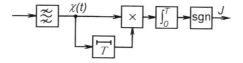

Figure 5.3 Optimum autocorrelated demodulator for binary PDM-1 signals.

demodulator input. This is a special feature of the autocorrelation method of reception, the noise immunity of which depends not only on the noise power spectral density, as with correlation methods, but also on the total noise power (that is on the noise frequency bandwidth).

On first sight it might seem that signal passband filtering is not stipulated in the autocorrelated processing algorithm (5.12). However if we return to the algorithm (5.9), from which algorithm (5.12) has been obtained, it is obvious that the signal $x(t)$ is submitted in the decomposition form by B harmonic components and, hence, is limited by the bandwidth $F = B/T$.

Actually the filter bandwidth in the demodulator of Figure 5.3 should be selected as a compromise: On the one hand, it should be as narrow as possible to reduce the noise power, and on the other hand, it should be sufficiently wide to reduce linear distortions of the signal and intersymbol interference. In narrowband PM autocorrelated demodulators, the system parameter FT, as a rule, is chosen within the limits from 1 to 2, and input filters with bandwidth $F \approx 1.5/T$ are mostly used.

The second important element of an autocorrelated demodulator of PDM-1 signals is a memory element or a duration T delay line. This unit should meet quite rigid requirements in the sense of accuracy of the delay duration. The fact is that in the systems with autocorrelated reception of first-order PDM signals the relation between the carrier frequency and chip duration cannot be arbitrary: Within the chip of duration T an integer of frequency carrier periods should be packed, that is, $\omega T = 2\pi k$, where k is the integer. In this case, the signal shape is repeated within two chips, which is a necessary condition of PDM-1 signal optimum processing.

Deviations from the equation $\omega T = 2\pi k$, caused, for example, by instability of the carrier frequency ω or time delay T, result in noise immunity decreases and then also in loss of demodulator serviceability.[1] The admissible deviations of the frequency $\Delta\omega$ and time delay ΔT can be determined from the following equation:

$$\Delta\varphi_{\text{ad}} = \Delta\omega T + \omega\Delta T + \Delta\omega\Delta T \qquad (5.13)$$

where $\Delta\varphi_{\text{ad}}$ is the admissible phase shift between the PDM-1 signal adjacent elements. To retain a high enough noise immunity, the value $\Delta\varphi_{\text{ad}}$ should not exceed a small fraction (5% to 10%) of the minimum permitted phase hop, which is π at two-level PDM, $\pi/2$ at four-level PDM, $\pi/4$ at eight-level PDM, and so on.

We should draw attention to the fact that, as can be seen from (5.13), the frequency and delay deviations can compensate one another if they have different signs. In particular, the methods of adaptive correction of delay instability are based on it [4]. Usually the admissible frequency and delay deviations are determined at

---

1. At the same time we should note that the considered autocorrelated demodulators are operable at any value of $\omega = 2\pi k/T$, where k is the integer. This property of the discrete invariance can be used for receiving the signals with hoplike (with a step $2\pi/T$) changes of a carrier.

the fixed and equal to the nominal value of one of these parameters; for example, $\Delta\omega_{ad}$ is determined at $\Delta T = 0$, and ΔT_{ad} is determined at $\Delta\omega = 0$. At the precise carrier frequency ($\Delta\omega = 0$), the admissible delay duration instability ΔT_{ad} follows from the simple relation

$$\Delta T_{ad} = \Delta\varphi_{ad}/\omega \tag{5.14}$$

Here we can see that it is useful to decrease the carrier frequency ω in order to reduce the requirements to the delay line stability. In connection with it in autocorrelated demodulators transfer of the received signal spectrum to the lower intermediate frequency is sometimes used. The extreme case of this conversion is "the transfer to zero frequency," that is, the selection of the signal quadrature envelopes by convolution with the reference oscillations, coinciding by a frequency with the carrier. The last approach results in a scheme that is equivalent to the optimum incoherent demodulator.

As is obvious from the scheme of Figure 5.3, the autocorrelated demodulator uses the same correlators as the coherent and optimum incoherent demodulators. The difference is that in the autocorrelated demodulator the previous chip of the received signal is used as a reference oscillation at the given chip processing. Therefore it is not effective to apply sign multipliers in autocorrelated demodulators—this results in appreciable energy losses. So the multiplier in the scheme of the autocorrelated receiver, if it is realized by the analog method, is a comparatively complex unit. For digital realization there are no peculiarities in this case.

As a rule, autocorrelated demodulators use low-pass filters (LPFs) instead of integrators. The appropriate modification of the main autocorrelated demodulator scheme is shown in Figure 5.4. This type of demodulator, in contrast to the active correlator-based schemes, generally speaking, can operate without timing recovery; however, it leads to lower noise immunity. To increase the noise immunity, the LPF output should be registered at the strictly definite moment. The optimum moment of signal sign registration depends on the channel frequency characteristics. It is determined by the synchronization pulse, which is generated by a lock unit (LU).

In addition to algorithm (5.12) and the scheme indicated in Figure 5.3, there are also other algorithms and PDM-1 autocorrelated processing schemes equivalent

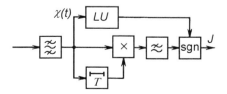

Figure 5.4 Modification of the autocorrelated demodulator for binary PDM signals.

to them, but they do not require the direct calculation of the received signal autocorrelation function. To find them let us return again to the general algorithms of (5.9) or (5.10). In square brackets in the left part of (5.10) there are sums of the decomposition coefficients of the signals received within two adjacent chips. Hence, the sum of squares of the square brackets is proportional to the energy of the sum of two adjacent chips. Similarly in square brackets in the right part of (5.10) there are the differences of the decomposition coefficients of the signal received within two adjacent chips; hence, the sum of squares of these differences is proportional to the energy of the difference of two adjacent chips. Thus, we can rewrite (5.10) in the following form:

$$\int_{T}^{2T} [x(t) + x(t - T)]^2 \, dt > \int_{T}^{2T} [x(t) - x(t - T)]^2 \, dt \qquad (5.15)$$

The equivalence of (5.15) and (5.12) is obvious. The inequality (5.15) can be presented in the form of the algorithm for calculating the transmitted binary symbol [similar to (5.12)]:

$$J = \mathrm{sgn}\left\{ \int_{T}^{2T} [x(t) + x(t - T)]^2 \, dt - \int_{T}^{2T} [x(t) - x(t - T)]^2 \, dt \right\} \qquad (5.16)$$

The demodulator corresponding to the algorithm (5.16) is shown in Figure 5.5. If we need to emphasize the difference of this scheme from the scheme of the autocorrelated demodulator shown in Figure 5.3, we will call it the energy demodulator of PDM-1 signals. From the viewpoint of noise immunity and requirements for carrier frequency and delay duration stability, the demodulators shown in Figures 5.3 and 5.5 are equivalent. However, in the demodulator of Figure 5.5 there is no a signal multiplier. Instead, the sum and the difference of adjacent chips are squared (S) and integrated. Then the results are compared in the subtraction scheme: If the energy of the signals' sum is more than the energy of the signals' difference, the zero phase difference and symbol +1 are fixed; the phase difference π and the

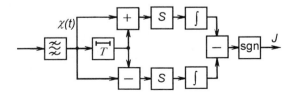

Figure 5.5 Optimum energy demodulator for binary PDM-1 signals.

symbol −1 are fixed in the opposite case. Squarers and integrators can be practically realized in the form of inertial squared detectors.

Thus, for analog realization of autocorrelated reception algorithm (5.16) and the scheme of Figure 5.5 appear to be more preferable. For digital realization it is better to use algorithm (5.12) and the scheme of Figure 5.3. However, the problem of realization is solved largely depending on the presented element base for the particular frequency range and transmission rate.

Among the multiphase systems most often in use are the autocorrelated modems with four-phase PDM-1, which can reduce the occupied frequency band almost twice at insignificant energy losses.

At four-level PDM, as a rule, two systems of phase differences are applied: (1) $\pi/4$, $3\pi/4$, $5\pi/4$, and $7\pi/4$ or (2) 0, $\pi/2$, π, and $3\pi/2$. The Gray code is used in both cases. The indicated systems are specified in Table 4.1 and in Figure 5.6 in vector form.

To synthesize the algorithms of autocorrelated demodulators of PDM-1 signals it is most convenient to use the general algorithms of PDM multiposition signals decoding, presenting the transmitted binary symbols through a sine and cosine of the received phase difference [see (2.79) through (2.86)]. At autocorrelated processing the sine and cosine included in these algorithms are determined by formulas (2.87) in the form of signal adjacent chips convolutions.

When the first signal system is used (Fig. 5.6), the algorithms of demodulation of the first and second binary subchannels are

$$J_{1n} = \text{sgn} \int_{(n-1)T}^{nT} x(t)\ x^*(t - T)\ dt$$

$$J_{2n} = \text{sgn} \int_{(n-1)T}^{nT} x(t)\ x(t - T)\ dt$$

(5.17)

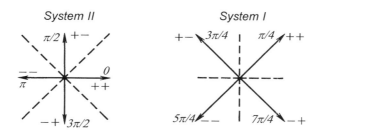

Figure 5.6 Systems of four-level PDM signals.

where Hilbert transformation is designated by an asterisk. The demodulator scheme that realizes algorithm (5.17) is illustrated in Figure 5.7.

When we use the second signal system (Fig. 5.6), the autocorrelated processing algorithm is as follows:

$$J_{1n} = \text{sgn}\left[\int_{(n-1)T}^{nT} x(t)\,x(t-T)\ dt + \int_{(n-1)T}^{nT} x(t)\,x^*(t-T)\ dt\right]$$

$$J_{2n} = \text{sgn}\left[\int_{(n-1)T}^{nT} x(t)\,x(t-T)\ dt - \int_{(n-1)T}^{nT} x(t)\ x^*(t-T)\ dt\right]$$

(5.18)

The corresponding demodulators are shown in Figure 5.8.

The considered autocorrelated demodulators of four-level PDM-1 signals contain a phase shifter by $\pi/2$, operating as a Hilbert converter. This phase shifter should be a broadband unit because it should carry out phase shifting for all spectral components of the received signal.

As is obvious from comparison schemes in Figures 5.7 and 5.8, the use of the phase differences $\pi/4$, $3\pi/4$, $5\pi/4$, and $7\pi/4$ results in a simpler demodulator. At the same time, to realize this advantage a more complex modulator is required because in this case a carrier has eight phase levels.

However, we can find a relation between ω and T such that the carrier will have only four phases (0, $\pi/2$, π, and $3\pi/2$) and the demodulator will be implemented by the simplest scheme, as indicated in Figure 5.7. For this purpose, as is obvious from the comparison of signal systems in Figure 5.6, we should shift every chip by $\pi/4$, which occurs automatically when fulfilling the relation

$$\omega T = 2\pi k + \pi/4 \tag{5.19}$$

where k is an integer.

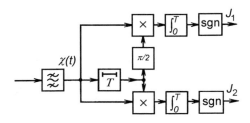

Figure 5.7 Autocorrelated demodulator for four-level PDM-1 signals with phase differences $\pi/4$, $3\pi/4$, $5\pi/4$, and $7\pi/4$.

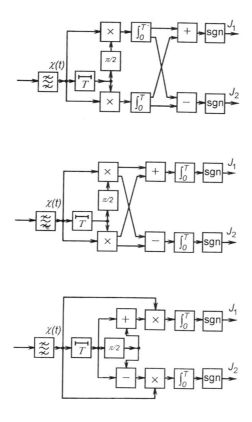

Figure 5.8 Modifications of the autocorrelated demodulator for four-level PDM-1 signals with the phase differences 0, $\pi/2$, π, and $3\pi/2$.

In general, note that autocorrelated demodulators allow us to receive signals with various minimum phase differences by changing the signal delay duration. The same effect can be reached by varying the carrier frequency, without any changes in the scheme of an autocorrelated demodulator.

For four-level PDM, as well as for two-level PDM, we can synthesize the algorithms and schemes of energy demodulators, similar to (5.16) and (5.15) and Figure 5.5. For example, in the case of phase differences 0, $\pi/2$, π, and $3\pi/2$, the demodulator scheme without signal multiplication units has the form shown in Figure 5.9. This energy demodulator contains four processing branches—according to a number of signal variants. The comparison scheme (CS) defines the strongest output signal: If the strongest signal is in branch 1, the zero phase difference is fixed as transmitted; if in branch 2, the phase difference is π; if in branches 3 and 4, the phase differences are $\pi/2$ and $3\pi/2$, respectively.

The scheme of an energy demodulator for four-level PDM signals can be made up in such a way that at its output there will be binary symbols, transmitted

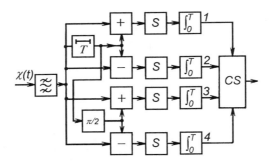

Figure 5.9 Energy (autocorrelated) demodulator for four-level PDM-1 signals with phase differences 0, $\pi/2$, π, and $3\pi/2$.

by subchannels. For example, the scheme of the autocorrelated demodulator in Figure 5.7 can be transformed to the form of the energy demodulator shown in Figure 5.10. Formally, by a number of units, this scheme is more complex, however, it does not contain signal multipliers, which in some cases could be more attractive.

We should mention that the considered autocorrelated demodulators of four-level PDM signals should meet the same requirements and use the same implementation approaches as the binary PDM demodulators. In particular, for analog realization, integrators with resets are replaced by LPFs that take samples at their outputs in moments, determined by clock locking pulses. For digital realization, memory digital registers are used instead of delay lines and digital correlators.

At eight-level PDM, as a rule, one of two phase-difference systems is used in combination with the Gray code, which in this case, as for four-level PDM, is the unique optimum keying code. These systems (see Table 4.2) can be written as follows:

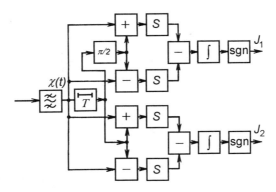

Figure 5.10 Energy (autocorrelated) demodulator for four-level PDM-1 signals with phase differences $\pi/4$, $3\pi/4$, $5\pi/4$, and $7\pi/4$.

$$\Delta\varphi_i = (2i - 1)\,\pi/8 \qquad \text{for } i = 1, 2, \ldots, 8 \qquad (5.20)$$

$$\Delta\varphi_i = (i - 1)\,\pi/4 \qquad \text{for } i = 1, 2, \ldots, 8 \qquad (5.21)$$

As in the case of four-level PDM, the demodulator algorithms at eight-level PDM are obtained by substituting expressions (2.87) for the sine and cosine of the first phase difference in the common algorithms of decoding (2.84).

When using the phase-difference system of (5.20), the algorithm of autocorrelated reception, determining binary symbols transmitted by three subchannels, has the following form:

$$J_{1n} = \text{sgn} \int_{(n-1)T}^{nT} x(t)\,x^*(t - T)\ dt$$

$$J_{2n} = \text{sgn} \int_{(n-1)T}^{nT} x(t)\,x(t - T)\ dt \qquad\qquad (5.22)$$

$$J_{3n} = \text{sgn}\left[\int_{(n-1)T}^{nT} x(t)\,x^*(t - T)\ dt + \int_{(n-1)T}^{nT} x(t)\,x(t - T)\ dt\right]$$

$$\times \text{sgn}\left[\int_{(n-1)T}^{nT} x(t)\,x(t - T)\ dt - \int_{(n-1)T}^{nT} x(t)\,x^*(t - T)\ dt\right]$$

The scheme corresponding to this algorithm is shown in Figure 5.11.

When using the phase-difference system of (5.21), the algorithm of eight-level PDM-1 signals demodulation is as follows:

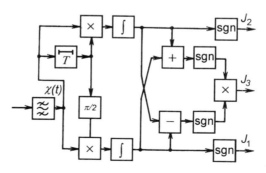

Figure 5.11 Autocorrelated demodulator for eight-level PDM-1 signals with phase differences $\Delta^1\varphi_i = (2i - 1)\,\pi/8$.

$$J_{1n} = \text{sgn}\left[\cos\frac{\pi}{8} \int\limits_{(n-1)T}^{nT} x(t)\,x^*(t-T)\,dt + \sin\frac{\pi}{8} \int\limits_{(n-1)T}^{nT} x(t)\,x(t-T)\,dt\right]$$

$$J_{2n} = \text{sgn}\left[\cos\frac{\pi}{8} \int\limits_{(n-1)T}^{nT} x(t)\,x(t-T)\,dt - \sin\frac{\pi}{8} \int\limits_{(n-1)T}^{nT} x(t)\,x^*(t-T)\,dt\right] \qquad (5.23)$$

$$J_{3n} = \text{sgn}\left[\cos\frac{3\pi}{8} \int\limits_{(n-1)T}^{nT} x(t)\,x^*(t-T)\,dt + \sin\frac{3\pi}{8} \int\limits_{(n-1)T}^{nT} x(t)\,x(t-T)\,dt\right]$$

$$\times\ \text{sgn}\left[\cos\frac{3\pi}{8} \int\limits_{(n-1)T}^{nT} x(t)\,x(t-T)\,dt - \sin\frac{3\pi}{8} \int\limits_{(n-1)T}^{nT} x(t)\,x^*(t-T)\,dt\right]$$

The appropriate (5.23) scheme of an autocorrelated demodulator of 8-ary PDM-1 signals is more complex than the scheme shown in Figure 5.11. Therefore, when phase differences (5.21) are used, it is also expedient to proceed to processing algorithm (5.22) by means of the additional $\pi/8$ phase shift at every chip. This can be accomplished, having chosen a carrier frequency and a chip duration that correspond to the following equality:

$$\omega T = 2\pi k + \pi/8 \qquad (5.24)$$

where k is the integer. In this case a chip contains $(k + 1/16)$ carrier periods.

The main practical issue of autocorrelated reception of PDM-1 signals is the implementation of the precise and stable unit for one-chip signal delay. This problem is similar to the problem of providing a precise and stable carrier phase at the output of the passive reference oscillation selector in the coherent demodulators. The given delay duration with the required accuracy, according to (5.13) and (5.14), should be provided at all temperature conditions and other destabilization factors. In some situations this requirement is difficult to fulfill. There are known cases because when it was impossible to realize a stable delay line we had to use PDM-2 instead of PDM-1. PDM-2 is insensitive to delay instability when using special invariant autocorrelated algorithms. These algorithms are considered in Section 5.4.

5.3 OPTIMUM AUTOCORRELATED (ENERGY) DEMODULATORS OF SECOND-ORDER PDM SIGNALS

The algorithms of demodulators of PDM-2 signals considered in this section can be obtained from the common optimum algorithm of unknown shape signal reception (see Section 5.1). These algorithms will be also optimum in the sense that they realize the generalized maximum likelihood rule [8].

The peculiarity of the synthesis of the optimum autocorrelated algorithm for PDM-2 signal processing, in comparison with PDM-1, lies in that the variants of PDM-2 signals are not orthogonal. Consequently, the subspaces of PDM-2 signal variants are crossed, and the energy of one of signal variants, being present in the subspace of the other, is not equal to zero. However, it is not difficult to be convinced that the intersection of subspaces of signal variants at binary PDM-2 is symmetric; that is, the energy portion of the first variant, being present in the subspace of the second one, is equal to the energy portion of the second variant, being present in the subspace of the first one. Therefore in this case, as well as at orthogonal signals, the common algorithm (5.4) can be used to synthesize the algorithm, corresponding to the generalized maximum likelihood rule.

Figure 5.12 shows four possible variants of an arbitrary antipodal signal within three chips when we have a fixed signal within the first chip: S_1, S_2, S_3, and S_4. Within

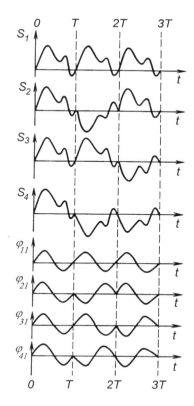

Figure 5.12 Variants of an arbitrary shaped signal within three chips and their basic functions.

every chip with duration T, the signal can have one of two forms, distinguished by a sign. If the variants S_1 or S_2 are transmitted, this corresponds to the second phase difference $\Delta^2\varphi = 0$; if S_3 or S_4 is transmitted, this corresponds to $\Delta^2\varphi = \pi$. The convolutions of the signal variants are equal to

$$(S_iS_j) = \int_0^{3T} S_i(t)\,S_j(t)\ dt = \begin{cases} 3E & \text{at } i = j \\ \pm E & \text{at } i \neq j \end{cases} \tag{5.25}$$

where E is the signal chip energy; $i,j =$1, 2, 3, and 4. The signals S_1 through S_4 are nonorthogonal, however their mutual scalar products are equal by the absolute value.

To use the common algorithm of arbitrary shaped signal processing, we should substitute in it the basic functions of signal variants:

$$
\begin{aligned}
&\text{for } S_1(t) \\
&\varphi_{1i} = \sin i\omega_0 t && 0 < t < 3T \\[1em]
&\text{for } S_2(t) \\
&\varphi_{2i} = \sin i\omega_0 t && \left\{\begin{array}{l} 0 < t < T \\ 2T < t < 3T \end{array}\right. \\[1em]
&\varphi_{2i} = -\sin i\omega_0 t && T < t < 2T \\[1em]
&\text{for } S_3(t) \\
&\varphi_{3i} = \sin i\omega_0 t && 0 < t < 2T \\
&\varphi_{3i} = -\sin i\omega_0 t && 2T < t < 3T \\[1em]
&\text{for } S_4(t) \\
&\varphi_{4i} = \sin i\omega_0 t && 0 < t < T \\
&\varphi_{4i} = -\sin i\omega_0 t && T < t < 3T
\end{aligned}
\tag{5.26}
$$

where $\omega_0 = 2\pi/T$. Figure 5.12 shows, as an example, the basic functions φ_{11}, φ_{21}, φ_{31}, and φ_{41}.

According to the common algorithm we should calculate the received signal energy in the subspaces of every of four transmitted signal variants, that is,

$$V_j = \sum_{i=1}^{B}\left\{\left[\int_0^{3T} x(t)\,\varphi_{ji}(t)\ dt\right]^2 + \left[\int_0^{3T} x(t)\,\varphi_{ji}^*(t)\ dt\right]^2\right\} \tag{5.27}$$

where $x(t)$ is the received signal; $j = $1, 2, 3, and 4. As an example we represent V_1 in more detail:

$$V_1 = \sum_{i=1}^{B} \left\{ \left[\int_0^T x(t) \cos i\omega_0 t \, dt + \int_T^{2T} x(t) \cos i\omega_0 t \, dt + \int_{2T}^{3T} x(t) \cos i\omega_0 t \, dt \right]^2 \right.$$

$$\left. + \left[\int_0^T x(t) \sin i\omega_0 t \, dt + \int_T^{2T} x(t) \sin i\omega_0 t \, dt + \int_{2T}^{3T} x(t) \sin i\omega_0 t \, dt \right]^2 \right\} \quad (5.28)$$

The integrals in square brackets are decomposition coefficients of the received signal within the first, second, and third chips. The sum of these coefficients is equal to the decomposition coefficient of the signal received within three chips and, consequently, V_1 is the energy of the sum of three series chips of the received signal.

Discussed in such a way, we realize that V_2 is the energy of the sum of the first and third chips minus the second one, and so on. Thus, we have

$$V_1 = \int_0^T [x(t) + x(t - T) + x(t - 2T)]^2 \, dt$$

$$V_2 = \int_0^T [x(t) - x(t - T) + x(t - 2T)]^2 \, dt \qquad (5.29)$$

$$V_3 = \int_0^T [x(t) + x(t - T) - x(t - 2T)]^2 \, dt$$

$$V_4 = \int_0^T [x(t) - x(t - T) - x(t - 2T)]^2 \, dt$$

The obtained optimum algorithm of receiving the unknown shape signals with PDM-2 is formulated as follows: If the value of V_1 or V_2 is maximum, the transmitted second phase difference is $\Delta^2 \varphi = 0$; if V_3 or V_4 is maximum, the transmitted phase difference is $\Delta^2 \varphi = \pi$.

The scheme of the optimum autocorrelated demodulator of two-level PDM-2 signals is shown in Figure 5.13. This demodulator, similar to the PDM-1 signal demodulator shown in Figure 5.5, can be called an energy demodulator. The values (5.29), proportional to the energy of corresponding combinations of received signal chips, are calculated in the demodulator by means of signal delay units, inverters (Inv), adders, squarers (S), and integrators. Then the maximum value is chosen in a comparator (Com).

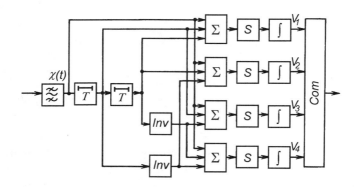

Figure 5.13 Optimum energy (autocorrelated) demodulator for two-level PDM-2 signals.

Comparing two schemes for optimum autocorrelated (energy) demodulators (Fig. 5.5 for PDM-1 and Fig. 5.13 for PDM-2), we note that the latter is almost twice as complex. However, the autocorrelated demodulator of PDM-2 signals (Fig. 5.13) has better noise immunity than the autocorrelated demodulator of PDM-1 signals (Fig. 5.5). Similarly, the optimum incoherent demodulator of PDM-2 signals provides better noise immunity than the optimum incoherent demodulator of PDM-1 signals.

Processing algorithm (5.29) can be easily presented through the usual scalar products of adjacent chips, if we open integrands and reduce the common components:

$$V'_{1n} = (x_n x_{n-1}) + (x_n x_{n-2}) + (x_{n-1} x_{n-2})$$

$$V'_{2n} = -(x_n x_{n-1}) + (x_n x_{n-2}) - (x_{n-1} x_{n-2})$$

$$V'_{3n} = (x_n x_{n-1}) - (x_n x_{n-2}) - (x_{n-1} x_{n-2})$$
(5.30)

$$V'_{4n} = -(x_n x_{n-1}) - (x_n x_{n-2}) + (x_{n-1} x_{n-2})$$

where

$$(x_n x_{n-1}) = \int_{(n-1)T}^{nT} x(t) x(t - T) \, dt$$

$$(x_n x_{n-2}) = \int_{(n-1)T}^{nT} x(t) x(t - 2T) \, dt$$
(5.31)

$$(x_{n-1} x_{n-2}) = \int_{(n-1)T}^{nT} x(t - T) x(t - 2T) \, dt$$

The demodulator corresponding to (5.30) is shown in Figure 5.14. It includes two autocorrelators and has the same noise immunity as the demodulator shown in Figure 5.13. This scheme is certainly more complex than the scheme of the elementary autocorrelated demodulator of PDM-1 signals (Fig. 5.3), however, it has greater noise immunity. The quantitative evaluation of a gain for the optimum autocorrelated reception of PDM-2 signals in comparison with similar reception for PDM-1 signals is considered in Chapter 6.

The demodulator of Figure 5.14 includes as a component the suboptimum PDM-2 signal demodulator (dotted line in Fig. 5.14), in which the scalar product $(x_n\, x_{n-2})$ of the received signal chips distant by two intervals is calculated. This part of the scheme, which is interesting by itself, is shown in Figure 5.15 as a demodulator of binary PDM-2 signals, realizing the following algorithm:

$$J = \text{sgn} \int_0^T x(t)\; x(t-2T)\; dt \tag{5.32}$$

From the comparison of (5.32) with (5.12) and Figure 5.15 with Figure 5.3 it follows that this suboptimum autocorrelated demodulator of PDM-2 signals has the same noise immunity as the optimum autocorrelated demodulator of PDM-1 signals.

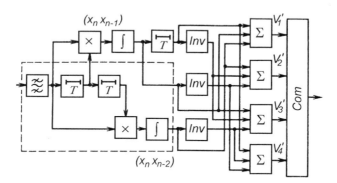

Figure 5.14 Modification of the optimum autocorrelated demodulator of two-level PDM-2 signals.

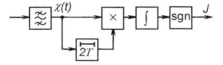

Figure 5.15 Suboptimum autocorrelated demodulator for binary PDM-2 signals.

The considered demodulators do not provide invariance to the carrier frequency of the PM signal in spite of the fact that they use second-order PDM. As a matter of fact, the algorithms of these demodulators are obtained directly from the general algorithm for receiving the unknown shape signal; at their synthesis it was assumed that, although the signal shape is unknown, it is repeated precisely at least within three adjacent chips. For sinusoidal signals it means that the carrier frequency and chip duration are connected in such a way that a chip contains an integer of carrier periods. Any deviation from this relation results at first in decreasing the noise immunity, and then in infringement of demodulator serviceability. Thus, the requirements in terms of the stability of the carrier frequency and delay duration for the considered demodulators of PDM-2 signals are the same as those for autocorrelated demodulators of PDM-1 signals (see Section 5.2). We can overcome the dependence of the noise immunity from the frequency and delay stability by means of special processing algorithms, which are considered in the next section.

5.4 CARRIER-FREQUENCY-INVARIANT AUTOCORRELATED DEMODULATORS OF PDM SIGNALS

As mentioned earlier, the discrete-difference signal conversion permits us to reduce or even completely remove the influence of spurious signal parameters. This conversion has a property of invariance to interfering effects. In particular, the discrete-difference conversion, consisting of modulating the second differences of signal initial phases, that is, PDM-2, allows us to implement the carrier frequency invariant digital transmission system, which is insensitive to random frequency deviations caused by instability of driving oscillators, by Doppler effects, and by other factors. However, to realize this potential property of invariance we should synthesize appropriate algorithms for processing the PDM-2 signals. The algorithms obtained earlier in this chapter for receiving PDM-2 signals are unsuitable for this purpose, because they operate only at the frequency values $f = k/T$, where k is an integer.[2]

To synthesize the reception algorithm, invariant to arbitrary frequency deviations (certainly, within the limits of a passband of the input demodulator filter), we use the general algorithm for demodulating the signals with two-phase PDM-2:

$$J_n = \text{sgn} \cos \Delta_n^2 \varphi \qquad (5.33)$$

where $\Delta_n^2 \varphi$ is the second phase difference of the signal received within the nth chip, and J_n is the binary symbol transmitted within the nth chip. Representing the cosine of the second phase difference $\Delta_n^2 \varphi$ through trigonometric functions of the first phase differences $\Delta_n^1 \varphi$ and $\Delta_{n-1}^1 \varphi$ within the nth and $(n-1)$th chips, we obtain

---

2. Therefore these algorithms can be considered invariant to the number k. All autocorrelated demodulators of signals with PDM-1 and PDM-2, considered in Sections 5.2 and 5.3, have the property of discrete invariance.

$$J_n = \text{sgn}\,(\cos \Delta_n^1 \varphi \, \cos \Delta_{n-1}^1 \varphi + \sin \Delta_n^1 \varphi \, \sin \Delta_{n-1}^1 \varphi) \qquad (5.34)$$

Included in (5.34) are the cosines and sines of the first phase differences that are equal to

$$\cos \Delta_{n-1}^1 \varphi = c \int_0^T x_{n-1}(t)\, x_{n-2}(t) \; dt$$

$$\sin \Delta_{n-1}^1 \varphi = c \int_0^T x_{n-1}(t)\, x_{n-2}^*(t) \; dt \qquad (5.35)$$

$$\cos \Delta_n^1 \varphi = c \int_0^T x_n(t)\, x_{n-1}(t) \; dt$$

$$\sin \Delta_n^1 \varphi = c \int_0^T x_n(t)\, x_{n-1}^*(t) \; dt$$

where $x_n(t)$, $x_{n-1}(t)$, and $x_{n-2}(t)$ are three successive in time signal chips, and c is the proportionality factor, depending on signal power and independent of its phase. Thus, we have the following algorithm of autocorrelated processing for two-level PDM-2 signals:

$$J_n = \text{sgn}\Bigg[\int_0^T x_n(t)\, x_{n-1}(t) \; dt \int_0^T x_{n-1}(t) \; x_{n-2}(t) \; dt$$

$$+ \int_0^T x_n(t)\, x_{n-1}^*(t) \; dt \int_0^T x_{n-1}(t)\, x_{n-2}^*(t) \; dt \Bigg] \qquad (5.36)$$

The scheme that realizes (5.36) is shown in Figure 5.16.

Let us consider the operator of signal conversion in square brackets of (5.36):

$$I[x(t)] = A \cos \Delta_n^2 \varphi = \int_0^T x_n(t)\, x_{n-1}(t) \; dt \int_0^T x_{n-1}(t)\, x_{n-2}(t) \; dt$$

$$+ \int_0^T x_n(t)\, x_{n-1}^*(t) \; dt \int_0^T x_{n-1}(t)\, x_{n-2}^*(t) \; dt \qquad (5.37)$$

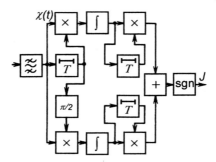

Figure 5.16 Absolutely carrier-frequency-invariant autocorrelated demodulator for binary PDM-2 signals.

The operator of (5.37) has a remarkable property: In the absence of noise, its value, by which the transmitted symbol is determined, does not depend on the carrier frequency of the received signal $x(t)$ and, consequently, when there is additive noise, the error probability does not depend on carrier frequency. Therefore, the demodulator corresponding to algorithm (5.36) is called the *absolutely carrier-frequency-invariant demodulator* (Fig. 5.16).

Let us consider the signal transformations in the carrier-frequency-invariant demodulator. In the absence of noise, the nth signal chip within the interval $(0, T)$ is equal to

$$x_n(t) = a \sin(\omega t + \varphi_n) \tag{5.38}$$

where ω is an arbitrary frequency, and φ_n is an initial phase of the nth chip. The $(n-1)$th and $(n-2)$th chips, delayed on time T and $2T$, respectively, can be represented within the interval $(0, T)$ as follows:

$$x_{n-1}(t) = a \sin(\omega t + \omega T + \omega\Delta T + \varphi_{n-1}) \tag{5.39}$$

$$x_{n-2}(t) = a \sin(\omega t + 2\omega T + 2\omega\Delta T + \varphi_{n-2}) \tag{5.40}$$

Here ΔT is the delay deviation from the nominal value T. Due to this deviation, we must take into account the interval T error, which, as indicated, leads to the same negative result as carrier frequency instability. Taking into account (5.38), (5.39), and (5.40), we can calculate one of the integrals in (5.37):

$$I_n = \int_0^T x_n(t)\,x_{n-1}(t)\ dt = \frac{a^2 T}{2} \cos(\varphi_n - \varphi_{n-1} - \omega T - \omega\Delta T)$$

$$- \frac{a^2}{4\omega} \sin(2\omega T + \varphi_n + \varphi_{n-1} + \omega T + \omega\Delta T)$$

$$+ \frac{a^2}{4\omega} \sin(\varphi_n - \varphi_{n-1} - \omega T - \omega\Delta T) \tag{5.41}$$

If the chip duration T and carrier frequency ω are such that $T/2 \gg 1/4\omega$, that is, $\omega T \gg 1/2$, and it is almost always fulfilled, then the second and the third components in (5.41) can be neglected. Then the required integral is approximately equal to

$$I_n \approx \frac{a^2 T}{2} \cos(\varphi_n - \varphi_{n-1} - \omega T - \omega \Delta T) \tag{5.42}$$

Similarly we obtain

$$I_{n-1} = \int_0^T x_{n-1}(t)\, x_{n-2}(t)\ dt \approx \frac{a^2 T}{2} \cos(\varphi_{n-2} - \varphi_{n-1} - \omega T - \omega \Delta T)$$

$$I_n^* = \int_0^T x_n(t)\, x_{n-1}^*(t)\ dt \approx \frac{a^2 T}{2} \sin(\varphi_n - \varphi_{n-1} - \omega T - \omega \Delta T) \tag{5.43}$$

$$I_{n-1}^* = \int_0^T x_{n-1}(t)\, x_{n-2}^*(t)\ dt \approx \frac{a^2 T}{2} \sin(\varphi_{n-2} - \varphi_{n-1} - \omega T - \omega \Delta T)$$

Substituting (5.43) and (5.42) in (5.37), we finally receive

$$\begin{aligned} I[x(t)] &= A(I_n I_{n-1} + I_n^* I_{n-1}^*) \\ &\approx \frac{a^4 T^2}{4} \cos(\varphi_n - 2\varphi_{n-1} + \varphi_{n-2}) \\ &= \frac{a^4 T^2}{4} \cos \Delta_n^2 \varphi \end{aligned} \tag{5.44}$$

The value of (5.44), corresponding to the demodulator output (Fig. 5.16), does not depend on both the carrier frequency ω and the delay duration error ΔT. In particular, we should emphasize the latter, because in some cases we cannot realize a stable delay line, which makes it impossible to use the autocorrelated demodulator of PDM-1 signals. [Compare (5.44) with expression (5.42), which corresponds to the output of the autocorrelated demodulator of PDM-1 signals (Fig. 5.3) and depends on both ω and ΔT.]

Thus, the useful signal at the output of the considered demodulator (Fig. 5.16) as well as its statistical performance—error probability—are the absolute invariants of the conversion, consisting of carrier frequency changing.

Analyzing the demodulator scheme, we notice that a large part of the scheme completely coincides with the corresponding part of the optimum incoherent demodulator of PDM-1 signals (see Fig. 4.1). The only difference is that one of them uses the local generator output as a reference oscillation, and the other uses a previous signal chip. Therefore the $\pi/2$ phase shifter, shown in Figure 5.16, should be wideband, in contrast to the phase shifter in Figure 4.1, which provides the $\pi/2$ phase shift for a single harmonic oscillation.

We can now state that the correlators of the PDM-2 autocorrelated demodulator convert the frequency uncertainty into phase uncertainty, which is then removed in the same way as in the optimum incoherent demodulator for PDM-1 signals (see Fig. 4.1). At the same time, the optimum incoherent demodulator of PDM-1 signals, removing the uncertainty of an initial signal phase, is transformed into a unit, removing the frequency uncertainty, by means of replacement of the harmonic reference oscillation by the previous signal chip.

Based on the last interpretation, we can make a number of modifications to the frequency-invariant autocorrelated demodulator for PDM-2 signals. One of them is shown in Figure 5.17. The part of this scheme up to the LPF$_1$ and LPF$_2$ outputs differs from the scheme of Figure 5.16 only in that the integrators are replaced by the LPFs. The LPF$_1$ and LPF$_2$ outputs, proportional to the sine and cosine of the first phase difference, modulate a carrier generated by a local oscillator. As a result oscillation with the first-order PDM is formed at the adder output. Then the latter is demodulated by the PDM-1 demodulator, consisting of a delay line, a multiplier, and an LPF. The locking pulses are required only in the output unit, fixing the signal sign; the other part of the scheme operates in asynchronous mode.

In the demodulators considered, the function of the input filter is the same as in the autocorrelated demodulators used for PDM-1 signals: It must limit the spectrum (power) of noise, coming to the demodulator input. The filter bandwidth should be as small as possible (see Section 5.2). At the same time these demodulators are intended mainly for channels with frequency deviations. Therefore, their input filter bandwidth should be extended in order to pass the signal at all possible changes of a carrier frequency. If we designate a difference between the maximum

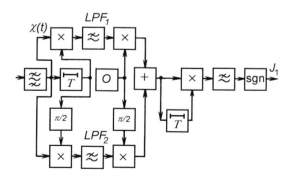

Figure 5.17 Modification of the frequency-invariant autocorrelated demodulator for binary PDM-2 signals.

and minimum carrier frequencies as δf, that is, $\delta f = f_{max} - f_{min}$, the filter should have the bandwidth $F'' = \delta f + F$, where F is the bandpass of the input filter at absolutely stable carrier frequency, usually equal to $1.5/T$. Thus, at frequency deviations, comparable or superior to the manipulation frequency $1/T$, the system parameter $F''T = 1.5 + \delta f/T$ is appreciably increased, which results in additional energy losses.

In the absence of carrier frequency deviations, the noise immunity of the autocorrelated PDM-1 demodulator (Fig. 5.3) is higher than the noise immunity of the absolutely invariant demodulator of PDM-2 signals (Fig. 5.16). The invariance to frequency deviations is achieved by decreasing the noise immunity. However, as the deviation of carrier frequency from a nominal value is increased, the advantage of PDM-1 decreases. The absolutely invariant PDM-2 demodulator begins to "win" against the autocorrelated PDM-1 demodulator at frequency deviations of $\Delta f > 1/6T$. So at carrier frequency deviations comparable to the manipulation frequency, the advantage of PDM-2 is not in doubt.

To find the algorithms of absolutely invariant autocorrelated demodulators of multiphase PDM-2 signals, let us use common algorithms for PM signal demodulation. If the second phase differences are $\Delta^2\varphi_1 = \pi/4$, $\Delta^2\varphi_2 = 3\pi/4$, $\Delta^2\varphi_3 = 5\pi/4$, and $\Delta^2\varphi_4 = 7\pi/4$, the common algorithm for binary subchannel demodulation has the following form ($\Delta_n^2\varphi$ is the received second phase difference):

$$J_{1n} = \text{sgn sin } \Delta_n^2\varphi$$
$$J_{2n} = \text{sgn cos } \Delta_n^2\varphi$$

(5.45)

Opening trigonometric functions, we obtain

$$J_{1n} = \text{sgn}(\sin \Delta_n^1\varphi \cos \Delta_{n-1}^1\varphi - \sin \Delta_{n-1}^1\varphi \cos \Delta_n^1\varphi)$$
$$J_{2n} = \text{sgn}(\sin \Delta_n^1\varphi \sin \Delta_{n-1}^1\varphi + \cos \Delta_n^1\varphi \cos \Delta_{n-1}^1\varphi)$$

(5.46)

The sines and cosines of the first phase differences of the received signal within the nth chip are determined by (5.35). Having substituted them into (5.46), we obtain the required algorithm, which is realized in the demodulator shown in Figure 5.18. The demodulator modification, operating according to algorithm (5.46) and similar to the scheme of Figure 5.17, is shown in Figure 5.19.

When using other variants of transmitted phase differences, the algorithms and schemes of demodulators for multilevel PDM-2 signals are similarly synthesized: A common demodulation algorithm is chosen and expressed in terms of sines and cosines of the second-order phase differences. Then sines and cosines of the second phase differences are expressed in terms of sines and cosines of the first differences, and the latter ones are expressed in terms of adjacent signal chip convolution according to (5.35). So, if the Gray code and phase differences (5.20) are used in

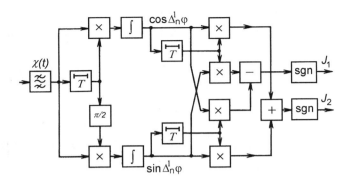

Figure 5.18 Carrier-frequency-invariant autocorrelated demodulator for four-level PDM-2 signals with the second phase differences $\pi/4$, $3\pi/4$, $5\pi/4$, and $7\pi/4$.

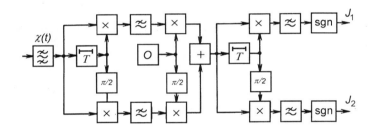

Figure 5.19 Modification of the carrier-frequency-invariant autocorrelated demodulator for four-level PDM-2 signals.

the system with eight-level PDM-2, the demodulation algorithm for three binary subchannels has the form:

$$J_{1n} = \text{sgn sin } \Delta_n^2\varphi$$

$$J_{2n} = \text{sgn cos } \Delta_n^2\varphi \qquad (5.47)$$

$$J_{3n} = \text{sgn}(\sin \Delta_n^2\varphi + \cos \Delta_n^2\varphi)\,\text{sgn}(\cos \Delta_n^2\varphi - \sin \Delta_n^2\varphi)$$

Opening (5.47), we obtain the carrier-frequency-invariant demodulator of eight-level PDM-2 signals, which is shown in Figure 5.20.

The demodulators considered in this section are absolutely invariant to a carrier frequency, and this property is a consequence of applying the processing operator (5.37). This operator can be used for making the system invariant to frequency not only at PDM-2, but also at other kinds of modulation, for example PDM-1 or FM.

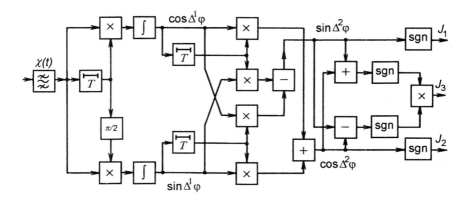

Figure 5.20 Carrier-frequency-invariant autocorrelated demodulator of eight-level PDM-2 signals with the second phase differences $\Delta^2\varphi_i = (2i - 1)\pi/8$.

In the case of PDM-1 signals the demodulation algorithm similar to (5.36) has the following form:

$$J_1 = \mathrm{sgn}\Bigg\{\Bigg[\int_{T/2}^{T} x_{n-1}(t)\,x_{n-1}(t - T/2)\ dt \int_{T}^{3T/2} x_n(t)\,x_{n-1}(t - T/2)\ dt$$

$$+ \int_{T/2}^{T} x_{n-1}(t)\,x^*_{n-1}(t - T/2)\ dt \int_{T}^{3T/2} x_n(t)\,x^*_{n-1}(t - T/2)\ dt\Bigg]$$

$$+ \Bigg[\int_{T}^{3T/2} x_n(t)\,x_{n-1}(t - T/2)\ dt \int_{3T/2}^{2T} x_n(t)\,x_{n-1}(t - T/2)\ dt$$

$$+ \int_{3T/2}^{2T} x_n(t)\,x^*_{n-1}(t - T/2)\ dt \int_{3T/2}^{2T} x_n(t)\,x^*_n(t - T/2)\ dt\Bigg]\Bigg\} \tag{5.48}$$

As in an autocorrelated PDM-2 demodulator three signal chips are processed in the appropriate PDM-1 demodulator and implemented according to (5.48). Three series halves of signal chips are processed (Fig. 5.21). These halves are designated on the time diagram in Figure 5.22, illustrating the principle of demodulator operation, in terms of $(n - 1) \cdot 1$, $(n - 1) \cdot 2$, $n \cdot 1$, and $n \cdot 2$.

At first we calculate the cosine of the second phase difference for the elements $(n - 1) \cdot 1$, $(n - 1) \cdot 2$, and $n \cdot 1$—the expression in the first square brackets of (5.48)—and then for the cosine of the second phases difference for the elements

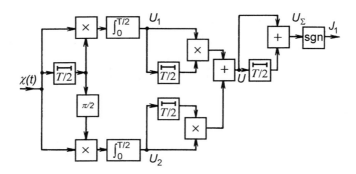

Figure 5.21 Carrier-frequency-invariant autocorrelated demodulator of binary PDM-1 signals.

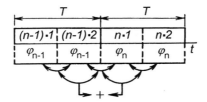

Figure 5.22 Time diagram of signal processing in the demodulator shown in Figure 5.21.

$(n-1) \cdot 2$, $n \cdot 1$, and $n \cdot 2$—the expression in the second square brackets of (5.48). The transmitted symbol is determined as sign of the sum of these cosines.

We explain the demodulator operation by means of an example of a harmonic signal with amplitude a, frequency ω, and phases φ_{n-1} and φ_n within adjacent chips. The signal samples at the integrator outputs in the moments T, $3T/2$, and $2T$ are

$$
\begin{aligned}
U_1(T) &= 0.25a^2 T \cos(0.5\omega T + \varphi_{n-1} - \varphi_{n-1}) \\
U_2(T) &= 0.25a^2 T \sin(0.5\omega T + \varphi_{n-1} - \varphi_{n-1}) \\
U_1(3T/2) &= 0.25a^2 T \cos(0.5\omega T + \varphi_n - \varphi_{n-1}) \\
U_2(3T/2) &= 0.25a^2 T \sin(0.5\omega T + \varphi_n - \varphi_{n-1}) \\
U_1(2T) &= 0.25a^2 T \cos(0.5\omega T + \varphi_n - \varphi_n) \\
U_2(2T) &= 0.25a^2 T \sin(0.5\omega T + \varphi_n - \varphi_n)
\end{aligned}
\tag{5.49}
$$

The signal samples at the output of the adder at the moments of $3T/2$ and $2T$ are

$$
\begin{aligned}
U(3T/2) &= U_1(T)\,U_1(3T/2) + U_2(T)\,U_2(3T/2) \\
&= 0.25a^2 T \cos(\varphi_n - \varphi_{n-1}) \\
U(2T) &= U_1(3T/2)\,U_1(2T) + U_2(3T/2)\,U_2(2T) \\
&= 0.25a^2 T \cos(\varphi_n - \varphi_{n-1})
\end{aligned}
\tag{5.50}
$$

and the sample at the output of the demodulator is

$$U_2 = U(3T/2) + U(2T) = 0.5a^2 T \cos(\varphi_n - \varphi_{n-1}) = 0.5a^2 T \cos \Delta_n^1 \varphi \quad (5.51)$$

Thus, the output signal of this autocorrelated demodulator is proportional to the cosine of the first phase difference of the transmitted signal chips and does not depend on the carrier frequency.

The autocorrelated PDM-1 demodulator (Fig. 5.21) has lower noise immunity than the autocorrelated PDM-2 demodulators, which is explained by losses at incoherent addition of signals (5.50). At the same time, the PDM-1 demodulator of Figure 5.21 requires the constancy of the frequency only within a 1.5-chip interval (Fig. 5.22) instead of a 3-chip interval in the case of an autocorrelated demodulator for PDM-2 signals.

On the basis of the convolution operator (5.37), we can similarly synthesize algorithms of carrier-frequency-invariant demodulators with FM and other kinds of modulation [2]. The corresponding demodulators have the property of absolute invariance to a carrier frequency; that is, the error probability caused by noise does not depend on a carrier frequency of the received signal if this signal does not leave the passband of the input filter. This property is achieved by means of some noise immunity decreases, stipulating both an algorithm for calculating the second phase difference and the input passband expansion. At the fixed carrier frequency the absolutely invariant autocorrelated PDM-2 demodulators concede to autocorrelated PDM-1 demodulators and to the optimum autocorrelated (energy) PDM-2 demodulators, and their energy loss reaches 2 to 3 dB.

The tendency to increase the noise immunity of autocorrelated PDM-2 demodulators, practically having inherited the property of invariance to carrier frequency, has resulted in the development of different modifications of relatively invariant autocorrelated PDM-2 demodulators. In these demodulators the error probability is not a strict invariant of carrier frequency and can be changed within given limits (from this we get the name *relatively invariant*), remaining on the average smaller than in absolutely invariant demodulators. Noise immunity increases are reached, as a rule, by complicating the demodulator scheme.

One of the relatively invariant demodulators is the so-called "multiphase" (or multichannel) autocorrelated demodulator of PDM-2 signals [2,9]. It is depicted in Figure 5.23. The received signal is processed in parallel by several autocorrelators, consisting of a phase shifter, a multiplier, and an integrator. The autocorrelators differ only by the phase shifters, that is, by phase shifts between received and delayed signals at multiplier inputs. Figure 5.23 shows the variant with six autocorrelators with the phase shifts 0, $\pi/6$, $2\pi/6$, $3\pi/6$, $4\pi/6$, and $5\pi/6$. The output signals of all autocorrelators are squared (S) and then come to the comparison circuit (CC), which determines the greatest of them. The CC controls the switch (Sw) in such a way that only one autocorrelator, having the greatest output signal, is connected to the output part of the demodulator. Consequently, we can calculate the sign of

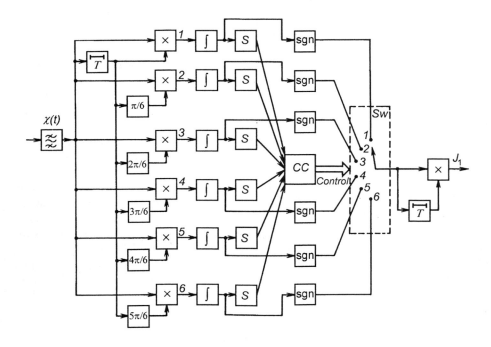

Figure 5.23 Multiphase autocorrelated demodulator for binary PDM-2 signals that is relatively invariant to a carrier frequency.

the second phase difference by a multiplying the signs of adjacent signal samples at the output of one autocorrelator.

As a result of this procedure, it is possible to increase the noise immunity in comparison with the absolutely invariant demodulators (Figs. 5.16 and 5.17), because, first, the demodulation is accomplished by only one autocorrelator with maximum signal-to-noise ratio at its output, and, second, the second difference is calculated by sample signs, as in the coherent demodulator. Thus, the potential noise immunity of the considered multichannel demodulator of PDM-2 signals should be close to the noise immunity of the autocorrelated demodulator of PDM-1 signals. This qualitative conclusion is confirmed by the corresponding calculations in Chapter 6. The real noise immunity of the given demodulator is decreased because of errors that result from the automatic choosing of the processing channel with the greatest signal, caused by both the equipment errors and external interferences. However, at comparatively slow changes of frequency, when we can use inert control, it is possible to considerably decrease or even practically reduce to zero the influence of these errors on the demodulator noise immunity.

The multiphase autocorrelated demodulator is relatively invariant to a carrier frequency, because the level of the maximum signal at the autocorrelator outputs is changed. To illustrate this, Figure 5.24 shows output signals of all six autocorrelators versus a carrier frequency.

Every one of the signals changes under the sinusoidal law; however, because the choosing of the maximum signal is automatic, the demodulator uses only the signals that appear in shaded areas. As is obvious from Figure 5.24, six autocorrelators are quite enough to keep the signal on the level, approximate to maximum. At this level, the output signal decreases not more than about 4% in comparison with the case for a fixed frequency. It is not difficult to determine from Figure 5.24 how much the useful signal decreases when a multiphase demodulator uses fewer processing channels, for example, three or two.

The multiphase autocorrelated demodulator for the case of four-level PDM-2 is shown in Figure 5.25 [9]. The output part of this scheme, consisting of delay elements and decoder (D), is connected to one of the autocorrelator pairs, which differ by the $\pi/2$ phase shift—the first and fourth autocorrelators, or the second and fifth autocorrelators, or the third and sixth autocorrelators. The choice of the pair is carried out by a unit, consisting of squares (S), three subtraction circuits (−), and a comparison circuit (CC). This unit chooses the pair with the least absolute value of the difference between squares of the autocorrelator output signals. The appropriate autochoosing algorithm can be formulated as follows: Autocorrelators should be chosen with an index i at an additional phase shift that is equal to

$$i = \arg \min_{j} | \cos^2(\Delta^1\varphi_\xi + \Delta\varphi_j) - \cos^2(\Delta^1\varphi_\xi + \Delta\varphi_j + \pi/2)| \qquad (5.52)$$

where $\Delta^1\varphi_\xi$ is the received first phase difference, and $\Delta\varphi_j$ is the additional phase shift, introduced by a phase shifter. In the considered case $j = 1$, 2, and 3,

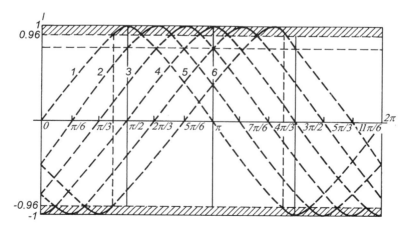

Figure 5.24 Output signals of autocorrelators in the multiphase demodulator versus the carrier frequency.

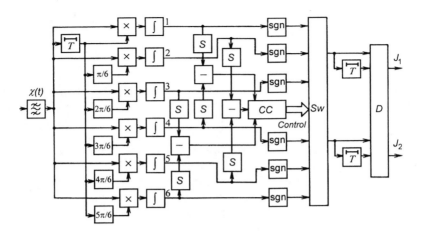

Figure 5.25 Multiphase autocorrelated demodulator for four-level PDM-2 signals that is relatively invariant to a carrier frequency.

$\Delta\varphi_1 = 0$; $\Delta\varphi_2 = \pi/6$; $\Delta\varphi_3 = \pi/3$. The autochoosing algorithm (5.52) corresponds to the PDM with phase differences $\Delta\varphi = \pi/4$, $3\pi/4$, $5\pi/4$, and $7\pi/4$. When the carrier frequency coincides with the nominal one, the cosines and sines of these differences are equal by an absolute value, and the second phase difference can be determined by cosine and sine signs, as in the coherent demodulator. If the carrier frequency does not coincide with the nominal value, we will be able to find two autocorrelators with minimum difference between the sine and cosine of the received phase difference, and the transmitted second phase difference can be also determined by signs of their output signals. The last procedure is executed by a decoder, operating according to the same algorithm as the decoder for the optimum incoherent demodulator of PDM-1 signals (see Section 4.2).

5.5 AUTOCORRELATED RECEPTION OF PDM SIGNALS AT FAST VARIATIONS OF A CARRIER FREQUENCY

By virtue of an invariance property of discrete-difference conversion, the PDM-2 method permits us to eliminate the influence of constant and linear components of signal phase deviation or a constant component of a signal frequency on the system noise immunity. Thus, the information is put into the second phase difference of three adjacent chips; the required condition of absolute system invariance is the constancy of a carrier frequency within the three-chip interval. However in communications channels with high-speed moving objects (Doppler communication channels) and in some other channels with variable parameters, we often come across situations in which the specified condition is not carried out and the signal frequency can be presented as a linear function of time:

$$\omega = \omega_0(1 + 2\alpha t) \tag{5.53}$$

where α takes positive or negative values depending on the sign of the frequency changing rate. Expression (5.53) corresponds, in particular, to various kinds of accelerated movement of the object. Let us evaluate the influence of frequency linear changes on performances of an autocorrelated PDM-2 demodulator, operating according to algorithm (5.36). For this purpose we substitute in (5.37) the expressions for three adjacent chips, which, taking into account (5.53), are as follows:

$$\begin{aligned}
x_n(t) &= a \sin[\omega_0 t(1 + \alpha t) + \varphi_n] \\
x_{n+1}(t) &= a \sin[\omega_0(t + T)(1 + \alpha t + \alpha T) + \varphi_{n+1}] \\
x_{n+2}(t) &= a \sin[\omega_0(t + 2T)(1 + \alpha t + 2\alpha T) + \varphi_{n+2}]
\end{aligned} \tag{5.54}$$

After substitution, calculation of integrals, and simplification we obtain

$$I = 0.25 a^4 T^2 (\sin \alpha\omega_0 T^2 / \alpha\omega_0 T^2)^2 \cos(\Delta_n^2\varphi + 2\alpha\omega_0 T^2) \tag{5.55}$$

From (5.55) follows (at $\alpha = 0$) the result of (5.44), having been obtained before. As is obvious from (5.55), the linear deviations of a signal frequency ($\alpha \neq 0$) can result in changing the sign of the demodulator output signal (the so-called "opposite operation"). They can also cause the lowering of an output signal level, which reduces the noise immunity. This result is physically similar to the effects that appear at autocorrelated reception of PDM-1 signals when the equality $\omega_0 T = 2k\pi$ (k is an integer) is not fulfilled.

To evaluate the influence of these factors in real conditions we analyze the part of expression (5.55) that is dependent on α, which (at $\Delta_n^2\varphi = 0$) is equal to

$$I(\gamma) = (\sin \gamma / \gamma)^2 \cos 2\gamma \tag{5.56}$$

where $\gamma = \alpha\omega_0 T^2$. Figure 5.26 shows the diagram for function (5.56). The shaded areas correspond to signal frequency changing rates at which the phenomenon of opposite operation takes place. On the abscissa, besides the γ value in radians, there is also the rate of changing the carrier frequency f', which according to (5.53) and (5.56) is equal to

$$f' = \frac{d\omega}{2\pi \, dt} = 2\alpha f_0 = \frac{\gamma F_M^2}{\pi} \tag{5.57}$$

This rate f' is represented in the diagram as a fraction of manipulation rate square $F_M^2 = 1/T^2$.

From the diagram there follows the conditions of expediently applying the autocorrelated reception of PDM-2 in communications channels with changing

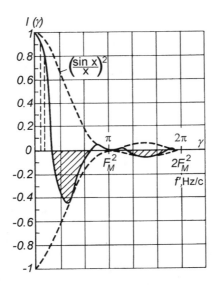

Figure 5.26 Output signal versus the frequency changing rate for an autocorrelated demodulator of PDM-2 signals.

carrier frequency. If the output signal level of a demodulator must be reduced at all frequency changes by not more than 20%, the following condition has to be fulfilled:

$$f' = 2\alpha f_0 < 2F_M^2/15 \tag{5.58}$$

From (5.58) we can make a conclusion: At autocorrelated reception according to algorithm (5.36), the system with PDM-2 practically keeps the invariance to signal frequency change, if the rate of this change does not exceed 13% of the square of manipulation rate.

In [10] on the basis of the preceding relations, an analysis has been made of PDM-2 system performance in different frequency ranges when communicating with high-speed moving objects. From this analysis it follows, in particular, that at communications with modern aircrafts the PDM-2 system does not practically reduce its noise immunity because of changing signal frequencies caused by transmitter and receiver movement.

At the same time, for communications with more high-speed objects or in the modes of information transmission with very low bit rates the condition of applying PDM-2 (5.58) can appear to be infringed. In these cases, third-order PDM should be used, which is actually invariant to all linear changes of a carrier frequency.

For PDM-3 adequate methods of demodulation should be developed that would allow the system to realize the specified property of invariance. The fact is that at an usual autocorrelated method the high-order finite phase differences are calculated by trigonometric functions of the first phase differences $\sin \Delta^1 \varphi$ and $\cos \Delta^1 \varphi$. The last ones are determined by scalar products of adjacent signal chips and actually coincide with the sine and cosine of these signal phase differences only when a carrier frequency is constant within the calculation interval. When the frequency is changing, the autocorrelated demodulator gives the averaged value of the trigonometric functions of the phase difference and, consequently, does not provide strict invariance of the system to these changes. We illustrate this statement using PDM-3 as an example. At binary PDM-3 with phase differences of 0 and π the algorithm of demodulating the appropriate signals has a form

$$J_{1n} = \text{sgn} \cos \Delta_n^3 \varphi \qquad (5.59)$$

Representing the cosine of the third phase difference through the cosine and sine of the first differences, and latter ones through scalar products of chips, we obtain

$$
\begin{aligned}
J_{1n} &= \text{sgn} \cos(\Delta_n^2 \varphi - \Delta_{n-1}^2 \varphi) \\
&= \text{sgn} \cos\left[(\Delta_n^1 \varphi - \Delta_{n-1}^1 \varphi) - (\Delta_{n-1}^1 \varphi - \Delta_{n-2}^1 \varphi) \right] \\
&= \text{sgn}\left[\cos(\Delta_n^1 \varphi - \Delta_{n-1}^1 \varphi) \cos(\Delta_{n-1}^1 \varphi - \Delta_{n-2}^1 \varphi) \right. \\
&\quad \left. + \sin(\Delta_n^1 \varphi - \Delta_{n-1}^1 \varphi) \sin(\Delta_{n-1}^1 \varphi - \Delta_{n-2}^1 \varphi) \right] \\
&= \text{sgn}\left\{ \left[(x_n x_{n-1})(x_{n-1} x_{n-2}) + (x_n x_{n-1}^*)(x_{n-1} x_{n-2}^*) \right] \right. \\
&\quad \times \left[(x_{n-1} x_{n-2})(x_{n-2} x_{n-3}) + (x_{n-1} x_{n-2}^*)(x_{n-2} x_{n-3}^*) \right] \\
&\quad + \left[(x_n x_{n-1}^*)(x_{n-1} x_{n-2}) - (x_n x_{n-1})(x_{n-1} x_{n-2}^*) \right] \\
&\quad \left. \times \left[(x_{n-1} x_{n-2}^*)(x_{n-2} x_{n-3}) - (x_{n-1} x_{n-2})(x_{n-2}^* x_{n-3}^*) \right] \right\} \qquad (5.60)
\end{aligned}
$$

where the round brackets designate the scalar products (convolutions) of the corresponding chips. Assuming that the signal frequency changes under the law of (5.53), we present the signals included in (5.60) by means of (5.54). As a result of calculations we obtain [10]:

$$J_{1n} = \text{sgn}\left[0.25 \alpha^4 T^2 \left(\frac{\sin \alpha \omega_0 T^2}{\alpha \omega_0 T^2} \right)^4 \cos \Delta_n^3 \varphi \right] \qquad (5.61)$$

It follows from (5.61) that at autocorrelated reception of PDM-3 signals according to algorithm (5.60), the sign of the demodulator output signal does not depend on the carrier frequency changing rate, however, the signal level is a function of a parameter

$$\beta = \left(\frac{\sin \, \alpha\omega_0 T^2}{\alpha\omega_0 T^2}\right)^4 \tag{5.62}$$

From (5.62) we can obtain the similar to (5.58) condition of applying the autocorrelated reception of PDM-3 signals at a high rate frequency changing. If we keep the demodulator output signal change to not more than 20%, this condition has the following form:

$$f' = df/dt = 2\alpha f_0 < 0.1 F_M^2 \tag{5.63}$$

that is, in this case the rate of signal frequency changing should not exceed 10% of a square of manipulation rate.

Condition (5.63), as well as (5.58), is usually carried out in real communications systems. At the same time, from the comparison of (5.63) with (5.58), it is obvious that at linear changing of a carrier frequency it is not expedient to apply PDM-3 with autocorrelated reception instead of PDM-2, because the restrictions on the frequency changing rate in both systems are practically identical. The difference is only that at linear changing of the frequency in the system with PDM-2 there can occur the phenomenon of opposite operation, but at PDM-3 it is impossible.

Thus, to realize strict invariance to linear changes of a signal frequency in the systems with the third-order PDM we should develop the algorithms for finding directly the finite phase differences, not connected with calculating the scalar products of elementary signals.

References

[1] Fink, L. M., *Discrete Message Transmission Theory*, 2nd ed., Moscow: Sov. Radio, 1970.

[2] Okunev, Yu. B., *Theory of Phase-Difference Modulation*, Moscow: Svyaz, 1979.

[3] Okunev, Yu. B., and L. A. Yakovlev, *Telecommunications Systems with Spread Spectrum Signals*, Moscow: Svyaz, 1968, p. 168.

[4] Shchelkunov, K. N., and E. S. Barbanel, "Application of the Phase Difference Modulation in Optical Communication Systems," *Telecom. Radioeng.*, Vol. 35, No. 4, 1979.

[5] Pent, M., "Double Differential PSK Scheme in the Presence of Doppler Shift," in *Digital Communications in Avionics*, AGARD Proceedings No. 239, 1978, pp. 43-1–43-11.

[6] Simon, M. K., S. M. Hinedi, and W. C. Lindsey, *Digital Communication Technique*, Englewood Cliffs, NJ: Prentice Hall, 1995.

[7] Petrovich, N. T., "New Methods of Phase Telegraphy," *Radiotechnika*, No. 10, 1957, pp. 47–54.

[8] Sidorov, N. M., "Reception of the Second-order PDM Signals by the Optimum Method of the Unknown Shape Signals Processing," *Commun. Tech.*, No. 9, 1984, pp. 7–16.

[9] Okunev, Yu. B., V. A. Pisarev, and V. K. Reshemkin, "The Design and Noise Immunity of Multiphase Autocorrelation Demodulators of Second-order DPSK Signals," *Telecom. Radioeng.*, Vol. 33/34, No. 6, 1979.

[10] Barbanel, E. S., Yu. B. Okunev, and V. K. Reshemkin, "PDM-2 Application in Channels with Fast-Changing Signal Frequency," *Radiotechnika*, No. 4, 1976, pp. 98–100.

CHAPTER 6

Noise Immunity of Phase and Phase-Difference Modulation Systems

6.1 NOISE IMMUNITY PERFORMANCE

Noise immmunity is a system property aimed at resisting the influence of noises and other interfering factors. It is quantitatively determined by a number of performance parameters.

In digital communication systems the most usable parameter of noise immunity is error probability, which is addressed either to a transmitted element set, called a *block* or *code combination,* or to one signal element. In the latter case, the performance of noise immunity is called the *bit error probability.* This probability is the basic characteristic of a modem as a unit of element-by-element reception.

If the discrete channel formed by a modem and a continuous channel is a binary, stationary, symmetric[1] channel with independent errors, the bit error probability is a comprehensive indicator of noise immunity. In this case all other indicators, for example, the probability of faulty decoding of a code combination, are unambiguously determined by the bit error probability.

The bit error probability, which will simply be called error probability, depends first of all on statistical characteristics and parameters of interference in the continuous communications channel. The most important statistical channel model, both theoretically and practically, is the AWGN channel. The AWGN channel contains the white noise with a uniform spectrum and with a normal distribution of instant values: Gaussian noise, which takes place at almost all communications channels, and in some of them is a unique or prevailing interference. Gaussian white noise

---

1. A binary channel is referred to as a *symmetric* channel if the probability of receiving the symbol 0 instead of the symbol 1 is equal to probability of receiving a 1 instead of a 0.

is completely determined by its spectral power density, that is, by the average noise power per one Hertz.

In this chapter we study the error probability (or bit error rate, BER) as a function of a ratio of signal energy to the spectral density of noise power. This ratio, which we shall call simply the *signal-to-noise ratio* (SNR), is designated as follows:

$$h^2 = E/N_0 = P_s T/N_0 \qquad (6.1)$$

where E is a signal element (chip) energy; P_s is signal power; T is chip duration; and N_0 is noise power spectral density. The SNR can be also presented in the another form:

$$h^2 = (P_s/P_n)FT \qquad (6.2)$$

where P_n is the noise power within frequency bandwidth F (F is the width of the noise spectrum at the demodulator input).

From (6.1) and (6.2), we can easily see that the value h^2 is a parameter of both the communications channel and the signal, because this parameter includes a chip duration and channel frequency bandwidth. The value h^2 is most often used as an argument of a function, presenting the error probability: $\rho = f(h^2)$.

Sometimes another SNR is used:

$$\Delta P = P_s/P_n \qquad (6.3)$$

which is also not a pure channel parameter, because the noise power P_n depends on the frequency band used at the demodulator input.

Therefore as a SNR performance, independent from signal chip duration as well as from the used channel band, alongside the dimensionless values h^2 and ΔP, the following value, measured in Hertz is used:

$$\delta = P_s/N_0 \qquad (6.4)$$

The mutual relations between (6.1) through (6.4) are obvious:

$$h^2 = \Delta PFT = \delta T$$
$$\Delta P = \delta/F \qquad (6.5)$$

Rather often h^2, ΔP, and δ are expressed in decibels:

$$h^2 = 10 \; \lg(P_s T/N_0) \; \text{dB}$$
$$\Delta P = 10 \; \lg(P_s/P_n) \; \text{dB} \qquad (6.6)$$
$$\delta = 10 \; \lg(P_s/N_0) \; \text{dB} \times \text{Hz}$$

In addition to the error probability, which is a function of SNR, as a noise immunity characteristic an energy loss (or gain) of the given system (or the given demodulator) to some other known system, for example, to the ideal system (or the perfect demodulator), is often used. Let some error probability p be reached in the reference (ideal) system at the SNR $h_i^2(p)$, and in the given system it is reached at $h^2(p)$. Then the energy loss (or gain) Δh of the given system to the reference one is

$$\Delta h(p) = 10 \lg \frac{h^2(p)}{h_i^2(p)} \tag{6.7}$$

Similarly we can calculate an energy loss by (6.3) and (6.4), and this loss will coincide with the loss calculated by (6.7) if chip durations are the same in compared systems.

In a common case energy losses are the function of the error probability, though sometimes energy losses, as will become obvious later, do not depend on the error probability.

When finding the relationship between the error probability and SNR, the function, which we traditionally call the *Laplace function,* is of great importance (Fig. 6.1):

$$\mathsf{F}(y) = \frac{1}{\sqrt{2\pi}} \int_{y}^{\infty} \exp\left(-\frac{x^2}{2}\right) dx \tag{6.8}$$

This function is an addition up to 1 of the integral distribution function $\Phi(y)$ of a normal random value with zero average value and dispersion equal to 1:

$$\mathsf{F}(y) = 1 - \Phi(y) \tag{6.9}$$

where

$$\Phi(y) = \frac{1}{\sqrt{2\pi}} \int_{-\infty}^{y} \exp\left(-\frac{x^2}{2}\right) dx \tag{6.10}$$

Integrals (6.8) and (6.10) are not expressed through the elementary functions, and the meanings of the Laplace function should be found by the tables, analytical approximating expressions, or numerical calculations under the approximated formulas. The tables of the integral distribution function of (6.10) have been published in many places, for example, in [1–3]. Using them and formula (6.9)

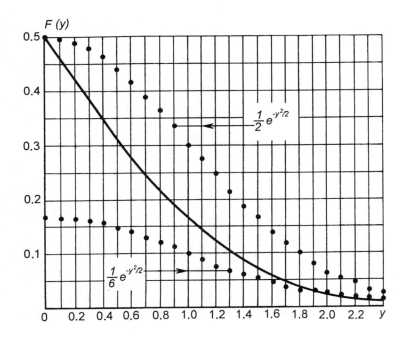

Figure 6.1 Laplace function.

we can determine (6.8). In addition, we often need to tabulate a function called the *probability integral* in mathematical literature:

$$\text{erf}(y) = \frac{2}{\sqrt{2\pi}} \int_0^y \exp\left(-\frac{x^2}{2}\right) dx \qquad (6.11)$$

Function (6.11) is sometimes called the *error integral* or *Cramp function*[2]. Through the tabulated function (6.11) it is not difficult to express Laplace function (6.8):

$$\mathsf{F}(y) = [1 - \text{erf}(y)]/2 \qquad (6.12)$$

Usually, in error probability calculations, we should use the values of the Laplace function with wide limits—from 5×10^{-1} up to 10^{-9} and less. Therefore, at the graphic representation of this function, it is convenient to use a logarithmic scale

---

2. The probability integral was found for the first time in Moivre's works (1730). Laplace in 1782 was the first to evaluate its meaning in terms of the tasks of probability theory (the Moivre–Laplace asymptotic, Laplace integrated theorem). The first tables of integrals were created by Cramp in 1799.

of the y coordinate as shown in Figure 6.2. On the x coordinate (above the arrows) are the arguments of the Laplace function, corresponding to integers of the logarithm of this function.

For numerical calculations it is convenient to represent function (6.10) by series (for small values of the argument) or the function (6.8) by asymptotic expres-

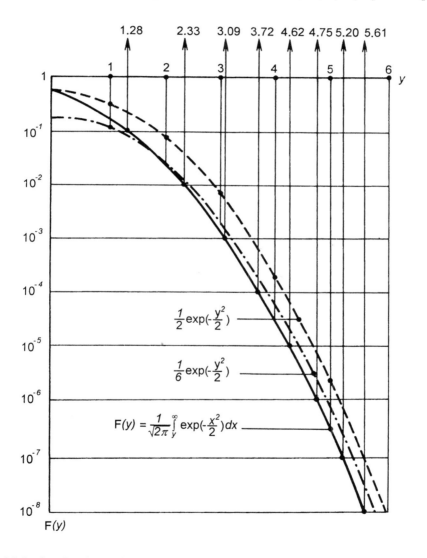

Figure 6.2 Laplace function and its simplest evaluations.

sions (for large values of the argument). The corresponding expressions are indicated, for example, in [4].

For analytical calculations that do not require great precision, different simple approximating expressions are used. In [4] it is recommended that the following approximate formula be used:

$$F(y) \approx 0.65 \exp[-0.443(y + 0.75)^2] \qquad (6.13a)$$

which provides an error of not more than 60% at $y \leq 5.5$. The following elementary approximation is also convenient:

$$F(y) \approx \frac{1}{6} \exp(-y^2/2) \qquad (6.13b)$$

This function is shown in Figure 6.2 by the dot-and-dash curve. In many cases the admissible upper bound of Laplace function is

$$F(y) < \frac{1}{2} \exp(-y^2/2) \qquad (6.14)$$

This upper bound is shown in Figure 6.2 by a dashed curve.

We emphasize that, as follows from the theory of the potential noise immunity (see later discussion), the curves shown in Figure 6.2 by the continuous and dashed lines are adequate for the potential noise immunity of coherent reception of binary PM signals and optimum incoherent reception of binary PDM signals, respectively. In addition, practical experience shows that if the noise immunity of a real coherent demodulator is located not higher than the dashed curve, the demodulator has been realized quite well, and if it is not higher than the dot-and-dash curve (at $p < 10^{-2}$) then the demodulator implementation is excellent.

6.2 FOUNDATIONS OF NOISE IMMUNITY THEORY: BINARY PM AND PDM SYSTEMS

Let us consider binary PM and PDM systems and find the potential noise immunity of receiving the appropriate two-phase signals when using coherent, optimum incoherent, and autocorrelated demodulators. The required expressions for error probabilities are fundamental in the optimum reception theory and form the basis for comparing the noise immunity of various processing algorithms and different real demodulators. Also, for beginners the procedure for deriving these expressions is a necessary introduction to the noise immunity calculation methods.

We begin with the simplest case: the coherent reception of two-phase PM signals (see Figs. 3.1 and 3.9). Thus according to (3.1) and (3.2) the demodulation algorithm is

$$J = \text{sgn} \int_0^T x(t) \, \sin(\omega t + \varphi) \, dt \tag{6.15}$$

Let the received signal $x(t)$ be equal to a sum of the useful signal, coinciding by a frequency and phase with a reference oscillation, and Gaussian noise $\xi(t)$:

$$x(t) = a \sin(\omega t + \varphi) + \xi(t) \tag{6.16}$$

Then, as follows from (6.15), the error will occur in the case, when

$$\int_0^T a \sin^2(\omega t + \varphi) \, dt + \int_0^T \xi(t) \, \sin(\omega t + \varphi) \, dt < 0 \tag{6.17}$$

that is, the error probability is the probability of

$$I_\xi < -aT/2 \tag{6.18}$$

where

$$I_\xi = \int_0^T \xi(t) \, \sin(\omega t + \varphi) \, dt \tag{6.19}$$

The stochastic integral (6.19) plays an important role in the potential noise immunity theory, and the method of its calculation can be found in textbooks [5,6].

A random value I_ξ is the linear conversion of the normal random process; therefore, it is the normal random value with zero mathematical expectation and dispersion:

$$D(I_\xi) = N_0 T/4 \tag{6.20}$$

We note that for the case of an arbitrary reference oscillation $S(t)$ with average power P_c the dispersion of the random value

$$I_\xi(S) = \int_0^T \xi(t) S(t) \, dt \tag{6.21}$$

is equal to

$$D[I_\xi(S)] = N_0 P_c T/2 = N_0 E_S/2 \tag{6.22}$$

where $E_S = \displaystyle\int_0^T S^2(t)\, dt$.

We give an elementary calculation of the dispersion of convolution of the signal $S(t)$ and the Gaussian interference $\xi(t)$ with the spectrum from zero up to F (in Hertz), that is, for a low-frequency equivalent of the received radio signal. Integral (6.21) can be replaced by a sum of products of signal and interference samples taken at intervals $\Delta t = 1/2F$:

$$I_\xi(S) = \sum_{n=1}^{N} \xi(n\Delta t)\, S(n\Delta t)\, \Delta t$$

where $N\Delta t = T$. The dispersion of the last sum is equal to a sum of summand dispersions, that is,

$$D(I_\xi) = \sum_{n=1}^{N} (\Delta t)^2\, S^2(n\Delta t)\, D[\xi(n\Delta t)]$$

The interference sample dispersion is equal to interference power, that is, $D[\xi(n\Delta t)] = P_i$; therefore,

$$D(I_\xi) = P_i \Delta t \sum_{n=1}^{N} S^2(n\Delta t)\, \Delta t$$

Because the last sum is equal to the energy E_s of the signal $S(t)$, finally we have

$$D[I_\xi(S)] = P_i \Delta t E_S = P_i E_S/2F = E_S N_0/2$$

Thus, the probability of fulfilling inequality (6.18) is equal to

$$p(I_\xi < -aT/2) = \int_{-\infty}^{-aT/2} \frac{1}{\sqrt{2\pi D(I_\xi)}} \exp\left[-\frac{z^2}{2D(I_\xi)}\right] dz = \mathsf{F}\left(\frac{aT}{2\sqrt{D(I_\xi)}}\right) \tag{6.23}$$

where F is the Laplace function of (6.8). The argument of the Laplace function in (6.23) can be transformed, using the designation of (6.1) and the formula of (6.20), as follows:

$$\frac{aT}{2\sqrt{D(I_\xi)}} = \frac{a\sqrt{T}}{\sqrt{N_0}} = \sqrt{2}\sqrt{\frac{E}{N_0}} = \sqrt{2}\, h \tag{6.24}$$

where, remember, h^2 is the ratio of signal energy to spectral density of noise power.

Thus, from (6.23) and (6.24) we obtain the required error probability p_0 of coherent reception of the two-phase PM signals as a function of the SNR h^2 [7]:

$$p_0 = \mathsf{F}(\sqrt{2}\,h) \qquad (6.25)$$

Formula (6.25) is the first fundamental expression of the noise immunity theory. It is considered to be the standard, when comparing different methods of digital transmission, and determines the minimum achievable error probability for element-by-element reception.

The last statement should be understood as follows: If some binary, equiprobable, and power P_s equivalent signals are transmitted and received within a time interval T, and the Gaussian white noise has the power spectrum density N_0, that is, the SNR $h^2 = P_s T/N_0$, it is impossible to distinguish the two specified signals with an error probability of less than (6.25).

In the diagrams of error probability versus SNR (Fig. 6.3), the curve, corresponding to (6.25), usually takes the extreme left-hand position (curve 1).

In real digital systems, error probability (6.25) is unattainable even at the most precise modem implementation because some part of the energy resources has to be spent, as shown earlier, to eliminate the uncertainty of the initial signal phase, that is, to establish the unique conformity between the received signals and the reference signals of the demodulator. One of the most effective ways to eliminate this ambiguity is the transition to PDM, which also has some losses in noise immunity compared with (6.25).

Let us determine the error probability at coherent reception of binary signals with first-order PDM (see Figs. 3.17 and 3.18). In this case, according to the results of Section 3.3, the transmitted binary symbol is defined as a product of the PM demodulator decisions on two adjacent chips:

$$J_n = \left[\operatorname{sgn} \int_{(n-1)T}^{nT} x(t)\,\sin(\omega t + \varphi)\,dt \right] \times \left[\operatorname{sgn} \int_{nT}^{(n+1)T} x(t)\,\sin(\omega t + \varphi)\,dt \right] \qquad (6.26)$$

As is obvious from (6.26), the error occurs in two cases: as a result of faulty registration of a symbol on the $(n-1)$th interval or a faulty registration of a symbol on the nth interval. At a faulty registration of symbols on both adjacent intervals, the error does not occur. The probability of one of these events occurring, that is, the probability that one of two adjacent symbols has been received correctly and the other erroneously, is equal to $p_0(1 - p_0)$, where p_0 is the error probability at the output of a coherent demodulator of PM signals. Because the specified events are independent, the required probability p_1 is equal to the sum of probabilities of each of them:

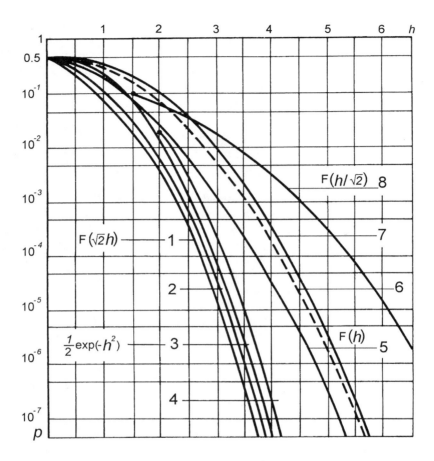

Figure 6.3 Error probability of binary systems at different methods of modulation and reception: (1) PM, coherent; (2) PDM-1 and PDM-2, coherent; (3) PDM-1, optimum incoherent; (4) PDM-1, autocorrelated $(FT = 2)$; (5) FM, coherent; (6) FM, optimum incoherent; (7) PDM-1, autocorrelated $(FT = 10)$; and (8) AM, coherent.

$$p_1 = 2p_0(1 - p_0) \qquad (6.27)$$

Substituting (6.25) in this formula, we obtain the second fundamental expression of the theory of PM signal reception—the error probability of the coherent demodulator of the two-phase PDM [8,9] (curve 2 in Fig. 6.3):

$$p_1 = 2F(\sqrt{2}\,h)\,[1 - F(\sqrt{2}\,h)] \qquad (6.28)$$

This is a beautiful formula, isn't it? An important conclusion follows from it: The error probability at coherent reception of two-phase PDM signals exceeds the error probability at coherent reception of two-phase PM signals by not more than two times. At small error probabilities ($p_1 < 0.1$) instead of (6.28) we can use the more simple approximated formula

$$p_1 \approx 2F(\sqrt{2}\,h) \tag{6.29a}$$

Comparing the diagrams of Figures 6.2 and 6.3 we can conclude that the error probability p_1 is approximated well by function (6.13b), which in this case takes the form

$$p_1 \approx \frac{1}{3} \exp(-h^2) \tag{6.29b}$$

Note that in contrast to the error probability (6.25), which is the ideal limit, the probability (6.28) is really achievable. Thus, the price of realizing the PM signal reception by PDM is a two-times increase in the error probability, which corresponds to energy losses of less than 1 dB.

This is not a serious price to pay (see Fig. 6.3), especially if we take into account that a two-times increase in the error probability at the transition from PM to PDM occurs not at the expense of additional independent errors, but at the expense of the groups of two errors. Actually, for every erroneous symbol at the output of a coherent PM demodulator one more erroneous symbol is added at the adjacent chip. This phenomenon of error duplication is generally typical for differential modulation, however, it is especially so for coherent reception of signals with PDM, when there are practically no single errors and almost only double (twin) errors take place.

The methods for twin error correction are known: The same codes are used that are used for correcting the single errors, and the difference lies in that the adjacent transmitted symbols are distributed in different code combinations, for example, all even and all odd symbols are separately coded. In addition, it is possible to use the special codes for correcting twin errors [10].

Relation (6.28) is also interesting in that it gives the error probability not only for the binary PDM-1, but also for coherent reception of binary signals with PDM-2 and, in general, for the coherent reception of binary signals with PDM-k, if k is an integer degree of number 2. This conclusion directly follows from the common algorithms (3.50), (3.51), and (3.55) because according to these algorithms at $k = 2^l$, where l is an integer, the transmitted symbol is determined by a product of two decisions at the output of the coherent demodulator of PM signals as well as in case (6.26).

Now we move on to the optimum incoherent reception of two-phase PDM-1 signals. The corresponding algorithm is represented by (4.10), and the scheme of the demodulator is shown in Figure 4.1. Algorithm (4.10) can be also presented in the form (4.12), which will be used for further calculations:

$$J_n = \text{sgn}[(X_{n-1} + X_n)^2 + (Y_{n-1} + Y_n)^2 - (X_n - X_{n-1})^2 - (Y_n - Y_{n-1})^2]$$
(6.30)

where X_n, X_{n-1}, Y_n, and Y_{n-1} are determined by (4.4). Let the signal, received within two chips, be equal to

$$x(t) = \begin{cases} a\sin(\omega t + \varphi) + \xi_{n-1}(t) & (n-1)T < t \le nT \\ a\sin(\omega t + \varphi) + \xi_n(t) & nT < t \le (n+1)T \end{cases}$$
(6.31)

where ξ_{n-1} and ξ_n are Gaussian noise within the intervals of the $(n-1)$th and nth chips. Then the values included in (6.30) are equal to

$$X_{n-1} = \int_{(n-1)T}^{nT} x(t)\sin\omega t\, dt = \frac{aT}{2}\cos\varphi + I_{\xi,n-1}$$

$$X_n = \int_{nT}^{(n+1)T} x(t)\sin\omega t\, dt = \frac{aT}{2}\cos\varphi + I_{\xi,n}$$
(6.32)

$$Y_{n-1} = \int_{(n-1)T}^{nT} x(t)\cos\omega t\, dt = \frac{aT}{2}\sin\varphi + I^*_{\xi,n-1}$$

$$Y_n = \int_{nT}^{(n+1)T} x(t)\cos\omega t\, dt = \frac{aT}{2}\sin\varphi + I^*_{\xi,n}$$

Stochastic integrals in (6.32) are similar to the integral (6.15), and random values $I_{\xi,n-1}$, $I_{\xi,n}$, $I^*_{\xi,n-1}$, and $I^*_{\xi,n}$ are similar to the value of (6.19). These random values are independent and have normal distribution with zero mean and dispersion (6.20). The random values, included in (6.30),

$$X_+ = X_{n-1} + X_n = aT\cos\varphi + I_{\xi,n-1} + I_{\xi,n}$$
$$Y_+ = Y_{n-1} + Y_n = aT\sin\varphi + I^*_{\xi,n-1} + I^*_{\xi,n}$$
$$X_- = X_n - X_{n-1} = I_{\xi,n} - I_{\xi,n-1}$$
$$Y_- = Y_n - Y_{n-1} = I^*_{\xi,n} - I^*_{\xi,n-1}$$
(6.33)

have normal distribution functions and are independent as the sums and differences of independent normal values, with X_- and Y_- having zero means, but X_+ and Y_+ having nonzero means:

$$m(X_+) = aT \cos \varphi \qquad m(Y_+) = aT \sin \varphi \tag{6.34}$$

The dispersions of all random values (6.33) are identical:

$$D(I) = N_0 T / 2 \tag{6.35}$$

Thus, the random value

$$\lambda_1 = \sqrt{(X_{n-1} + X_n)^2 + (Y_{n-1} + Y_n)^2}$$

has Ricean distribution function W_1; and the random value

$$\lambda_2 = \sqrt{(X_n - X_{n-1})^2 + (Y_n - Y_{n-1})^2}$$

has Rayleigh distribution function W_2:

$$W_1(\lambda_1) = \frac{\lambda_1}{D} \exp\left(-\frac{\lambda_1^2 + B^2}{2D}\right) I_0\left(\frac{B\lambda_1}{D}\right) \tag{6.36}$$

$$W_2(\lambda_2) = \frac{\lambda_2}{D} \exp\left(-\frac{\lambda_2^2}{2D}\right) \tag{6.37}$$

$$B = \sqrt{m^2(X_+) + m^2(Y_+)} = aT \tag{6.38}$$

where I_0 is the modified first kind Bessel function of the zero order. Obviously, an error at transmission of the phase difference $\Delta\varphi_1 = 0$ will occur, according to (6.30), if $\lambda_1 < \lambda_2$. The probability of this event is equal to

$$p = \int_0^\infty W_1(\lambda_1) \, d\lambda_1 \int_{\lambda_1}^\infty W_2(\lambda_2) \, d\lambda_2 = \int_0^\infty \frac{\lambda_1}{D} \exp\left(-\frac{\lambda_1^2 + B^2}{2D}\right)$$

$$\times I_0\left(\frac{B\lambda_1}{D}\right) d\lambda_1 \int_{\lambda_1}^\infty \frac{\lambda_2}{D} \exp\left(-\frac{\lambda_2^2}{2D}\right) d\lambda_2 \tag{6.39}$$

Calculating the integrals in (6.39), we obtain [9]:

$$p_1 = 0.5 \exp(-B^2/4D) \tag{6.40}$$

Substituting the values for B and D from (6.35) and (6.38) into (6.40), we finally get the error probability of the optimum incoherent reception of binary PDM signals [11,12] in the following form [taking into account the designations (6.1)]:

$$p_1 = \frac{1}{2}\exp(-h^2) \qquad (6.41)$$

This formula determines the maximum possible noise immunity of receiving the binary PDM-1 signals in the channel with an uncertain, uniformly distributed signal phase, when processing within the two chips interval.

Expression (6.41) is illustrated in Figure 6.3 by curve 3, which is the third of the fundamental relations of PM signal reception theory considered here.

We should mention that the result (6.41) can be obtained in another way, if we take into consideration the fact that variants of a binary PDM-1 signal

$$S_1(t) = a\sin(\omega t + \varphi) \qquad 0 < t \le 2T$$

$$S_2(t) = \begin{cases} a\sin(\omega t + \varphi) & 0 < t \le T \\ -a\sin(\omega t + \varphi) & T < t \le 2T \end{cases}$$

are orthogonal in a strong sense [9,12] within the interval $2T$. As for orthogonal signals within the T time interval the error probability at the optimum incoherent reception is equal to

$$p_{ort} = 0.5\exp(-h^2/2)$$

where $h^2 = P_sT/N_0$, for signals with PDM we immediately obtain (6.41).

Relation (6.41), having the external grace inherent to all fundamental physical laws, plays an important role in the theory of PDM. First, (6.41) shows that the optimum incoherent processing of binary PDM signals is close to the coherent reception of these signals for noise immunity. As can be seen from the diagrams and calculations of (6.41) and (6.28), at binary PDM-1 the energy loss of the optimum incoherent reception to the coherent one at small error probabilities is less than 1 dB and decreases in accordance with the increase of h^2. Therefore in the systems with two-phase PDM-1 there is no need to use coherent processing in attempts to increase noise immunity. This property of the optimum incoherent reception quantitatively confirms the thesis important in PDM theory: For high-quality reception of two-phase signals, we do not have to have the coherent reference oscillations.

Second, (6.41) is a lower bound for error probability of autocorrelated demodulators for PDM-1 and PDM-2 signals. As shown later, at relatively invariant methods of signal processing relation (6.41) is a good estimate for the noise immunity of autocorrelated demodulators. It is widely used in analytical studies.

As has been mentioned before, at the coherent reception of binary PDM signals there occur almost exclusively twin errors. At the optimum incoherent reception of the same signals both twin and single errors occur, with a fraction of the last ones being more than 50% of all errors.

The other difference of optimum incoherent reception from coherent reception lies in the relationship between the noise immunity of PDM-1 and PDM-2. When coherent processing is used, the binary PDM-1 and PDM-2 systems have identical noise immunity; when optimum incoherent processing is used, the system with PDM-2 has higher noise immunity than the system with PDM-1. Because the problem of comparative noise immunity of the incoherent reception of PDM-1 and PDM-2 signals is very important it is considered separately in Section 6.4.

Here we consider the third of the main methods of PDM signal reception based on an autocorrelated processing algorithm. At two-phase PDM-1, as was shown in Chapter 5, the appropriate algorithm consists of calculating the scalar product of the adjacent chips $x_1(t)$ and $x_2(t)$:

$$J = \text{sgn} \int_0^T x_1(t)\, x_2(t)\ dt \qquad (6.42)$$

Let us assume that within the considered two chips the identical signals $S(t)$ are transmitted, that is, the phase difference is equal to zero, and the signals included in (6.42) can be presented in the following form:

$$\begin{aligned} x_1(t) &= S(t) + \xi_1(t) \\ x_2(t) &= S(t) + \xi_2(t) \end{aligned} \qquad (6.43)$$

where ξ_1 and ξ_2 are the realizations of the normal random process within the chips. Having substituted (6.43) into (6.42), we obtain

$$J = \text{sgn}\left[\int_0^T S^2(t)\ dt + \int_0^T S(t)\, \xi_1(t)\ dt + \int_0^T S(t)\, \xi_2(t)\ dt + \int_0^T \xi_1(t)\, \xi_2(t)\ dt \right] \qquad (6.44)$$

The first integral in (6.44) is equal to the energy of the useful signal, the second and third ones [see (6.21)] are equal to the random values with the normal distribution.

The peculiarities of noise immunity calculations for autocorrelated reception are combined with the fourth integral in (6.44), which is not a normal random value. To overcome the difficulties connected with it, we represent the signal $S(t)$ and interferences $\xi_1(t)$ and $\xi_2(t)$, included in (6.44), by means of Fourier series expansions within the interval of the chip duration T. Thus Fourier coefficients are equal to

$$a_k = \frac{2}{T} \int_0^T S(t) \, \sin\frac{2\pi k}{T} t \, dt$$

$$b_k = \frac{2}{T} \int_0^T S(t) \, \cos\frac{2\pi k}{T} t \, dt$$

$$\alpha_{1k} = \frac{2}{T} \int_0^T \xi_1(t) \, \sin\frac{2\pi k}{T} t \, dt \qquad (6.45)$$

$$\beta_{1k} = \frac{2}{T} \int_0^T \xi_1(t) \, \cos\frac{2\pi k}{T} t \, dt$$

$$\alpha_{2k} = \frac{2}{T} \int_0^T \xi_2(t) \, \sin\frac{2\pi k}{T} t \, dt$$

$$\beta_{2k} = \frac{2}{T} \int_0^T \xi_2(t) \, \cos\frac{2\pi k}{T} t \, dt$$

A number of coefficients (6.45) not equal to zero in the signal and interference expansions is determined by bandwidth F of a bandpass filter at the input of the autocorrelated demodulator and is equal to $2FT$. Replacing the integrals in (6.44) by the sums of products of the appropriate expansion coefficients, we obtain

$$J = \text{sgn} \sum_{k=k_1}^{k_2} (a_k^2 + b_k^2 + a_k\alpha_{1k} + a_k\alpha_{2k} + b_k\beta_{1k} + b_k\beta_{2k} + \alpha_{1k}\alpha_{2k} + \beta_{1k}\beta_{2k})$$

$$(6.46)$$

where $k_2 - k_1 + 1 = FT$.

To calculate the error probability, it is convenient to represent (6.46) in the form of squares sum; for this purpose we express the random values α and β through a sum and a difference of independent random values γ and δ:

$$\alpha_{1k} = \gamma_{1k} + \gamma_{2k} \qquad \alpha_{2k} = \gamma_{1k} - \gamma_{2k}$$
$$\beta_{1k} = \delta_{1k} + \delta_{2k} \qquad \beta_{2k} = \delta_{1k} - \delta_{2k}$$

Then (6.46) can be written in the following form:

$$J = \operatorname{sgn} \sum_{k=k_1}^{k_2} [(a_k + \gamma_{1k})^2 + (b_k + \delta_{1k})^2 - (\gamma_{2k}^2 + \delta_{2k}^2)] \tag{6.47}$$

An error will occur if the value in square brackets is negative, namely,

$$\sum_{k=k_1}^{k_2} (a_k + \gamma_{1k})^2 + (b_k + \delta_{1k})^2 < \sum_{k=k_1}^{k_2} (\gamma_{2k}^2 + \delta_{2k}^2) \tag{6.48}$$

The probability of the event (6.48) is equal to

$$p_1 = \int_0^\infty W_1(\lambda_1) \, d\lambda_1 \int_{\lambda_1}^\infty W_2(\lambda_2) \, d\lambda_2 \tag{6.49}$$

where $W_1(\lambda_1)$ and $W_2(\lambda_2)$ are the probability densities of the left and right parts of inequality (6.48), respectively. To determine $W_1(\lambda_1)$ and $W_2(\lambda_2)$ we note that the right part of (6.48) contains the squares of independent normal random values with zero means, and the left part (6.48) contains the squares of independent normal random values with nonzero means. Consequently, $W_2(\lambda_2)$ is the probability density of a χ^2 distribution with $2q$ degrees of freedom [13]:

$$W_2(\lambda_2) = \frac{1}{2^q \Gamma(q) D} \left(\frac{\lambda_2}{D}\right)^{q-1} \exp\left(-\frac{\lambda_2}{2D}\right) \tag{6.50}$$

and $W_1(\lambda_1)$ is the probability density of the noncentral χ^2 distribution with $2q$ degrees of freedom [14]:

$$W_1(\lambda_1) = \frac{1}{2D} \left(\frac{\lambda_1}{\Delta}\right)^{(q-1)/2} \exp\left(-\frac{\lambda_1 + \Delta}{2D}\right) I_{q-1}\left(\frac{\sqrt{\lambda_1 \Delta}}{D}\right) \tag{6.51}$$

where $q = FT$; D is the dispersion of the random values γ_{1k}, γ_{2k}, δ_{1k}, and δ_{2k} (at white noise the dispersion does not depend on k);

$$\Delta = \sum_{k=k_1}^{k_2} a_k^2 + b_k^2 = 2P_s \tag{6.52}$$

is the parameter of noncentrality, proportional to signal power P_s; I_{q-1} is the modified first kind Bessel function of the $(q-1)$th order; and $\Gamma(q)$ is the gamma function.

The dispersion of the random values γ and δ can be expressed through the dispersion of the random values α and β using the result (6.20) and taking into account the formulas (6.45):

$$D = \frac{D(\alpha, \beta)}{2} = \frac{1}{2}\frac{4D(I_{\xi})}{T^2} = \frac{N_0}{2T} \tag{6.53}$$

The relation between the SNR h^2 and the parameters Δ and D of the distributions (5.50) and (6.51) follows from (6.52) and (6.53):

$$h^2 = P_s T/N_0 = \Delta/D \tag{6.54}$$

Now it remains to calculate (6.49) using (6.50) and (6.51). The internal integral in (6.49) can be expressed, using the representation of the incomplete gamma function [1], in the following form:

$$\int\limits_{\lambda_1}^{\infty} W_2(\lambda_2)\ d\lambda_2 = \int\limits_{\lambda_1}^{\infty}\frac{1}{2^q\Gamma(q)D}\left(\frac{\lambda_2}{D}\right)^{q-1}\exp\left(-\frac{\lambda_2}{2D}\right)$$

$$= \int\limits_{\lambda_1/2D}^{\infty}\frac{z^{q-1}e^{-z}}{\Gamma(q)}\ dz = \exp\left(-\frac{\lambda_1}{2D}\right)\sum_{k=0}^{q-1}\frac{1}{k!}\left(\frac{\lambda_1}{2D}\right)^k \tag{6.55}$$

Then the required error probability is equal to

$$p_1 = \int\limits_0^{\infty}\frac{1}{2D}\left(\frac{\lambda_1}{\Delta}\right)^{(q-1)/2}\exp\left(-\frac{\lambda_1 + \Delta}{2D}\right)I_{q-1}\left(\frac{\sqrt{\lambda_1\Delta}}{D}\right)$$

$$\times \exp\left(-\frac{\lambda_1}{2D}\right)\sum_{k=0}^{q-1}\frac{1}{k!}\left(\frac{\lambda_1}{2D}\right)^k\ d\lambda_1 = e^{-2h^2}\sum_{k=0}^{q-1}\frac{1}{k!}\left(\frac{1}{2h^2}\right)^{(q-1)/2}$$

$$\times \int\limits_0^{\infty}x^{k+(q-1)/2}e^{-2x}I_{q-1}\left(2\sqrt{2}xh\right)\ dx \tag{6.56}$$

Using the tabular integral [3]

$$\int\limits_0^{\infty}x^{k+\frac{1}{2}\nu}e^{-ax}J_{\nu}(2\beta\sqrt{x})\ dx = k!\beta^{\nu}e^{-\beta^2/\alpha}\alpha^{-k-\nu-1}L_k^{\nu}\left(\frac{\beta^2}{\alpha}\right)$$

where J_ν is the Bessel function of the order ν, and L_k^ν is the Laguerre polynomial of the degree k, as well as that at integers n

$$I_n(x) = i^{-n}J_n(ix)$$

we shall calculate the integral in (6.56):

$$\int_0^\infty x^{k+(q-1)/2}e^{-2x}I_{q-1}(2\sqrt{2x}h)\ dx = k!\,(\sqrt{2}h)^{q-1}e^{-h^2}2^{-k-q}L_k^{q-1}(-h^2) \qquad (6.57)$$

Having substituted (6.57) into (6.56), after conversions we obtain

$$p_1 = \frac{e^{-h^2}}{2^q}\sum_{k=0}^{q-1}\frac{1}{2^k}L_k^{q-1}(-h^2)$$

At last, using the representation of the Laguerre polynomial in the form of a sum, we find [15,16]:

$$p_1 = \frac{e^{-h^2}}{2^q}\sum_{l=0}^{q-1}\frac{(h^2)}{l!}\sum_{k=l}^{q-1}\frac{1}{2^k}C_{k+q-1}^{k-l} \qquad (6.58)$$

where $q = FT$; C_{k+q-1}^{k-l} is a number of combinations from $(k + q - 1)$ by $k - l$.

It is convenient to present formula (6.58) in the following form:

$$p_1 = \frac{1}{2}e^{-h^2} + \frac{e^{-h^2}}{2^q}\sum_{l=1}^{q-1}\frac{(h^2)^l}{l!}\sum_{k=l}^{q-1}\frac{1}{2^k}C_{k+q-1}^{k-l} \qquad (6.59)$$

In this relation the first member is the error probability of the optimum incoherent reception of binary PDM signals [compare with (6.41)], and the second, strictly non-negative member shows how the autocorrelated reception error probability is more than the optimum incoherent reception error probability.

Formula (6.59) is the fourth of the fundamental relations considered here for the theory of PM signal reception noise immunity. From this relation it follows that, in contrast to the correlation methods of processing, coherent and optimum incoherent, at autocorrelated processing the error probability depends not only on the SNR h^2, but also on the system parameter $2FT$. This thought can also be expressed as follows: At the correlation reception, noise immunity depends on the noise spectral density, and at autocorrelated reception on noise power. The latter one increases proportionally to the passband width at the input of the demodulator in the case of the white noise.

At $q = FT = 1$, (6.59) results in the fundamental relation (6.41) obtained earlier; in this idealized case the bandpass filter is becoming a filter matched with a harmonic signal, and the autocorrelated demodulator acts as though it has been transformed into the optimum incoherent one. However, at once we should emphasize that it is practically impossible to transform the autocorrelated demodulator to the optimum incoherent by simply narrowing the passband of the input filter because the intersymbol interference will cause a sharp degradation in noise immunity.

In fact, (6.59) can be used at the integers of $q = FT$, which are more than one, when the impact of intersymbol interference can be neglected. For the most important cases ($FT = 2$, 3, 4, and 5), we obtain from (6.59) [15]:

$$p_1(FT = 2) = 0.5e^{-h^2}(1 + h^2/4) \tag{6.60}$$

$$p_1(FT = 3) = p_1(FT = 2) + \frac{e^{-h^2}}{2^6}(h^4 + 4h^2) \tag{6.61}$$

$$p_1(FT = 4) = p_1(FT = 3) + \frac{e^{-h^2}}{3!2^7}(h^6 + 12h^4 + 30h^2) \tag{6.62}$$

$$p_1(FT = 5) = p_1(FT = 4) + \frac{e^{-h^2}}{4!2^9}(h^8 + 24h^6 + 167h^4 + 336h^2) \tag{6.63}$$

The higher FT, the more error probability.

Figure 6.3 shows the noise immunity curve of autocorrelated reception for two-phase PDM-1 signals at $FT = 2$ [curve 4, made up according to (6.60)]. At $FT = 2$ the energy loss Δh of the autocorrelated demodulator to the optimum incoherent one is insignificant and is equal to about 1 dB. At increasing FT this loss increases and at $FT = 10$ it reaches about 3 dB (curve 7).

At large FT, (6.59) is inconvenient for calculations. In this case we can use the approximated relation, obtained as a result of Gaussian approximation of the random value, which is in square brackets of (6.44). From this formula it follows that the error for transmitting the phase difference $\Delta\varphi = 0$ will occur if

$$\theta = \theta_1 + \theta_2 + \theta_3 < -E \tag{6.64}$$

where $E = \displaystyle\int_0^T S^2(t)\, dt$;

$$\theta_1 = \int_0^T S_1(t) \, \xi_1(t) \, dt$$

(6.65a)

$$\theta_2 = \int_0^T S(t) \, \xi_2(t) \, dt$$

$$\theta_3 = \int_0^T \xi_1(t) \, \xi_2(t) \, dt$$

(6.65b)

Random values θ_1 and θ_2 have a normal distribution with the zero mean and dispersion [see (6.22)]:

$$D(\theta_1) = D(\theta_2) = EN_0/2$$

(6.66)

To find the statistical characteristics of the random value θ_3, we present interferences ξ_1 and ξ_2 in the form of the expansion by the orthonormalized basis

$$\xi_1(t) = \sum_{k=k_1}^{k_2} \alpha_{1k} \sin k\frac{2\pi}{T}t + \beta_{1k} \cos k\frac{2\pi}{T}t$$

(6.67)

$$\xi_2(t) = \sum_{k=k_1}^{k_2} \alpha_{2k} \sin k\frac{2\pi}{T}t + \beta_{2k} \cos k\frac{2\pi}{T}t$$

where α_{1k}, α_{2k}, β_{1k}, and β_{2k} are determined according to (6.45). Having substituted (6.67) into (6.65) we obtain

$$\theta_3 = \frac{T}{2} \sum_{k=k_1}^{k_2} \alpha_{1k}\alpha_{2k} + \beta_{1k}\beta_{2k}$$

(6.68)

The coefficients in (6.68) are independent random values and have the dispersion N_0/T, therefore the dispersion of the random value θ_3 is

$$D(\theta_3) = 0.25T^2(k_2 - k_1 + 1)2N_0^2/T^2 = 0.5N_0^2FT$$

(6.69)

If the number of summands in (6.68), equal to the integer part of $2FT$, is quite great, the random value θ_3 can be considered an approximate normal random value with the zero mean and dispersion (6.69).

Thus, coming back to (6.64), we see that the random value θ is approximately Gaussian with zero mean and dispersion

$$D(\theta) = D(\theta_1) + D(\theta_2) + D(\theta_3) = EN_0 + 0.5N_0^2 FT \qquad (6.70)$$

and, consequently, the required error probability

$$p_1 \approx \frac{1}{\sqrt{2\pi D(\theta)}} \int_E^\infty \exp\left[-\frac{z^2}{2D(\theta)}\right] dz = \mathsf{F}\left[\frac{E}{\sqrt{D(\theta)}}\right] \qquad (6.71)$$

As

$$\frac{E}{\sqrt{D(\theta)}} = \frac{h}{\sqrt{1 + FT/2h^2}} \qquad (6.72)$$

we finally obtain the approximated formula [17]

$$p_1 \approx \mathsf{F}\left(\frac{h}{\sqrt{1 + FT/2h^2}}\right) < \frac{1}{2} \exp\left[-\frac{h^2}{2}\left(1 + \frac{FT}{2h^2}\right)^{-1}\right] \qquad (6.73)$$

Formula (6.73) is the fifth of the fundamental relations considered here for the theory of noise immunity of PM signal reception: the error probability at autocorrelated reception of binary PDM-1 signals in systems with high FT.

As is obvious from (6.73), with increasing FT and constant SNR the error probability is increasing, with this increase being especially expressed at relatively small h^2.

At larger h^2, when $2h^2 \gg FT$ or, just the same, when the signal power is much more than the interference power $P_s \gg P_i$, the error probability of autocorrelated reception slightly depends on the parameter FT and, as is obvious from (6.73), approximates to the error probability at coherent reception of orthogonal signals

$$p = \mathsf{F}(h) \qquad (6.74)$$

In Figure 6.3, curve 5 corresponds to formula (6.74) and curve 7 to formula (6.73) at $FT = 10$. We notice that at $FT = 10$ the noise immunity of autocorrelated reception of binary PDM signals practically coincides with the noise immunity of the optimum incoherent reception of FM signals, which is calculated by the formula

$$p = \frac{1}{2} \exp(-h^2/2) \qquad (6.75)$$

Curve 6 corresponds to formula (6.75).

Remember that expression (6.73) is approximated. Two assumptions have been made: The random value θ_3 is considered to be Gaussian, and the random values θ_1, θ_2, and θ_3 are considered to be independent. Actually, θ_3 is not Gaussian, and θ_1, θ_2 and θ_3 are dependent. However, as system parameter FT increases, the random value θ_3 is "normalized," and the statistical dependence between θ_1, θ_2, and θ_3 becomes weak. The analysis shows that at $FT \gg 10$ both assumptions are practically acceptable, and (6.73) gives quite precise results. This conclusion can be confirmed by quantitative comparison of results, obtained at the calculations according to (6.73) and (6.59). The calculations show that at $FT \gg 10$ the energy error does not exceed 1 dB using the approximated formula [15].

So, as indicated, the fundamental relations of the theory of noise immunity of the phase-modulated signals reception—the formulas (6.25), (6.28), (6.41), (6.59), and (6.73)—determine the potential noise immunity of the binary signals with PM and PDM at different processing algorithms in the AWGN channel.

The obtained results are illustrated by Table 6.1. In addition to the results for phase-modulated signals, the table also shows comparison error probabilities of coherent and incoherent reception for signals with frequency modulation (columns 8 and 10). In addition, column 5 gives a function $[\exp(-h^2)]/6$, approximating the error probability of the coherent reception of binary PDM-1 and PDM-2 signals.

6.3 COHERENT RECEPTION OF HIGH-ORDER PDM SIGNALS

Returning to coherent processing, we consider PDM signals for an arbitrary high order and find the expression for error probability, generalizing the fundamental relation (6.28).

We designate the error probability of the two-position system with PDM of the kth order by $p_1^{(k)}$, and the error probability of the two-position ideal system with PM (that is, PDM-0), as before by p_0.

The general formula for $p_1^{(k)}$ at an arbitrary k can be determined on the basis of the common algorithm (3.55) of coherent reception of signals with binary PDM-k. As long as in the last algorithm the products of signs are included, an error in determining the transmitted symbol will occur only in that case, when a number of erroneous cofactors (that is, a number of errors at the output of a coherent demodulator of PM signals) will appear odd. The total number R of cofactors in algorithm (3.55) is equal, as was shown in Section 3.4, to a number of odd integers in the kth line of the Pascal triangle [see the explanations given for algorithm (3.50)]. It can be analytically expressed by the Hamming weight $V(k)$ of the binary integer k according to the formula

$$R = 2^{V(k)} \tag{6.76}$$

Table 6.1
Error Probabilities for Binary PM and PDM Systems

h^2 (dB)	$h^2 = \frac{P_sT}{N_0}$	Coherent, PM $F(\sqrt{2}h)$	Coherent, PDM-1, PDM-2 $2F(\sqrt{2}h) \times [1 - F(\sqrt{2}h)]$	$\frac{1}{6}\exp(-h^2)$	Optimum Incoherent, PDM-1 $\frac{1}{2}\exp(-h^2)$	Autocorrelated, PDM-1 (FT=2) $\frac{1}{2}(1+h^2/4) \times \exp(-h^2)$	Coherent, FM $F(h)$	Autocorrelated, PDM-1 (FT=10) $F(h/\sqrt{1+5/h^2})$	Optimum Incoherent, FM $\frac{1}{2}\exp(-h^2/2)$
1	2	3	4	5	6	7	8	9	10
	0	5.00×10^{-1}	5.00×10^{-1}	1.67×10^{-1}	5.00×10^{-1}	5.00×10^{-1}	5.00×10^{-1}	5.00×10^{-1}	5.00×10^{-1}
0.00	1.0	7.87×10^{-2}	1.45×10^{-1}	6.13×10^{-2}	1.84×10^{-1}	2.30×10^{-1}	1.59×10^{-1}	3.42×10^{-1}	3.03×10^{-1}
3.01	2.0	2.28×10^{-2}	4.46×10^{-2}	2.26×10^{-2}	6.77×10^{-2}	1.02×10^{-1}	7.87×10^{-2}	2.25×10^{-1}	1.84×10^{-1}
4.77	3.0	7.15×10^{-3}	1.43×10^{-2}	8.30×10^{-3}	2.49×10^{-2}	4.36×10^{-2}	4.16×10^{-2}	1.44×10^{-1}	1.12×10^{-1}
6.02	4.0	2.34×10^{-3}	4.67×10^{-3}	3.05×10^{-3}	9.16×10^{-3}	1.83×10^{-2}	2.28×10^{-2}	9.13×10^{-2}	6.77×10^{-2}
6.99	5.0	7.83×10^{-4}	1.56×10^{-3}	1.12×10^{-3}	3.37×10^{-3}	7.58×10^{-3}	1.27×10^{-2}	5.69×10^{-2}	4.10×10^{-2}
7.78	6.0	2.66×10^{-4}	5.32×10^{-4}	4.13×10^{-4}	1.24×10^{-3}	3.10×10^{-3}	7.15×10^{-3}	3.52×10^{-2}	2.49×10^{-2}
8.45	7.0	9.10×10^{-5}	1.89×10^{-4}	1.52×10^{-4}	4.56×10^{-4}	1.25×10^{-3}	4.08×10^{-3}	2.16×10^{-2}	1.51×10^{-2}
9.03	8.0	3.17×10^{-5}	6.35×10^{-5}	5.59×10^{-5}	1.68×10^{-4}	5.03×10^{-4}	2.34×10^{-3}	1.32×10^{-2}	9.16×10^{-3}
9.54	9.0	1.10×10^{-5}	2.21×10^{-5}	2.06×10^{-5}	6.17×10^{-5}	2.01×10^{-4}	1.35×10^{-3}	8.08×10^{-3}	5.55×10^{-3}
10.00	10.0	3.87×10^{-6}	7.74×10^{-6}	7.57×10^{-6}	2.27×10^{-5}	7.94×10^{-5}	7.83×10^{-4}	4.91×10^{-3}	3.37×10^{-3}
10.41	11.0	1.36×10^{-6}	2.73×10^{-6}	2.78×10^{-6}	8.35×10^{-6}	3.13×10^{-5}	4.56×10^{-4}	2.98×10^{-3}	2.04×10^{-3}
10.79	12.0	4.82×10^{-7}	9.63×10^{-7}	1.02×10^{-6}	3.07×10^{-6}	1.23×10^{-5}	2.66×10^{-4}	1.80×10^{-3}	1.24×10^{-3}
11.14	13.0	1.71×10^{-7}	3.41×10^{-7}	3.77×10^{-7}	1.13×10^{-6}	4.80×10^{-6}	1.55×10^{-4}	1.09×10^{-3}	7.52×10^{-4}
11.46	14.0	6.07×10^{-8}	1.21×10^{-7}	1.39×10^{-7}	4.16×10^{-7}	1.87×10^{-6}	9.10×10^{-5}	6.59×10^{-4}	4.56×10^{-4}
11.76	15.0	2.16×10^{-8}	4.32×10^{-8}	5.10×10^{-8}	1.53×10^{-7}	7.26×10^{-7}	5.12×10^{-5}	3.98×10^{-4}	2.77×10^{-4}
12.04	16.0	7.71×10^{-9}	1.54×10^{-8}	1.88×10^{-8}	5.63×10^{-8}	2.81×10^{-7}	3.17×10^{-5}	2.40×10^{-4}	1.68×10^{-4}
12.30	17.0	2.76×10^{-9}	5.51×10^{-9}	6.90×10^{-9}	2.07×10^{-8}	1.09×10^{-7}	1.87×10^{-5}	1.44×10^{-4}	1.02×10^{-4}
12.55	18.0	9.87×10^{-10}	1.97×10^{-9}	2.54×10^{-9}	7.62×10^{-9}	4.19×10^{-8}	1.10×10^{-5}	8.66×10^{-5}	6.17×10^{-5}
12.79	19.0	3.54×10^{-10}	7.07×10^{-10}	9.34×10^{-10}	2.80×10^{-9}	1.61×10^{-8}	6.54×10^{-6}	5.17×10^{-5}	3.74×10^{-5}
13.01	20.0	1.27×10^{-10}	2.54×10^{-10}	3.44×10^{-10}	1.03×10^{-9}	6.18×10^{-9}	3.87×10^{-6}	3.17×10^{-5}	2.27×10^{-5}
13.98	25.0	7.69×10^{-13}	1.54×10^{-12}	2.32×10^{-12}	6.94×10^{-12}	5.03×10^{-11}	2.87×10^{-7}	2.50×10^{-6}	1.86×10^{-6}
14.77	30.0	4.74×10^{-15}	9.49×10^{-15}	1.56×10^{-14}	4.68×10^{-14}	3.98×10^{-13}	2.16×10^{-8}	1.98×10^{-7}	1.53×10^{-7}
15.44	35.0	2.96×10^{-17}	5.93×10^{-17}	1.05×10^{-16}	3.15×10^{-16}	3.07×10^{-15}	1.65×10^{-9}	1.56×10^{-8}	1.25×10^{-8}
16.02	40.0	1.87×10^{-19}	3.74×10^{-19}	7.08×10^{-19}	2.12×10^{-18}	2.34×10^{-17}	1.27×10^{-10}	1.24×10^{-9}	1.03×10^{-9}

Error Probability at Receiving Two-Position Signals

The probability of the event, that from total number R of the cofactors there appear to be erroneous $(2i + 1)$ cofactors, is equal to

$$p_0^{2i+1}(1 - p_0)^{R-(2i+1)}$$

and a number similar events is equal to a number of combinations from R by $(2i + 1)$, designated through C_R^{2i+1}. Thus, the required error probability is

$$p_1^{(k)} = \sum_{i=0}^{R/2-1} C_R^{2i+1} p_0^{2i+1}(1 - p_0)^{R-2i-1} \tag{6.77}$$

Taking into account (6.76) and using the Gallager formula [19] for the sums of the form (6.77)

$$\sum_{\substack{j \\ \text{odd}}}^{R} C_R^j p_0^j (1 - p_0)^{R-j} = \frac{1 - (1 - 2p_0)^R}{2}$$

we convert (6.77) to the following compact form:

$$p_1^{(k)} = \frac{1}{2}\left[1 - (1 - 2p_0)^{2^{V(k)}}\right] \tag{6.78}$$

The obtained formula has a fundamental meaning in the difference methods theory. First of all, note the wider application field of relation (6.78), than that for which it was particularly derived. Generally speaking, formula (6.78) determines how often errors appear when calculating the binary finite differences according to the initial samples with independent errors rate p_0. It determines the degree of multiplying the independent errors when calculating the final differences of the kth order and gives quantitative evaluation of the "error propagation coefficient" depending on interference parameters.

We should then emphasize that formula (6.78) generalizes fundamental relations (6.25), (6.27), and (6.28). In fact, for $k = 0, 1, 2, 3$, taking into account that $V(0) = 0$, $V(1) = 1$, $V(2) = 1$, $V(3) = 2$, from (6.78) we obtain:

$$p_1^{(0)} = p_0 \tag{6.79}$$

$$p_1^{(1)} = p_1^{(2)} = 2p_0(1 - p_0) \tag{6.80}$$

$$p_1^{(3)} = 4p_0(1 - p_0)(1 - 2p_0 + 2p_0^2) \tag{6.81}$$

For the AWGN channel, p_0 is determined by formula (6.25), and from (6.80) we obtain (6.28).

Formula (6.80) is correct when calculating the binary finite differences of any order k, equal to an integer degree of 2:

$$p_1^{(k=2^l)} = 2p_0(1 - p_0) \tag{6.82}$$

The general expression (6.78) for error probability at coherent reception of signals with two-phase PDM of the kth order in the AWGN channel takes the form

$$p_1^{(k)} = \frac{1}{2}\left\{1 - [1 - 2F(\sqrt{2}h)]^{2^{V(k)}}\right\} \tag{6.83}$$

from which (6.25) follows at $k = 0$, and (6.28) follows at $k = 1, 2, 4$, and so on.

At small error probabilities ($p_0 < 0.1$), expression (6.78) is approximated well by the following function:

$$p_1^{(k)} \approx 2^{V(k)}p_0 \tag{6.84}$$

The coefficient, which is before p_0, is equal to a number of errors, appearing at the output of the coherent PDM-k demodulator at one single error at the output of the coherent PM demodulator. This coefficient is called the *error propagation coefficient*. Thus, the error propagation coefficient at coherent reception of signals with binary PDM-k depends only on a number of 1 (Hamming weight) in the binary integer k:

$$R_1^{(k)} \approx 2^{V(k)} \tag{6.85}$$

The result is both simple and unexpected. The diagram of dependence (6.85) is shown in Figure 6.4. It is quite interesting to note that the propagation coefficient is not, as might be expected, the monotone increasing function of the used difference order; instead, it has minimums at the points $k = 2^l$ and maximums at the points $k = 2^l - 1$.

The approximated expressions for error probability by the error propagation coefficient can be also made up for the coherent reception of multiphase signals with high-order PDM. For this purpose remember that the scheme of a coherent demodulator of signals with phase-difference modulation of the kth order can be presented as a coherent demodulator of a signal phase and k series included discrete phase-difference calculators (see Fig. 3.35). The error probability $p_N^{(k)}$ of such a demodulator in the system with N-valued PDM-k can be expressed by the error probability at the PDM-0 demodulator output and by the error propagation coefficient $R_N^{(k)}$ according to the formula

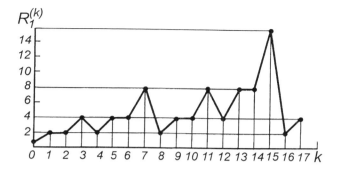

Figure 6.4 The error propagation coefficient at the coherent reception of binary PDM-k signals.

$$p_N^{(k)} \approx R_N^{(k)} p_N^{(0)} \qquad (6.86)$$

where $p_N^{(0)}$ is the error probability of a coherent demodulator of signals with N-valued PM. The coefficient $R_N^{(k)}$ is equal to a number of errors at the demodulator output of the system with N-valued PDM-k at a single error at the demodulator output of the system with N-valued PDM-0. The smaller $p_N^{(0)}$ is, the more precise formula (6.86) is.

To find $R_N^{(k)}$ as a function of N and k we should create and analyze the error multiplication tables resembling the shape of the known Pascal triangle, which serves to help us find binomial coefficients. In the first line of such a table, the received sequence of signal phases with a single error is written, and in the second and all the remaining lines the finite differences of the first order, second order, and so on are written. For convenience when creating the tables, it is expedient to use two methods.

First, the discrete set of phases at the coherent demodulator output forms an algebraic ring; this ring should be replaced by an isomorphic ring of positive integers. For example, a set of phase $\{0;\ \pi/2;\ \pi;\ 3\pi/2\}$ is replaced by set of numbers $\{0;\ 1;\ 2;\ 3\}$ with the operation of addition modulo 4, and a set of phases $\{0;\ \pi/4;\ \pi/2;\ 3\pi/4;\ \pi;\ 5\pi/4;\ 3\pi/2;\ 7\pi/4\}$ is replaced by a set of numbers $\{0;\ 1;\ 2;\ 3;\ 4;\ 5;\ 6;\ 7\}$ with the operation of addition modulo 8. Second, without loss of generality, we can associate 0 with correctly received elements. Then the error propagation coefficient will be equal to a number of nonzero elements in the appropriate line of the Pascal triangle.

For the case of two-phase PDM-k the table of error propagation is shown in Table 6.2. The error propagation coefficient (6.85), represented in Figure 6.4, follows from this table.

The example of for creating the Pascal triangle for the algebraic ring $\{0;\ 1;\ 2;\ 3\}$ is shown in Table 6.3, which illustrates the multiplication of an error corresponding to changing the information phase by $\pi/2$. It is easy to determine

Table 6.2
Error Propagation in Two-Level PDM-k Systems

PDM-0 (PM)	0 1 0
PDM-1	0 1 1 0
PDM-2	0 1 0 1 0
PDM-3	0 1 1 1 1 0
PDM-4	0 1 0 0 0 1 0
PDM-5	0 1 1 0 0 1 1 0
PDM-6	0 1 0 1 0 1 0 1 0
PDM-7	0 1 1 1 1 1 1 1 0
PDM-8	0 1 0 0 0 0 0 0 1 0
PDM-9	0 1 1 0 0 0 0 0 1 1 0
PDM-10	0 1 0 1 0 0 0 0 0 1 0 1 0
PDM-11	0 1 1 1 1 0 0 0 0 1 1 1 1 0
PDM-12	0 1 0 0 0 1 0 0 0 1 0 0 0 1 0
PDM-13	0 1 1 0 0 1 1 0 0 1 1 0 0 1 1 0
PDM-14	0 1 0 1 0 1 0 1 0 1 0 1 0 1 0 1 0
PDM-15	0 1 1 1 1 1 1 1 1 1 1 1 1 1 1 1 0
PDM-16	0 1 0 0 0 0 0 0 0 0 0 0 0 0 0 0 1 0

that the phase change by $3\pi/2$ results in the same values $R^{(k)}$, and the phase change by π results in the $R^{(k)} = 2^{V(k)}$ as in the case with two-phase PDM. Thus, at four-phase PDM ($\Delta^k \varphi = 0, \pi/2, \pi, 3\pi/2$) the error propagation coefficient takes, depending on the initial error, the values either on the lower boundary for $R^{(k)}$, or on the dashed curve (see Fig. 6.5).

Table 6.4 shows an example for creating the Pascal triangle for the algebraic ring {0; 1; 2; 3; 4; 5; 6; 7}. The table illustrates the error propagation at the phase change by $\pi/4$. According to this table, in cases of eight-phase PDM ($\Delta^k \varphi = 0, \pi/4, \pi/2, 3\pi/4, \pi, 5\pi/4, 3\pi/2, 7\pi/4$) or four-phase PDM ($\Delta^k \varphi = \pi/4, 3\pi/4, 5\pi/4, 7\pi/4$), the error propagation coefficient takes the position either on the lower bound for $R^{(k)}$, or on the dashed or dashed-dotted curves (see Fig. 6.5).

In the general case we can formulate the following rules to evaluate the error propagation coefficient:

1. The values of $R_N^{(k)}$ at any N are within the limits

$$R_1^{(k)} = 2^{V(k)} \le R_N^{(k)} \le k + 1 \tag{6.87}$$

and at $k = 2^m - 1$, where m is an integer, the upper and lower boundaries coincide, that is,

$$R_N^{(2^m-1)} = 2^{V(k)} = k + 1 = 2^m \tag{6.88}$$

Table 6.3
Error Propagation in Four-Level PDM-k Systems

$k = 0$	0 1 0
1	0 3 1 0
2	0 1 2 1 0
3	0 3 3 1 1 0
4	0 1 0 2 0 1 0
5	0 3 1 2 2 3 1 0
6	0 1 2 3 0 3 2 1 0
7	0 3 3 3 3 1 1 1 1 0
8	0 1 0 0 0 2 0 0 0 1 0
9	0 3 1 0 0 2 2 0 0 3 1 0
10	0 1 2 1 0 2 0 2 0 1 2 1 0
11	0 3 3 1 1 2 2 2 2 3 3 1 1 0
12	0 1 0 2 0 3 0 0 0 3 0 2 0 1 0
13	0 3 1 2 2 1 3 0 0 1 3 2 2 3 1 0
14	0 1 2 3 0 1 2 3 0 3 2 1 0 3 2 1 0
15	0 3 3 3 3 3 3 3 3 1 1 1 1 1 1 1 1 0
16	0 1 0 0 0 0 0 0 2 0 0 0 0 0 0 1 0
17	0 3 1 0 0 0 0 0 2 2 0 0 0 0 0 3 1 0

and at $k = 2^m$, the $R_N^{(k)}$ values have local minima.

2. The more N, the more close the values $R_N^{(k)}$ to the upper boundary. The maximum $R_N^{(k)}$ takes place at the minimum initial error. In the 2^N phase system with PDM-k ($\Delta^k \varphi_j = j\pi/2^{N-1}$; $j = 0, 1, \ldots, 2^{N-1}$) these maximum values are less than $(k + 1)$ by a number of binomial coefficients C_k^i ($i = 0, 1, \ldots, k$), divisible by 2^N.

3. At multiphase PDM, the error propagation coefficient depends on the initial phase error in a PM demodulator. Its values at $\pm\pi/4$, $\pm\pi/2$, and $\pm\pi$ are given in Table 6.5.

The obtained estimates of the error propagation coefficient allow us to find the relations between error probabilities of coherent demodulators of signals with different order PDMs. Having found the error probability at coherent reception of a multiposition PM signal and having multiplied it by the appropriate propagation coefficient, we can evaluate error probability when receiving high-order PDM signals.

6.4 OPTIMUM INCOHERENT RECEPTION OF SECOND-ORDER PDM SIGNALS

The noise immunity of binary systems for different methods of processing was studied in Section 6.2. However, the problem of noise immunity of PDM-2 signal

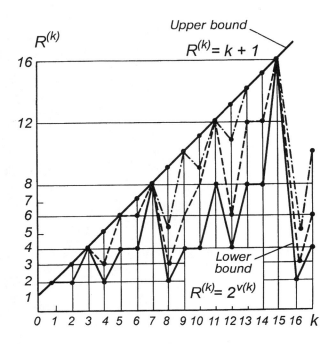

Figure 6.5 The error propagation coefficient versus the order of phase differences. Dashed line, $\pi/2$; dashed-dotted line, $\pi/4$.

incoherent reception has not been considered, and the discussion of PDM-1 signal incoherent reception has been limited to the case of two-chip processing. We decided to consider these themes in a separate section because of their importance.

The purpose of the following analysis is, however, not only to find error probability at optimum incoherent reception for PDM-2 signals, but also to determine the important fact that at an increasing interval for PDM signal processing the noise immunity of the optimum incoherent reception approximates the noise immunity of the ideal coherent reception.

As an object for analysis, let us use algorithms for the optimum incoherent reception of binary PDM-2 signals (4.40) or (4.41) and the corresponding demodulators (see Fig. 4.13). Remember that at PDM-2, processing is fulfilled within three adjacent chips, which contain one of four variants of a signal S_1, S_2, S_3, or S_4 [see Fig. 4.12 and (4.38)]. By means of convolutions of the received signal and these variants, the demodulator calculates values V_1, V_2, V_3, and V_4 by the formulas (4.40) or (4.41); the maximum number of them determines the transmitted binary symbol.

In contrast to PDM-1 signals, the signals of (4.38) are not orthogonal; therefore, to calculate noise immunity for the considered method of reception we should use the upper bound for error probability of the optimum incoherent reception

Table 6.4
Error Propagation in Eight-Level PDM-*k* Systems

$k = 0$	0 1 0
1	0 7 1 0
2	0 1 6 1 0
3	0 7 3 5 1 0
4	0 1 4 6 4 1 0
5	0 7 5 6 2 3 1 0
6	0 1 2 7 4 7 2 1 0
7	0 7 7 3 3 5 5 1 1 0
8	0 1 0 4 0 6 0 4 0 1 0
9	0 7 1 4 4 2 6 4 4 7 1 0
10	0 1 6 5 0 2 4 2 0 5 6 1 0
11	0 7 3 1 5 6 6 2 2 3 7 5 1 0
12	0 1 4 2 4 7 0 4 0 7 4 2 4 1 0
13	0 7 5 2 6 5 7 4 4 1 3 2 6 3 1 0
14	0 1 2 3 4 1 6 3 0 3 6 1 4 3 2 1 0
15	0 7 7 7 7 3 3 3 3 5 5 5 5 1 1 1 1 0
16	0 1 0 0 0 4 0 0 0 6 0 0 0 4 0 0 0 1 0
17	0 7 1 0 0 4 4 0 0 3 6 0 0 4 4 0 0 7 1 0

Table 6.5
Error Propagation Coefficient in Multiphase PDM-*k* Systems

$\Delta\varphi$	Multiplication Coefficient at k, Equal to																	
	0	*1*	*2*	*3*	*4*	*5*	*6*	*7*	*8*	*9*	*10*	*11*	*12*	*13*	*14*	*15*	*16*	*17*
$\pm\,\pi/4$	1	2	3	4	5	6	7	8	5	10	9	12	11	14	15	16	5	10
$\pm\,\pi/2$	1	2	3	4	3	6	6	8	3	6	8	12	6	8	8	16	3	6
$\pm\,\pi$	1	2	2	4	2	4	4	8	2	4	4	8	4	8	8	16	2	4

of arbitrary *m*-ary equipower signals [4]. The corresponding estimate is formulated as follows.

On transmission of one of *m* equipower signals $S_i(t)$, the error probability, that is, the probability that instead of S_i there will be fixed any other signal S_j ($j = 1, 2, \ldots, m, j \neq i$), is equal to

$$
\begin{aligned}
p_i \leq \frac{1}{2} \exp\left(-\frac{h_o^2}{2}\right) \sum_{j=1}^{m} & \left[I_0\left(\frac{R_{ij} h_o^2}{2}\right) \right] \\
& + 2 \sum_{n=1}^{\infty} \left[\left(\frac{1 - \sqrt{1 - R_{ij}^2}}{R_{ij}}\right)^n I_n\left(\frac{R_{ij} h_o^2}{2}\right) \right]
\end{aligned}
\tag{6.89}
$$

where

$$R_{ij} = \sqrt{\rho_{ij}^2 + \rho_{ij}^{*2}}$$

$$\rho_{ij} = \frac{1}{E} \int_0^\tau S_i(t)\, S_j(t)\ dt$$

$$\rho_{ij}^* = \frac{1}{E} \int_0^\tau S_i(t)\, S_j^*(t)\ dt \qquad (6.90)$$

$$E = \int_0^\tau S_j^2(t)\ dt$$

$$h_o^2 = \frac{P_s \tau}{N_0} = \frac{E}{N_0}$$

and I_n is the modified Bessel function of the first kind and nth order; τ is the signal processing duration; and P_s is the signal power.

Let the signal S_1 have been transmitted (see Fig. 4.12). Then it is not difficult to calculate according to the formulas (6.90) at $j = 2, 3, 4$:

$$|\rho_{1j}| = 1/3$$
$$\rho_{1j}^* = 0 \qquad (6.91)$$
$$R_{1j} = 1/3$$

Furthermore, as far as the processing is carried out within three chips, that is, $\tau = 3T$, then according to (6.90) and (6.91)

$$h_o^2 = 3P_s T / N_0 = 3h^2 \qquad (6.92)$$

Finally, we need to take into account the fact that according to algorithm (4.40), when receiving signal S_2 instead of transmitted signal S_1 the error does not occur (see Fig. 4.12). Therefore in the first sum in (6.89) we have to keep only two ($j = 3, 4$) of three components ($j = 2, 3, 4$). Taking into account the last circumstance and using (6.91) and (6.92), we obtain from (6.89) that the error probability on transmission of signal S_1 does not exceed the value [20]

$$p_1^{(2)} \le \exp\left(-\frac{3h^2}{2}\right)\left[I_0\left(\frac{h^2}{2}\right) + 2\sum_{n=1}^{\infty} (3 - \sqrt{8})^n I_n\left(\frac{h^2}{2}\right) \right] \qquad (6.93)$$

Similarly it is easy to determine that on transmission of other variants of a signal the error probability is defined by the same relation. That is why (6.93) is the required estimate of the upper bound of the error probability for the optimum incoherent reception of binary PDM-2 signals.

The calculation of (6.93) is submitted in Table 6.6 and shown by curve 3 in Figures 6.6 and 6.7. (Figure 6.7 shows in larger scale a part of Figure 6.6 at small h^2. Table 6.6 indicates the numbers of the formulas, according to which the calculation in the appropriate column has been done.)

As can be seen from these results, the upper bound of error probability $p_1^{(2)}$ is less than $0.5 \exp(-h^2)$, at error probabilities of less than 0.05. In addition, the results of statistical modeling of algorithm (4.40) [21], indicated in Table 6.6 and in Figures 6.7 and 6.6 (curve 2), show that for any SNR h^2 the following fundamental relation is true:

Table 6.6
Error Probabilities for Coherent and Incoherent PDM-2 Demodulators

h^2 (dB)	$h^2 = \dfrac{P_S T}{N_0}$	Coherent, PDM-1 and PDM-2 (6.28)	PDM-2 (6.93)	PDM-2, Modeling	PDM-1 (6.41)
			Optimum Incoherent		
	0	5.00×10^{-1}	1.0	5.00×10^{-1}	5.00×10^{-1}
0	1.0	1.45×10^{-1}	2.57×10^{-1}	1.80×10^{-1}	1.84×10^{-1}
1.58	1.44	8.56×10^{-2}	1.46×10^{-1}	1.10×10^{-1}	1.18×10^{-1}
2.92	1.96	4.66×10^{-2}	7.67×10^{-2}	6.20×10^{-2}	7.04×10^{-2}
3.01	2.00	4.46×10^{-2}	7.31×10^{-2}		6.77×10^{-2}
4.08	2.56	2.70×10^{-2}	3.73×10^{-2}	3.30×10^{-2}	3.86×10^{-2}
4.77	3.00	1.42×10^{-2}	2.23×10^{-2}		2.49×10^{-2}
5.11	3.24	1.08×10^{-2}	1.69×10^{-2}	1.60×10^{-2}	1.96×10^{-2}
6.02	4.00	4.67×10^{-3}	7.11×10^{-3}	6.40×10^{-3}	9.16×10^{-3}
6.85	4.84	1.86×10^{-3}	2.79×10^{-3}	2.40×10^{-3}	3.95×10^{-3}
6.99	5.00	1.56×10^{-3}	2.34×10^{-3}		3.37×10^{-3}
7.60	5.76	6.88×10^{-4}	1.02×10^{-3}	9.50×10^{-4}	1.58×10^{-3}
7.78	6.00	5.32×10^{-4}	7.87×10^{-4}		1.24×10^{-3}
7.96	6.25	4.06×10^{-4}	6.01×10^{-4}	5.60×10^{-4}	9.65×10^{-4}
8.45	7.00	1.82×10^{-4}	2.69×10^{-4}		4.56×10^{-4}
9.03	8.00	6.35×10^{-5}	9.26×10^{-5}		1.68×10^{-4}
9.54	9.00	2.21×10^{-5}	3.22×10^{-5}		6.17×10^{-5}
10.00	10.00	7.74×10^{-6}	1.12×10^{-5}		2.27×10^{-5}
10.40	11.00	2.73×10^{-6}	3.95×10^{-6}		8.35×10^{-6}
10.80	12.00	9.63×10^{-7}	1.39×10^{-6}		3.07×10^{-6}
11.10	13.00	3.41×10^{-7}	4.92×10^{-7}		1.13×10^{-6}
11.50	14.00	1.21×10^{-7}	1.75×10^{-7}		4.16×10^{-7}
11.80	15.00	4.32×10^{-8}	6.21×10^{-8}		1.53×10^{-7}
12.00	16.00	1.54×10^{-8}	2.22×10^{-8}		5.63×10^{-8}

Error Probability at Reception of Two-Phase Signals

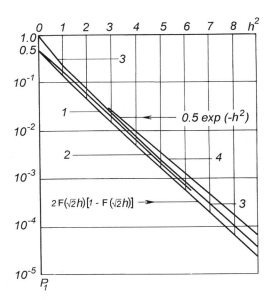

Figure 6.6 Comparative noise immunity of two-phase systems with PDM at the optimum methods of reception: (1) PDM-1 and PDM-2, coherent; (2 and 3) PDM-2, optimum incoherent (modeling and the upper bound estimate); and (4) PDM-1, optimum incoherent.

$$p_1^{(2)} < p_1^{(1)} = 0.5 \exp(-h^2) \tag{6.94}$$

Thus, at optimum incoherent reception of signals with known frequency but indefinite initial phase, the system with PDM-2 has a higher noise immunity than the system with PDM-1. The gain of PDM-2, however, is insignificant; for example, at $p = 10^{-2}$ it is equal to about 0.5 dB, and at $p = 10^{-4}$ it is close to 0.4 dB.

However, it is conceptually important to realize that in the channel with constant parameters at optimum incoherent reception the prolonging of the processing interval (at transition from PDM-1 to PDM-2 it increases by 50%) results in a decrease in the error probability, and the noise immunity of incoherent reception approximates the noise immunity of coherent one.

Table 6.6 and Figures 6.6 and 6.7 (curve 1) indicate the error probabilities of coherent reception of PDM-1 and PDM-2 signals. The comparison of these data with the obtained results shows that at PDM-1 the loss of the optimum incoherent reception to the coherent reception is 0.7 dB at $p = 10^{-2}$ and 0.6 dB at $p = 10^{-4}$, and at PDM-2 the loss of the optimum incoherent reception to the coherent reception in the range of error probabilities 10^{-2} to 10^{-4} does not exceed 0.2 dB.

This comparison allows us to make an important conclusion: Algorithms (4.40) or (4.41) of the optimum incoherent reception of binary PDM-2 signals practically solve the problem of achieving the potential noise immunity of two-phase signals

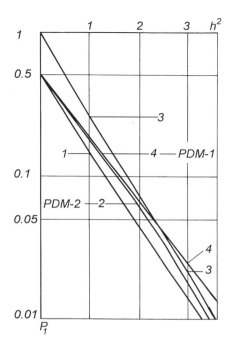

Figure 6.7 Fragment of the comparative noise immunity diagram in Figure 6.6.

without using any information about the initial carrier phase. This effect has been achieved by prolonging the signal processing interval.

Obviously, the same effect is also achieved at incoherent reception of the PDM-1 signals, if the processing interval is increased up to three chips. The corresponding algorithms were discussed in Section 4.5; they were constructed on the basis of calculating the same components (4.40) or (4.41), which were used for optimum incoherent reception of the PDM-2 signals. The noise immunity of incoherent algorithms for binary PDM-1 signals processing within three chips is determined, as well as in the case of PDM-2, by expression (6.93), which is the upper bound for the error probability.

The preceding results reflect features of the comparatively new algorithms for PDM signal processing, namely, optimum multisymbol incoherent processing. When increasing the number of processing signal elements (symbols), the error probability of the multisymbol incoherent processing decreases and in the limit tends to the noise immunity of the perfect coherent processing.

The dependence of the appropriate error probability from the processing interval can be estimated by means of formulas (6.205), (6.206), and (6.207), obtained in Section 6.8, which studies the noise immunity of quasicoherent reception of signals. For quasicoherent reception the decision about the transmitted symbol is made on the basis of processing a "training signal" of fixed length. As long as the algorithms considered in this section are optimum for any fixed processing interval, formulas (6.205), (6.206), and (6.207) can be the upper bound evaluations for the appropriate algorithms of multisymbol processing. The corresponding diagrams for the case of four-phase PDM are shown in Figure 4.20.

The multisymbol incoherent processing of phase-modulated signals is one of the most promising directions for improving digital communication systems. It allows us to realize practically the maximum possible noise immunity on the basis of incoherent processing, which does not require signal phase adjustment.

6.5 FOUR-PHASE PM AND PDM SYSTEMS

Of all nonbinary (multiposition) systems of digital transmission of messages by phase-modulated signals it is expedient to select and consider separately four-phase systems with PM and PDM, which are of the greatest practical interest (QPSK systems). Systems with four-phase PM and PDM provide us with the possibility to double the transmission rate in the same frequency band in comparison with the binary PM and PDM systems. However, even more important, they can provide the same transmission rate in a twice narrower band without losses or with insignificant losses of noise immunity.

Taking into account these properties of four-phase systems, we consider separately their noise immunity for different processing methods similar to what was done in Section 6.2 for the case of binary systems.

Let us begin with a consideration of coherent reception of four-phase PM signals. There are four signals in this case:

$$S_1(t) = a \sin \omega t$$
$$S_2(t) = a \cos \omega t \qquad (6.95)$$
$$S_3(t) = -a \cos \omega t$$
$$S_4(t) = -a \sin \omega t$$

On transmission of signal $S_1(t)$, a coherent receiver according to (3.1) will make a correct decision when the following inequalities take place:

$$\int_0^T x(t)\, S_1(t)\, dt > \int_0^T x(t)\, S_2(t)\, dt \qquad (6.96a)$$

$$\int_0^T x(t)\, S_1(t)\, dt > \int_0^T x(t)\, S_3(t)\, dt \qquad (6.96b)$$

$$\int\limits_0^T x(t)\, S_1(t)\ dt > \int\limits_0^T x(t)\, S_4(t)\ dt \tag{6.96c}$$

Let us designate the probability of fulfilling inequality (6.96a) as q_2, and (6.96b) as q_3. The random events (6.96a) and (6.96b) are compatible and independent, because the signal $S_1(t)$ orthogonal to the signals $S_2(t)$ and $S_3(t)$, and the signals $S_2(t)$ and $S_3(t)$ are antipodal.

To prove the independence of the depicted events and to calculate further the error probability, we substitute the received mixture $x(t)$ of the signal $S_1(t)$ with noise $\xi(t)$ into inequalities (6.96a) and (6.96b). From (6.96a) we obtain the inequality

$$E > -\theta_1 + \theta_2 \tag{6.97}$$

and from (6.96b) we obtain the inequality

$$E > -\theta_1 - \theta_2 \tag{6.98}$$

where

$$E = \int\limits_0^T S_1^2(t)\ dt \qquad \theta_1 = \int\limits_0^T S_1(t)\, \xi(t)\ dt \qquad \theta_2 = \int\limits_0^T S_2(t)\, \xi(t)\ dt$$

The random values θ_1 and θ_2 are independent (by virtue of orthogonality of signals S_1 and S_2) normal random values. Therefore the right parts of inequalities (6.97) and (6.98) as the sum and difference of independent normal random values are also independent normal random values. The probability of simultaneous fulfillment of inequalities (6.96a) and (6.96b) is equal to the product of probabilities of fulfilling every one of these inequalities separately, that is, $q_2 q_3$. In addition, if inequalities (6.96a) and (6.96b) are carried out, (6.96c) is carried out as well. In fact, (6.96c) can be presented in the form

$$E > -\theta_1 \tag{6.99}$$

This inequality is the simple sum of inequalities (6.97) and (6.98). From here we know that probability of simultaneous fulfillment of all three inequalities (6.96), that is, the probability of correct reception, is equal to

$$p_c = q_2\, q_3 \tag{6.100}$$

Now let us calculate q_2 and q_3, using inequalities (6.97) and (6.98). The values θ_1 and θ_2 are independent normal random values with zero means and dispersions $D(\theta_1) = D(\theta_2) = N_0E/2$ according to (6.22). Therefore,

$$D(-\theta_1 + \theta_2) = D(-\theta_1 - \theta_2) = N_0E$$

Thus,

$$q_2 = q_3 = \frac{1}{\sqrt{2\pi N_0 E}} \int_{-\infty}^{E} \exp\left(-\frac{y^2}{2N_0 E}\right) dy = 1 - \mathsf{F}(h) \qquad (6.101)$$

where F, as before, is the Laplace function (6.8). The required error probability is equal to

$$p_2 = 1 - p_c = 1 - q_2^2 \qquad (6.102)$$

Substituting (6.101) in (6.102), we obtain the following expression for the error probability at coherent reception of four-phase PM signals:

$$p_2 = 2\mathsf{F}(h)\left[1 - \frac{1}{2}\mathsf{F}(h)\right] \qquad (6.103)$$

This remarkable formula is similar in its structure to the formula for the error probability at coherent reception of binary PDM signals [see (6.28)]. It demonstrates that the error probability at coherent reception of the four-phase PM signals is about two times more than the error probability at coherent reception of two orthogonal signals (at identical h^2!), that is, the system with four-phase PM, which is almost four times more frequency-efficient than the system with FM, does not practically concede it by noise immunity.

Having compared (6.103) with expression (6.25) for the error probability at binary PM, we can make a conclusion that an increase in the transmission rate by two times, achieved by increasing a phase number from two up to four, results in energy losses of about two times, that is, about 3 dB.

However, it is very interesting that at a constant transmission rate, the transition from two-phase PM to four-phase PM does not practically require the SNR to increase. In fact, from (6.103) it is obvious that a good estimation of error probability at its small values is the formula

$$p_2 \approx 2\mathsf{F}(h) \qquad (6.104)$$

Let $h_1^2 = P_sT_1/N_0$ be the SNR in a two-phase PM system at chip duration T_1. The error probability in this system according to (6.25) is $p_1 = \mathsf{F}(\sqrt{2}h_1)$. In the system with four-phase PM at the same information transmission rate, the chip duration will be two times more, that is, $T_2 = 2T_1$. Hence, in system with four-phase PM the SNR is $h_2^2 = 2P_sT_1/N_0 = 2h^2$, and according to (6.104) the error probability is

$$p_2 \approx 2\mathsf{F}(\sqrt{2}h_1) = 2p_1$$

The difference in the error probability of about two times is certainly insignificant, but a two-times increase in the chip duration is a very important argument for the benefit of four-phase PM in comparison with two-phase PM. Due to chip lengthening, it is possible to reduce the occupied frequency band, to reduce out-of-band radiation, to increase the resistance to multibeam propagation, and so on. We pay attention to these simple facts because they often slip away from the engineers' attention.

The error probability at the coherent reception of four-phase PM signals as a function of SNR h^2, calculated according to (6.103), is shown in Figure 6.8 (curve 4) in comparison with the similar dependencies at other modulation and demodulation methods.

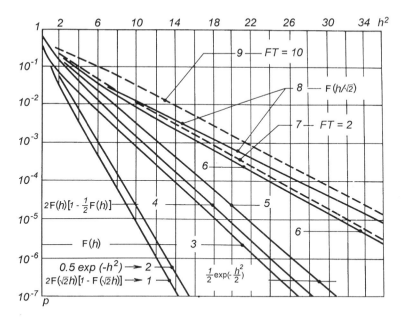

Figure 6.8 Comparative noise immunity of two- and four-phase systems at various modulation and reception methods: (1) 2PDM (BDPSK), coherent; (2) 2PDM, optimum incoherent; (3) FM, coherent; (4) 4PM (QPSK), coherent; (5) FM, optimum incoherent; (6) 4PDM (QDPSK), optimum incoherent; (7) 4PDM, autocorrelated (*FT* = 2); (8) AM, coherent; (9) 4PDM, autocorrelated (*FT* = 10).

The numerical results of the calculation for formula (6.103) are given in Table 6.7 (column 4). Notice that at $h^2 = 0$ the probability $p_2 = 3/4$ because in this case one of four equiprobable signal variants can be received.

Formula (6.103) determines the probability of faulty reception of a 4-ary symbol. When using the Gray code, binary subchannels of four-phase systems are completely symmetric and have the identical probability of faulty binary symbol reception. Therefore, at small error probabilities, the probability of simultaneous errors in both subchannels (double bit error) is negligibly small when compared with the error probability in only one of the subchannels (single bit error). Thus, formula (6.104) approximately determines the probability that one of a pair of transmitted binary symbols will be incorrect. As a result, at the coherent reception of four-position PM signals the bit error probability or BER, is equal to

$$p_2(\text{BER}) \approx \mathsf{F}(h) \qquad (6.105)$$

This dependence is depicted by curve 3 in Figure 6.8, which corresponds to the potential noise immunity of two orthogonal signal coherent receptions. The numerical results of the calculation according to (6.105) are shown in Table 6.7 (column 3).

At coherent reception of four-phase PDM signals, a phase-difference discrete calculator–decoder is added to a coherent demodulator of four-phase PM signals. The use of this unit leads to error propagation, the mechanism of which was considered in Section 6.3. At the first-order PDM every error at the coherent PM demodulator output leads (with the probability close to 1) to two errors following each other at the PDM demodulator output. There are twin errors in binary subchannels of the coherent demodulator of four-phase PDM signals as well as in the similar demodulator of binary PDM signals, and the probability of these error pairs is approximately equal to (6.105).

Thus, the bit error probability at coherent reception of four-phase PDM signals is determined by the approximate formula (6.104). As a result of this, curve 4 in Figure 6.8 can serve as an estimation for the BER at coherent reception of the four-phase PDM signals with the only remark that at small h^2 it should tend to 0.5, but not to 0.75, as shown in the figure. With the same remark for the specified case, the numerical data of column 4, Table 6.7 could serve as estimates.

In the case of coherent reception of PDM signals as well as at PM, the opportunity arises to change the transmission rate for the noise immunity. At equal bit rates the four-phase PDM system has practically the same noise immunity as the two-phase system. On the other hand, at equal chip durations the four-phase PDM system provides a two times higher bit rate than the two-phase PDM system, however it twice concedes the latter by the noise immunity (the required SNR).

We proceed to calculate noise immunity of the optimum incoherent reception of four-phase PDM signals. For this purpose let us address algorithm (4.15), corresponding to reception of a signal with phase difference variants of $\pi/4$, $3\pi/4$,

Table 6.7
Error Probabilities for Four-Phase PM and PDM Systems

		Coherent		Error Probability at Reception — Optimum Incoherent			Autocorrelated	
h^2 (dB)	h^2	Binary FM $F(h)$	Four-Phase PM (6.103)	Binary FM $0.5\exp(-h^2/2)$	Four-Phase PDM (6.125)	Four-Phase PDM (6.126)	Four-Phase PDM (6.133) $FT=2$	$FT=10$
1	2	3	4	5	6	7	8	9
	0	5.00×10^{-1}	7.50×10^{-1}	5.00×10^{-1}	5.00×10^{-1}	5.50×10^{-1}	5.00×10^{-1}	5.00×10^{-1}
0.00	1.0	1.59×10^{-1}	2.92×10^{-1}	3.03×10^{-1}	2.69×10^{-1}	2.44×10^{-1}	3.04×10^{-1}	7.78×10^{-1}
3.01	2.0	7.87×10^{-2}	1.51×10^{-1}	1.84×10^{-1}	1.64×10^{-1}	1.54×10^{-1}	1.94×10^{-1}	3.10×10^{-1}
4.77	3.0	4.16×10^{-2}	8.15×10^{-2}	1.12×10^{-1}	1.07×10^{-1}	1.02×10^{-1}	1.56×10^{-1}	2.61×10^{-1}
6.02	4.0	2.28×10^{-2}	4.51×10^{-2}	6.77×10^{-2}	7.16×10^{-2}	6.93×10^{-2}	8.71×10^{-2}	1.80×10^{-1}
6.99	5.0	1.27×10^{-2}	2.52×10^{-2}	4.10×10^{-2}	4.92×10^{-2}	4.79×10^{-2}	5.99×10^{-2}	1.36×10^{-1}
7.78	6.0	7.15×10^{-3}	1.43×10^{-2}	2.49×10^{-2}	3.42×10^{-2}	3.35×10^{-2}	4.18×10^{-2}	1.02×10^{-1}
8.45	7.0	4.08×10^{-3}	8.13×10^{-3}	1.51×10^{-2}	2.40×10^{-2}	2.36×10^{-2}	2.93×10^{-2}	7.60×10^{-2}
9.03	8.0	2.34×10^{-3}	4.68×10^{-3}	9.16×10^{-3}	1.70×10^{-2}	1.68×10^{-2}	2.08×10^{-2}	5.66×10^{-2}
9.54	9.0	1.35×10^{-3}	2.70×10^{-3}	5.55×10^{-3}	1.21×10^{-2}	1.20×10^{-2}	1.48×10^{-2}	4.20×10^{-2}
10.00	10.0	7.84×10^{-4}	1.57×10^{-3}	3.37×10^{-3}	8.65×10^{-3}	8.56×10^{-3}	1.05×10^{-2}	3.12×10^{-2}
10.40	11.0	4.56×10^{-4}	9.12×10^{-4}	2.04×10^{-3}	6.20×10^{-3}	6.15×10^{-3}	7.56×10^{-3}	2.31×10^{-2}
10.80	12.0	2.66×10^{-4}	5.32×10^{-4}	1.24×10^{-3}	4.46×10^{-3}	4.43×10^{-3}	5.43×10^{-3}	1.71×10^{-2}
11.10	13.0	1.55×10^{-4}	3.10×10^{-4}	7.52×10^{-4}	3.22×10^{-3}	3.20×10^{-3}	3.92×10^{-3}	1.26×10^{-2}
11.50	14.0	9.10×10^{-5}	1.82×10^{-4}	4.56×10^{-4}	2.33×10^{-3}	2.32×10^{-3}	2.83×10^{-3}	9.33×10^{-3}
11.80	15.0	5.42×10^{-5}	1.08×10^{-4}	2.76×10^{-4}	1.74×10^{-3}	1.68×10^{-3}	2.10×10^{-3}	6.96×10^{-3}
12.00	16.0	3.17×10^{-5}	6.34×10^{-5}	1.68×10^{-4}	1.22×10^{-3}	1.22×10^{-3}	1.49×10^{-3}	5.08×10^{-3}
12.30	17.0	1.87×10^{-5}	3.74×10^{-5}	1.02×10^{-4}	8.88×10^{-4}	8.86×10^{-4}	1.08×10^{-3}	3.76×10^{-3}
12.60	18.0	1.10×10^{-5}	2.21×10^{-5}	6.17×10^{-5}	6.45×10^{-4}	6.44×10^{-4}	7.85×10^{-4}	2.78×10^{-3}
12.80	19.0	6.54×10^{-6}	1.31×10^{-5}	3.74×10^{-5}	4.71×10^{-4}	4.70×10^{-4}	5.72×10^{-4}	2.05×10^{-3}
13.00	20.0	3.87×10^{-6}	7.74×10^{-6}	2.27×10^{-5}	3.44×10^{-4}	3.43×10^{-4}	4.17×10^{-4}	1.51×10^{-3}
13.20	21.0	2.30×10^{-6}	4.60×10^{-6}	1.38×10^{-5}	2.50×10^{-4}	2.50×10^{-4}	3.04×10^{-4}	1.12×10^{-3}
13.40	22.0	1.36×10^{-6}	2.72×10^{-6}	8.35×10^{-6}	1.83×10^{-4}	1.81×10^{-4}	2.22×10^{-4}	8.29×10^{-4}
13.60	23.0	8.10×10^{-7}	1.62×10^{-6}	5.07×10^{-6}	1.34×10^{-4}	1.33×10^{-4}	1.63×10^{-4}	6.10×10^{-4}
13.80	24.0	4.82×10^{-7}	9.64×10^{-7}	3.07×10^{-6}	9.80×10^{-5}	9.62×10^{-5}	1.19×10^{-4}	4.51×10^{-4}
14.00	25.0	2.87×10^{-7}	5.74×10^{-7}	1.86×10^{-6}	7.17×10^{-5}	6.99×10^{-5}	8.71×10^{-5}	3.34×10^{-4}
14.80	30.0	2.16×10^{-8}	4.32×10^{-8}	1.53×10^{-7}	1.53×10^{-5}	1.53×10^{-5}	1.85×10^{-5}	7.33×10^{-5}
15.50	35.0	1.65×10^{-9}	3.30×10^{-9}	1.28×10^{-8}	3.29×10^{-6}	3.30×10^{-6}	3.98×10^{-6}	1.63×10^{-5}
16.00	40.0	1.27×10^{-10}	2.54×10^{-10}	1.03×10^{-9}	7.15×10^{-7}	7.15×10^{-7}	8.66×10^{-7}	3.59×10^{-6}

$5\pi/4$, and $7\pi/4$. Because the binary subchannels of this 4-ary system are symmetric, it is sufficient to determine the error probability in one of them.

Within the nth and $(n-1)$th chips let the signals with amplitude a and initial phases φ_{n-1} and φ_n be transmitted, with the phase difference being

$$\Delta\varphi = \varphi_n - \varphi_{n-1} = \pi/4 \tag{6.106}$$

Then, as follows from Table 4.1 and (4.15), an error will occur in the case when

$$Z = X_n X_{n-1} + Y_n Y_{n-1} < 0 \tag{6.107}$$

The values X_n, X_{n-1}, Y_n, Y_{n-1} are, as follows from (4.4), independent normal random values with the dispersions [see (6.20)]

$$D_1 = N_0 T/4 \tag{6.108}$$

and the means

$$m(X_{n-1}) = 0.5aT \cos \varphi_{n-1}$$
$$m(X_n) = 0.5aT \cos \varphi_n \tag{6.109}$$

$$m(Y_{n-1}) = 0.5aT \sin \varphi_{n-1}$$
$$m(Y_n) = 0.5aT \sin \varphi_n \tag{6.110}$$

For further calculations we present (6.107) in the squared form:

$$Z = X_1^2 - X_2^2 + Y_1^2 - Y_2^2 \tag{6.111}$$

where

$$X_1 = 0.5(X_n + X_{n-1})$$
$$X_2 = 0.5(X_n - X_{n-1})$$
$$Y_1 = 0.5(Y_n + Y_{n-1}) \tag{6.112}$$
$$Y_2 = 0.5(Y_n - Y_{n-1})$$

The random values X_1, X_2, Y_1 and Y_2 are independent normal random values. According to (6.109), (6.110), and (6.112) their dispersions are

$$D = N_0 T/8 \tag{6.113}$$

and the means are equal to

$$m(X_1) = 0.25aT(\cos \varphi_n + \cos \varphi_{n-1})$$
$$m(Y_1) = 0.25aT(\sin \varphi_n + \sin \varphi_{n-1}) \qquad (6.114)$$
$$m(X_2) = 0.25aT(\cos \varphi_n - \cos \varphi_{n-1})$$
$$m(Y_2) = 0.25aT(\sin \varphi_n - \sin \varphi_{n-1})$$

Using representation (6.111), we can write the initial inequality (6.107) as

$$\lambda_1 = X_1^2 + Y_1^2 < X_2^2 + Y_2^2 = \lambda_2 \qquad (6.115)$$

Because the left and right parts of this inequality contain squared sums of the independent random values with the nonzero means [see (6.114)], the probability of satisfying this inequality is equal to

$$p_{\Delta\varphi}(Z < 0) = \int_0^\infty W(\lambda_1)\, d\lambda_1 \int_{\lambda_1}^\infty W(\lambda_2)\, d\lambda_2 \qquad (6.116)$$

where $W(\lambda_1)$ and $W(\lambda_2)$ are the probability densities of the left and right parts of (6.115), with both densities corresponding to the noncentral χ^2 distribution with two degrees of freedom.

The index $\Delta\varphi$ in (6.116) emphasizes that this probability depends on the transmitted phase difference $\Delta\varphi = \varphi_n - \varphi_{n-1}$ because trigonometric functions of adjacent chips phases are included in (6.114).

The density calculations for the noncentral χ^2 distribution with $2q$ degrees of freedom were performed in Section 6.2 when calculating the noise immunity of autocorrelated reception [see (6.51)]. At $q = 1$ from (6.51) we can obtain the probability density required for the calculation according to (6.116):

$$W(\lambda) = \frac{1}{2D} \exp\left(-\frac{\lambda + \Delta}{2D}\right) I_0\left(\frac{\sqrt{\lambda\Delta}}{D}\right) \qquad (6.117)$$

where D is the dispersion (6.113), and Δ is the noncentrality parameter, equal to the sum of mathematical expectation squares of random values (6.112) and in this case taking two values according to (6.114):

$$\Delta_1 = m^2(X_1) + m^2(Y_1) = 0.25a^2T^2 \cos^2(\Delta\varphi/2)$$
$$\Delta_2 = m^2(X_2) + m^2(Y_2) = 0.25a^2T^2 \sin^2(\Delta\varphi/2) \qquad (6.118)$$

Thus, the required probability (6.116) is expressed as the integral

$$p_{\Delta\varphi}(Z<0) = \int\limits_{0}^{\infty} \frac{1}{2D} \exp\left(-\frac{\lambda_1 + \Delta_1}{2D}\right) I_0\left(\frac{\sqrt{\lambda_1\Delta_1}}{D}\right) d\lambda_1$$

$$\times \int\limits_{\lambda_1}^{\infty} \frac{1}{2D} \exp\left(-\frac{\lambda_2 + \Delta_2}{2D}\right) I_0\left(\frac{\sqrt{\lambda_2\Delta_2}}{D}\right) d\lambda_2 \qquad (6.119)$$

Taking into account the great importance of the required result for noise immunity theory, we can make the calculation of the integral (6.119), following [16,22]. Having entered the normalization of noncentrality parameters

$$\delta_1 = \Delta_1/D \qquad \delta_2 = \Delta_2/D$$

and obtained new variables by means of monotonic functional transformation

$$x = \sqrt{\lambda_1/D} \qquad y = \sqrt{\lambda_2/D}$$

we can present (6.119) in the following form:

$$p_{\Delta\varphi}(Z<0) = J = \int\limits_{0}^{\infty} x \exp\left(-\frac{\delta_1 + x^2}{2}\right) I_0\left(\sqrt{\delta_1}x\right) dx$$

$$\times \int\limits_{x}^{\infty} y \exp\left(-\frac{\delta_2 + y^2}{2}\right) I_0\left(\sqrt{\delta_2}y\right) dy \qquad (6.120)$$

The interior integral in (6.120) is the Marcum Q function of two variables $Q(\alpha,\beta)$, representing the probability that the point of two-dimensional space with independent coordinates, having normal distribution with the zero means and dispersions equal to 1, will appear outside the circle of radius β with the center, standing by α from the beginning of coordinates:

$$Q(\alpha,\beta) = \int\limits_{\beta}^{\infty} t \exp\left(-\frac{\alpha^2 + t^2}{2}\right) I_0(\alpha t) dt \qquad (6.121)$$

Using the property of the Q function [23]

$$Q(\alpha,\beta) + Q(\beta,\alpha) = 1 + \exp\left(-\frac{\alpha^2 + \beta^2}{2}\right) I_0(\alpha\beta)$$

we can present (6.120) as a sum of the following three integrals:

$$J = J_1 + J_2 + J_3 = \int_0^\infty x \exp\left(-\frac{\delta_1 + x^2}{2}\right) I_0\left(\sqrt{\delta_1}x\right) dx$$

$$+ \int_0^\infty x \exp\left(-\frac{\delta_1 + x^2}{2}\right) I_0\left(\sqrt{\delta_1}x\right) \exp\left(-\frac{\delta_2 + x^2}{2}\right) I_0\left(\sqrt{\delta_2}x\right) dx$$

$$- \int_0^\infty x \exp\left(-\frac{\delta_1 + x^2}{2}\right) I_0\left(\sqrt{\delta_1}x\right) dx \int_{\sqrt{\delta_2}}^\infty y \exp\left(-\frac{x_2 + y^2}{2}\right) I_0\left(xy\right) dy \qquad (6.122)$$

The first integral in (6.122) is equal to 1 as the area under a curve of probability density: $J_1 = 1$. The second integral in (6.122), with the use of the tabular integral [3]

$$\int_0^\infty x \exp(-px^2) I_\nu(bx) I_\nu(cx) \, dx = \frac{1}{2p} \exp\left(\frac{b^2 + c^2}{4p}\right) I_\nu\left(\frac{bc}{2p}\right)$$

is reduced to the form

$$J_2 = 0.5 \exp[0.25(-\delta_1 - \delta_2)] \, I_0(0.5\sqrt{\delta_1 \delta_2})$$

In the third integral, localizing the variables and changing the order of integration, we obtain

$$J_3 = -\int_{\sqrt{\delta_2}}^\infty y \exp\left(-\frac{y^2}{2} - \frac{\delta_1}{2}\right) dy \int_0^\infty x \exp(-x^2) I_0\left(\sqrt{\delta_1}x\right) I_0\left(yx\right) dx$$

$$= -\frac{1}{2} \int_{\sqrt{\delta_2}}^\infty y \exp\left(-\frac{y^2}{2} - \frac{\delta_1}{2}\right) \exp\left(\frac{\delta_1 + y^2}{4}\right) I_0\left(\frac{\sqrt{\delta_1}}{2}\right) dy$$

$$= -Q\left(\sqrt{\frac{\delta_1}{2}}, \sqrt{\frac{\delta_2}{2}}\right)$$

$$= Q\left(\sqrt{\frac{\delta_2}{2}}, \sqrt{\frac{\delta_1}{2}}\right) - 1 - \exp\left(-\frac{\delta_1 + \delta_2}{4}\right) I_0\left(\frac{1}{2}\sqrt{\delta_1 \delta_2}\right)$$

Substituting the received results into (6.122), we finally obtain

$$J = Q\left(\sqrt{\frac{\delta_2}{2}}, \sqrt{\frac{\delta_1}{2}}\right) - 0.5 \exp[0.25(-\delta_1 - \delta_2)] I_0 (0.5\sqrt{\delta_1 \delta_2}) \qquad (6.123)$$

Note that a similar formula was obtained earlier by another way in [23].

Using (6.123) as well as the dispersions (6.113) and noncentrality parameters (6.118), having been calculated before, we present the required probability (6.119) through the signal parameters, the SNR h^2 and the transmitted phase difference $\Delta\varphi$:

$$p_{\Delta\varphi}(Z < 0) = Q(\alpha,\beta) - 0.5e^{-h^2} I_0(h^2 \sin \Delta\varphi) \qquad (6.124)$$

where

$$\alpha = \sqrt{2}h \sin(\Delta\varphi/2)$$
$$\beta = \sqrt{2}h \cos(\Delta\varphi/2)$$

Remember that here I_0 is the modified Bessel function of the first kind and zero order, and $Q(\alpha,\beta)$ is the Marcum function [see (6.121)]. The tables of Q functions can be found in [1,4,23]; Q-function representations also take place in the form of series and asymptotic expressions [4].

We should emphasize that (6.124) is actually the conditional probability that the received phase difference cosine, proportional to Z [see (6.107)], is less than zero if the phase difference $\Delta\varphi$ has been transmitted.

In the four-phase system considered, as determined by condition (6.106), $\Delta\varphi = \pi/4$. Consequently, the required error probability in a binary subchannel of the optimum incoherent demodulator of four-phase PDM-1 signals is equal to

$$p_2^{(1)} = Q(\alpha,\beta) - 0.5e^{-h^2} I_0 (h^2/\sqrt{2}) \qquad (6.125)$$

where

$$\alpha = \sqrt{2}h \sin(\pi/8)$$
$$\beta = \sqrt{2}h \cos(\pi/8)$$

Probability (6.125) is submitted as curve 6 in Figure 6.8, and the numerical calculations under this formula are depicted in Table 6.7. The approximate relation for probability (6.125) is recommended in [23]; this relation is expressed by Laplace function (6.8) as follows:

$$p_2^{(1)} \approx 1.1 \; \mathsf{F} \, (0.765 h) \tag{6.126}$$

The calculations show that formula (6.126) gives a rather good estimation (see Table 6.7) of probability (6.125). From (6.126), in particular, it follows that $p_2^{(1)}$ is less than the error probability at the coherent reception of signals with amplitude modulation

$$p_{\mathrm{AM}} = \mathsf{F} \, (0.707 h)$$

and is greater than the error probability at the optimum incoherent reception of signals with frequency modulation

$$p_{\mathrm{FM}} = 0.5 \; \exp \, (-h^2/2)$$

Compare curves 5, 6, and 8 in Figure 6.8.

The analysis of the obtained formulas (6.125) and (6.126) and the diagrams in Figure 6.8 demonstrates that for 4-ary PDM, the loss of optimum incoherent reception to the coherent reception at the same chip is larger than in the case of binary PDM. It is quite sufficient to compare a pair of curves 1 and 2 (see Fig. 6.8), relating to binary PDM, with a pair of curves 3 and 6, relating to 4-ary PDM. At binary modulation this loss does not exceed 1 dB and with error probability decreasing it also quickly decreases. At the same time at 4-ary modulation, the loss of the incoherent reception to the coherent one changes a little and is about 2 dB.

For incoherent reception, the transition from 2-ary to 4-ary PDM at a constant bit rate, that is, at twice the increase in chip duration, is accompanied by some noise immunity decrease that does not take place for coherent reception. For example, the incoherent reception of two-phase signals requires the SNR $h_1^2 = 10.8$ to achieve the error probability 10^{-5}. At the four-phase signals and the same bit rate, the chip duration will be doubled, and the SNR will become equal to $h_2^2 = 21.6$. However, to achieve the former error probability 10^{-5} in this case, it is necessary to have $h_2^2 = 31.4$. The value

$$\Delta h = 10 \; \log \; 31.4/21.6 = 1.6 \; \mathrm{dB}$$

constitutes the energy losses of four-phase PDM when compared to two-phase PDM at the incoherent reception of signals with fixed rate.

Thus, for incoherent reception in contrast to coherent reception, the two times increase chip duration, when going from two-phase to four-phase PDM, does not completely compensate for the energy losses caused by increasing the phase variant number by two times. This defect can be overcome by going to PDM signal processing within more than two chips. Because when the processing interval increases, the noise immunity of optimum incoherent processing also increases (see Sections 4.5 and 6.4), the losses at transition from two-phase to four-phase PDM (at a constant bit rate) will decrease, tending to zero as in the case of perfect coherent processing.

The same effect, that is, the error probability approximation of incoherent processing to the error probability of coherent processing, will also take place at transition to PDM-2. As mentioned earlier, for two-phase PDM-2 the loss of optimum incoherent reception to perfect coherent reception is extremely small. At four-phase PDM-2 it is also less than at the four-phase PDM-1.

Now let us address autocorrelated reception of signals with four-phase PDM-1. When using the phase differences 0, $\pi/2$, π, and $3\pi/2$ and the Gray code, the appropriate algorithm is determined by expressions (5.18).

Taking into account the symmetry of binary subchannels in the four-phase demodulator, we can determine the required error probability as the error probability, for example, in the first subchannel. This probability, in its turn, is equal to the probability that the value in square brackets of the first relation of (5.18) is negative at transmission of the phase difference $\Delta\varphi = 0$:

$$I = \int_0^T x_n(t)\, x_{n-1}(t)\ dt + \int_0^T x_n(t)\, x_{n-1}^*(t)\ dt < 0 \tag{6.127}$$

where

$$\begin{aligned} x_n(t) &= S(t) + \xi_2(t) \\ x_{n-1}(t) &= S(t) + \xi_1(t) \end{aligned} \tag{6.128}$$

where $\xi_1(t)$ and $\xi_2(t)$ are the realizations of Gaussian noise within adjacent signal chips.

To calculate further we use the same method used in Section 6.2 for analyzing the noise immunity for autocorrelated reception of binary PDM signals. Let us present the signal and the interferences in (6.128) as orthogonal function decompositions:

$$\begin{aligned} S(t) &= \sum_{k=k_1}^{k_2} a_k \sin k\omega t + b_k \cos k\omega t \\ \xi_1(t) &= \sum_{k=k_1}^{k_2} \alpha_{1k} \sin k\omega t + \beta_{1k} \cos k\omega t \\ \xi_2(t) &= \sum_{k=k_1}^{k_2} \alpha_{2k} \sin k\omega t + \beta_{2k} \cos k\omega t \end{aligned} \tag{6.129}$$

where a_k, b_k, α_{1k}, α_{2k}, β_{1k}, and β_{2k} are the decomposition coefficients, calculated by (6.45). Then taking into account the Hilbert transformation of the signal and noise within the $(n-1)$th chip, inequality (6.127) looks like

$$I = \sum_{k=k_1}^{k_2} [a_k^2 + b_k^2 + a_k(\alpha_{1k} + \alpha_{2k}) + b_k(\beta_{1k} + \beta_{2k}) + \alpha_{1k}\alpha_{2k} + \beta_{1k}\beta_{2k}$$

$$+ a_k(\beta_{1k} - \beta_{2k}) + b_k(\alpha_{2k} - \alpha_{1k}) + \beta_{1k}\alpha_{2k} - \alpha_{1k}\beta_{2k}] < 0 \qquad (6.130)$$

Here $k_2 - k_1 + 1 = FT$, where F is a bandwidth of the input filter of the autocorrelated demodulator, and α and β are independent normal random values with zero mean and dispersion N_0/T [see (6.53)].

If we reduce the squared form in the left part of (6.130) to the canonical form, we can use the well-known probability distribution of a sum of squares of independent random values. Having made the appropriate transformations by replacing the variables, we can present inequality (6.130) as

$$\sum_{k=k_1}^{k_2} (U_{1k}^2 + V_{2k}^2) < \sum_{k=k_1}^{k_2} (U_{2k}^2 + V_{1k}^2) \qquad (6.131)$$

where U_{1k}, U_{2k}, V_{1k}, and V_{2k} are independent normal random values with the same dispersion as the random values in (6.130), and with the means:

$$m(U_{1k}) = [a_k(\sqrt{2} + 1) + b_k]/2$$

$$m(U_{2k}) = [a_k(\sqrt{2} - 1) - b_k]/2$$

$$m(V_{1k}) = [a_k + b_k(\sqrt{2} - 1)]/2$$

$$m(V_{2k}) = [-a_k + b_k(\sqrt{2} + 1)]/2$$

Thus, the sums in the left and right parts of (6.131) have *noncentral χ^2* distributions with $2q = 2FT$ degrees of freedom that are represented by formula (6.51). Hence, the probability of satisfying inequality (6.131), that is, the required error probability, is equal to[3]

---

3. Having compared (6.132) with (6.56), note that in this section we have come to another form of representation of the error probability of autocorrelation reception. This fact has been caused by the fact that both parts of inequality (6.131), in contrast to inequality (6.48), are noncentral χ^2 distributed. Consequently, further calculations differ from the method of Section 6.2, and the final formula (6.133) differs by its structure from (6.59).

$$p_2 = \int\limits_0^\infty \frac{1}{2D}\left(\frac{x_1}{\Delta_1}\right)^{\frac{q-1}{2}} \exp\left(-\frac{x_1 + \Delta_1}{2D}\right) I_{q-1}\left(\frac{\sqrt{x_1\Delta_1}}{D}\right) dx_1$$

$$\times \int\limits_{x_1}^\infty \frac{1}{2D}\left(\frac{x_2}{\Delta_2}\right)^{\frac{q-1}{2}} \exp\left(-\frac{x_2 + \Delta_2}{2D}\right) I_{q-1}\left(\frac{\sqrt{x_2\Delta_2}}{D}\right) dx_2 \qquad (6.132)$$

where D is the dispersion of random values U and V in (6.131), and the following are the noncentrality parameters:

$$\Delta_1 = \left(1 + \frac{1}{\sqrt{2}}\right) \sum_{k=k_1}^{k_2} (a_k^2 + b_k^2)$$

$$\Delta_2 = \left(1 - \frac{1}{\sqrt{2}}\right) \sum_{k=k_1}^{k_2} (a_k^2 + b_k^2)$$

The calculation of integral (6.132) is similar to the calculation of (6.119). As a result we obtain [16]:

$$p_2 = \left[Q\left(\sqrt{2}h \sin\frac{\pi}{8}, \sqrt{2}h \cos\frac{\pi}{8}\right) - \frac{1}{2}e^{-h^2} I_0\left(\frac{h^2}{\sqrt{2}}\right) \right]$$

$$+ \left\{ \frac{1}{2}e^{-h^2} \sum_{k=1}^{q-1} C_{k,q}[(\sqrt{2} + 1)^k - (\sqrt{2} - 1)^k] I_k\left(\frac{h^2}{\sqrt{2}}\right) \right\} \qquad (6.133)$$

where Q is the Marcum function [see (6.121)]; I_k is the modified Bessel function of the kth order; $C_{k,q} = 2B_{1/2}(k + q, q - k)/B(k + q, q - k)$; $B_{1/2}$ is the incomplete beta function; and B is the beta function [1].

The expression in square brackets of (6.133) is the error probability in a binary subchannel of the optimum incoherent demodulator of four-phase PDM signals [see (6.125)]. Hence, the expression in curly brackets of (6.133) shows how much the error probability is increased at the transition from the optimum incoherent reception to the autocorrelated one when $FT = q$. At $q = 1$ formula (6.133) converts into (6.125).

In Figure 6.8 formula (6.133) at $FT = 2$ is presented by curve 7, and at $F = 10$ by curve 9. At small FT the autocorrelated demodulator concedes a little to the optimum incoherent demodulator by the noise immunity.

The calculation of the noise immunity of autocorrelated reception of signals under formulas (6.59), (6.73), and (6.133) is depicted as diagrams in Figure 6.9.

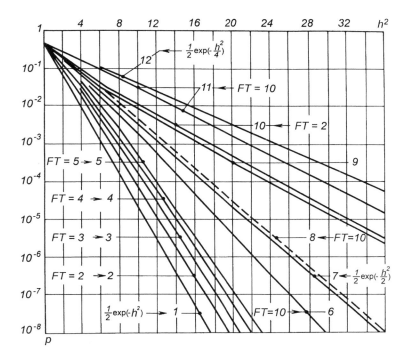

Figure 6.9 Comparative noise immunity of autocorrelated reception of two- and four-phase PDM signals: (1) two-phase PDM-1, optimum incoherent; (2–5) two-phase PDM-1, autocorrelated; (6) two-phase PDM-1, autocorrelated (exact formula); (7) FM, optimum incoherent; (8) two-phase PDM-1 (approximate formula); (9) four-phase PDM-1, optimum incoherent; (10,11) four-phase PDM-1, autocorrelated; (12) FM, autocorrelated ($FT = 2$).

This figure also shows the reference exponential curves 1, 7, and 12. Curve 1 corresponds to the noise immunity of the optimum incoherent reception of binary PDM-1 signals, curve 7 corresponds to the same for FM signals, and curve 12 corresponds to autocorrelated reception of the FM signal at $FT = 2$. Curve 9 corresponds to the optimum incoherent reception of four-phase PDM-1 and is constructed according to (6.125). The remainder of the curves in Figure 6.9 correspond to autocorrelated reception: Curves 2 through 6 and 8 correspond to two-phase PDM-1; 10 and 11 correspond to four-phase PDM-1 with system parameter FT equal to 2 and 10.

We should pay attention to the fact that energy losses of autocorrelated reception in comparison with optimum incoherent reception at four-phase PDM are insignificant and appreciably less than at two-phase PDM. This circumstance is an additional argument for the benefit of autocorrelated processing of multiposition signals with PDM, at least, at small system parameter FT.

6.6 MULTIPHASE PM AND PDM SYSTEMS

For multiphase PM and PDM systems there are no graceful and precise expressions for error probability as was the case for two- and four-phase PM and PDM.

Let us begin with consideration of coherent reception of m-ary PM signals. The corresponding general algorithm is represented as the inequalities (3.1), in which the variants of the signal S_i have the form (3.2). Let the received signal $x(t)$ be a sum of a desired signal and a Gaussian noise

$$x(t) = a \sin(\omega t + \varphi_i) + \xi(t) \tag{6.134}$$

Having substituted (6.134) into (3.1), we can write this inequalities set in the following form:

$$\int_0^T a^2 \sin^2(\omega t + \varphi_i) \, dt + a \cos \varphi_i \int_0^T \xi(t) \sin \omega t \, dt + a \sin \varphi_i \int_0^T \xi(t) \cos \omega t \, dt$$

$$> \int_0^T a^2 \sin(\omega t + \varphi_i) \sin(\omega t + \varphi_j) \, dt + a \cos \varphi_j \int_0^T \xi(t) \sin \omega t \, dt$$

$$+ a \sin \varphi_j \int_0^T \xi(t) \cos \omega t \, dt \tag{6.135}$$

The integrals included in the expression have already been considered in the previous sections. Designating

$$\theta_s = \int_0^T \xi(t) \sin \omega t \, dt \tag{6.136}$$

$$\theta_c = \int_0^T \xi(t) \cos \omega t \, dt$$

we obtain from (6.135)

$$0.5a^2 T + a \cos \varphi_i \, \theta_s + \sin \varphi_i \, \theta_c$$
$$> 0.5a^2 T \cos(\varphi_j - \varphi_i) + a \cos \varphi_j \, \theta_s + a \sin \varphi_j \, \theta_c \tag{6.137}$$

Having put the components that contain random values in the right part of the inequality, and the other ones in the left part, we obtain

$$0.5a^2 T \sin^2 0.5(\varphi_j - \varphi_i) > \theta \tag{6.138}$$

where $j = 1, 2, \ldots, m; \; j \neq i$:

$$\theta = (\cos \varphi_j - \cos \varphi_i)\theta_s + (\sin \varphi_j - \sin \varphi_i)\theta_c \tag{6.139}$$

The random value θ has a Gaussian distribution with zero average value and a dispersion

$$
\begin{aligned}
D(\theta) &= (\cos \varphi_j - \cos \varphi_i)^2 D(\theta_s) + (\sin \varphi_j - \sin \varphi_i)^2 D(\theta_c)\\
&= N_0 T \sin^2 0.5(\varphi_j - \varphi_i)
\end{aligned}
\tag{6.140}
$$

Therefore the probability of fulfilling the jth inequality of (6.138) is equal to

$$p_j = \frac{1}{\sqrt{2\pi D(\theta)}} \int_{-\infty}^{R} \exp\left[-\frac{x^2}{2D(\theta)}\right] dx \tag{6.141}$$

where $R = (aT/2) \sin^2 0.5(\varphi_j - \varphi_i)$, that is,

$$p_j = 1 - \mathsf{F}\left[\sqrt{2}\,h|\sin 0.5(\varphi_j - \varphi_i)|\right] \tag{6.142}$$

Let us return to the initial assumption that the signal S_i has been transmitted. It will be received without an error if inequality (6.138) is correct for all $j \neq i$. It takes place if two inequalities from (6.138), corresponding to the nearest to φ_i phases φ_j, are correct. Really, the sine argument in (6.138) does not exceed $\pi/2$, therefore the left part of the inequality is a monotonically increasing function of the phase difference $(\varphi_j - \varphi_i)$. Consequently, if this inequality is correct at the minimum phase difference equal to $\pm 2\pi/m$, it is all the more correct at larger differences.

Thus, the signal will be received without an error if (6.138) takes place at

$$\Delta\varphi_1 = \varphi_j - \varphi_i = 2\pi/m \qquad \text{and} \qquad \Delta\varphi_2 = \varphi_j - \varphi_i = -2\pi/m$$

These two events are compatible and dependent. That is why the probability of correct reception p can be written as

$$p = p_{2\pi/m}\,p_{-2\pi/m}(2\pi/m) \tag{6.143}$$

where $p_{2\pi/m}$ is the unconditional probability of fulfilling (6.138) at $\varphi_j - \varphi_i = 2\pi/m$, equal to

$$p_{2\pi/m} = 1 - F\left[\sqrt{2}\,h\,\sin(\pi/m)\right] \tag{6.144}$$

and $p_{-2\pi/m}(2\pi/m)$ is the conditional probability of fulfilling inequality (6.138) at $\varphi_j - \varphi_i = -2\pi/m$ and an additional condition that this inequality is fulfilled at $\varphi_j - \varphi_i = 2\pi/m$. The last probability is within the limits

$$p_{2\pi/m} \le p_{-2\pi/m}(2\pi/m) < 1 \tag{6.145}$$

The left part of this double inequality turns to the equality for four-phase systems when $2\pi/m = \pi/2$. In this case, the events, consisting of fulfilling (6.138) at $\varphi_j - \varphi_i = \pm\pi/2$ become independent, as was proved in Section 6.5. The 1 in the right part of (6.145) is the limit for modulation multiplicity increasing, when the dependence between the specified events is increasing. Multiplying (6.145) by $p_{2\pi/m}$ and taking into account (6.143) and (6.144), we have

$$\left[1 - F(\sqrt{2}\,h\,\sin(\pi/m))\right]^2 \le p < 1 - F\left[\sqrt{2}\,h\,\sin(\pi/m)\right] \tag{6.146}$$

At last, from (6.146) we obtain the following estimation of the error probability $p_N = 1 - p$ at the coherent reception of signals with m-phase (N-valued, $m = 2^N$) PM:

$$F\left(\sqrt{2}\,h\,\sin\frac{\pi}{2^N}\right) < p_N \le 2F\left(\sqrt{2}\,h\,\sin\frac{\pi}{2^N}\right)\left[1 - \frac{1}{2}F\left(\sqrt{2}\,h\,\sin\frac{\pi}{2^N}\right)\right] \tag{6.147}$$

From the right part of (6.147) we obtain the precise expression for noise immunity of the four-phase ($N = 2$) system with PM, which has been obtained before [see (6.103)].

Analyzing the result of (6.147), note that the obtained estimate of probability p_N is rather precise for practice; it is within the limits, which differ from each other by less than two times. This estimate determines the error probability of the m-position symbol. To obtain from it the estimate of the bit error probability we have to keep in mind that at small error probabilities and when using the Gray code, the probability of getting errors in several binary subchannels simultaneously is considerably less than the error probability in one subchannel. Therefore the equivalent bit error probability $p_{N(BER)}$ can be estimated by the approximate formula

$$p_{N(BER)} \approx (2/N)F\left[\sqrt{2}\,h\,\sin\frac{\pi}{2^N}\right] \tag{6.148}$$

from which, by the way, formula (6.105) follows for four-phase ($N = 2$) PM.

For N-valued PDM-1, taking into account the multiplicity coefficient, equal to two (see Section 6.3), we can write [similar to (6.148)] the following approximate formula for the equivalent bit error probability:

$$p_{N(\text{BER})}^{(1)} \approx (4/N) \, \mathsf{F}\left[\sqrt{2}\, h \, \sin\frac{\pi}{2^N}\right] \qquad (6.149)$$

The diagrams of the error probabilities versus SNR at coherent reception of signals with multiphase PM, constructed according to formula (6.148), are shown in Figure 6.10. These diagrams illustrate the comparative noise immunity of the coherent reception of multiposition PM signals with the same chip duration and different transmission rate, which in this case is proportional to modulation multiplicity N. It is obvious that when improving the bit rate by means of increasing a number of signal phase levels, the noise immunity rapidly decreases. For example, when doubling the bit rate by means of transition from 2-phase ($N = 1$) to 4-phase ($N = 2$) PM, we are required to increase the SNR by 3 dB to keep the same BER. However, when doubling the bit rate by transition from 4-phase ($N = 2$) PM to 16-phase ($N = 4$) PM, we are required to increase the SNR by more than 10 dB to keep the same BER.

In this connection we should emphasize that at a constant bit rate only the systems with $N = 1$ and 2 (two- and four-phase systems) are equivalent in terms of noise immunity; at $N > 2$, the increase in chip duration does not compensate noise immunity losses, which are caused by increasing a number of phase positions. To

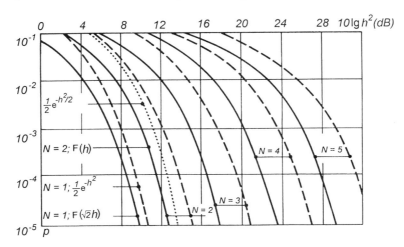

Figure 6.10 Comparative noise immunity of the systems with $m = 2^N$-phase (N-valued) PM and PDM. Solid line, coherent reception of PM signals; dashed line, optimum incoherent reception of PDM signals.

illustrate this aspect, we find the required energy per one bit of the transmitted information as a function of modulation multiplicity at coherent signal processing. Let us assume that the error probabilities in binary and m-ary systems are equal. According to (6.147) this means that

$$h_m \sin(\pi/m) = h_2 \qquad (6.150)$$

where h_2^2, h_m^2 is the signal energy to noise power density ratio necessary for achieving the given BER in binary and m-ary systems respectively.

Now the coefficient, showing how many times the signal power should be increased at transition from two variants of phase differences to m variants, that is, at $\log_2 m$ times increasing the bit rate, to keep the constant BER, is equal to

$$\Delta h(m) = h_m^2/h_2^2 = 1/\sin^2(\pi/m) \qquad (6.151)$$

The diagram of energy loss dependence $\Delta h(m)$ is shown in Figure 6.11.

By means of dependence (6.151) we can determine whether the transition to signals of higher multiplicity is preferable from the energy point of view of keeping a constant information transmission rate. To do this, let us make up the relation

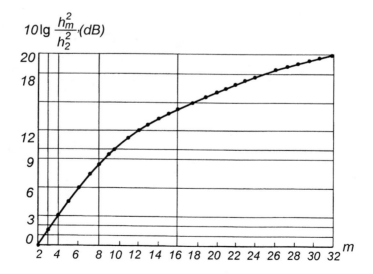

Figure 6.11 Energy loss versus a number of phases at constant BER (coherent processing).

$$\eta = \frac{\Delta h(m)}{\log_2 m} = \frac{1}{\log_2 m \sin^2(\pi/m)} \tag{6.152}$$

illustrating the relative signal energy expenditure per information bit.

From the diagram of relation (6.152), shown in Figure 6.12, it is obvious that the optimum system in terms of energy expenditure per a transmitted information bit is the three-phase system: $\Delta\varphi_1 = 0$, $\Delta\varphi_2 = 2\pi/3$, and $\Delta\varphi_3 = 4\pi/3$ ($\eta = 0.84$). By this criterion the four-phase PDM systems do not concede the binary PDM systems. When further increasing the number of phases, energy expenditure per a transmitted information bit rapidly increases.

In studying the noise immunity of coherent reception in this section, we first determined the error probability of an *m*-ary symbol, and then the equivalent BER was calculated. However, we can do it differently—we can first find the error probability of separate binary subchannels. This way is convenient if the structure of the keying code is known. In multilevel systems with PDM, as a rule, the keying code used is the Gray code, for which algorithms have been determined for binary subchannel demodulation as well as for the relations between subchannel noise immunity. For example, the noise immunity of the first two binary subchannels at any modulation multiplicity is identical; the error probability of the third subchannel is about twice that of the first or second one; the error probability of the fourth subchannel is about four times that of the first or the second one, and so on. Thus,

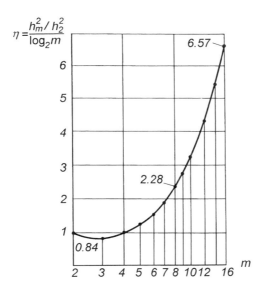

Figure 6.12 Relative energy expenditure per an information bit for coherent reception of *m*-ary PM signals.

when using the Gray code we can determine the error probability in every binary subchannel separately and estimate the average BER.

We explain the noise immunity calculation method, considered later, by using an example of a system with eight-level PDM that has the following phase differences: $\Delta\varphi_i = (2i - 1)\,\pi/8$, where $i = 1, 2, \ldots$ (Fig. 6.13).

The conformities between these phase differences and binary symbols in each of the three binary subchannels are given in Table 4.2. In particular, for the second binary subchannel these conformities are shown in Figure 6.13 by the signs "+" and "−."

The dashed vertical line in Figure 6.13 divides the areas of correct reception of binary symbols "+" and "−". As is obvious, the symbol transmitted by the second subchannel coincides with the cosine sign of the transmitted phase difference $\Delta\varphi_i$. That is why in the demodulator this symbol is determined by the cosine sign of the received phase difference $\Delta\varphi$ [see (4.22)]. Obviously, the error at the demodulation of the second subchannel will occur, when the cosine signs of the transmitted and received phase differences do not coincide. Hence, the error probability p_{32} in the second subchannel of the system with the 8-ary PDM is equal to

$$p_{32} = \sum_{i=1}^{8} p(\Delta\varphi_i)\, p_{\Delta\varphi_i}(\mathrm{sgn}\cos\Delta\varphi_\xi \neq \mathrm{sgn}\cos\Delta\varphi_i) \tag{6.153}$$

where $p(\Delta\varphi_i)$ is the probability of transmitting the phase difference $\Delta\varphi_i$, and $p_{\Delta\varphi_i}(\cdot)$ is the conditional probability that the cosine sign of the received phase difference does not coincide with the cosine sign of the transmitted phase difference, if the phase difference $\Delta\varphi_i$ has been transmitted. From Figure 6.13, we see that the specified conditional probabilities are identical for $\Delta\varphi_1, \Delta\varphi_4, \Delta\varphi_5, \Delta\varphi_8$ as

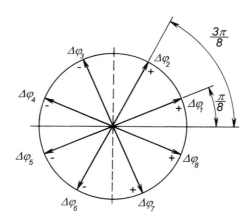

Figure 6.13 Areas of correct reception of binary symbols in the eight-phase PDM system.

well as for $\Delta\varphi_2$, $\Delta\varphi_3$, $\Delta\varphi_6$, $\Delta\varphi_7$ because the phase differences, combined in blocks of these four differences, have the same distance from the boundary of the correct reception areas. Thus, at equiprobable transmitted phase differences, we obtain from (6.153)

$$p_{32} = \frac{1}{8}\left[4p_{\Delta\varphi_1}(\text{sgn cos }\Delta\varphi_\xi \neq \text{sgn cos }\Delta\varphi_1) + 4p_{\Delta\varphi_2}(\text{sgn cos }\Delta\varphi_\xi \neq \text{sgn cos }\Delta\varphi_2)\right]$$

or finally

$$p_{32} = \frac{1}{2}\left[p_{\pi/8}(\cos \Delta\varphi_\xi < 0) + p_{3\pi/8}(\cos \Delta\varphi_\xi < 0)\right] \qquad (6.154)$$

For two other binary subchannels of the eight-phase PDM system, we have $p_{31} = p_{32}$, $p_{32} < p_{33} < 2p_{32}$, and the average bit error probability p_3 is within the limits of $p_{32} < p_3 < 1.33p_{32}$, that is, formula (6.154) is also a good estimation for the average (by subchannels) bit error probability in the 8-ary PDM system. Similarly we can present the error probability in the systems with higher multiplicity.

For example, in the 16-ary PDM system the error probability in the first two subchannels is equal to

$$p_{41} = p_{42} = \frac{1}{4}\left[p_{\pi/16}(\cos \Delta\varphi_\xi < 0) + p_{3\pi/16}(\cos \Delta\varphi_\xi < 0)\right.$$

$$\left. + p_{5\pi/16}(\cos \Delta\varphi_\xi < 0) + p_{7\pi/16}(\cos \Delta\varphi_\xi < 0)\right] \qquad (6.155)$$

When using the Gray code, we have $p_{43} \approx 2p_{42}$, and $p_{44} \approx 4p_{42}$. Therefore in this system the average error probability is $p_4 \approx 2p_{42}$.

In the 32-ary PDM system the error probability in the first two subchannels is equal to

$$p_{51} = p_{52} = \frac{1}{8}\sum_{i=1}^{8} p_{(2i-1)\pi/32}(\cos \Delta\varphi_\xi < 0) \qquad (6.156)$$

In the general case, for the $m = 2^N$-ary (N-valued) system the error probability in the first two subchannels is equal to

$$p_{N1} = p_{N2} = \frac{1}{2^{N-2}}\sum_{i=1}^{2^{N-2}} p_{(2i-1)\pi/2^N}(\cos \Delta\varphi_\xi < 0) \qquad (6.157)$$

Thus, the estimation of the error probability in binary subchannels of multiposition systems with PDM, as is obvious from (6.154) through (6.157), amounts to calculating the conditional probability $p_{\Delta\varphi}$ (cos $\Delta\varphi_\xi < 0$) of the event that the cosine of the received phase difference $\Delta\varphi_\xi$ is less than zero at transmitting this or that phase difference $\Delta\varphi$.

At the optimum incoherent reception, when cos $\Delta\varphi_\xi$ is proportional to $Z = X_n X_{n-1} + Y_n Y_{n-1}$, where X and Y are the projections of the received signal onto orthogonal reference signals, determined by (4.4), this probability is determined by integral (6.116) and according to the results of Section 6.5 is equal to (6.124).

Having substituted the appropriate values $\Delta\varphi$ into (6.124) we can get any of the required error probabilities (6.154) through (6.157). The corresponding formulas contain Q functions and Bessel functions and are rather bulky and inconvenient for analytical studies, though there are not any principal difficulties at numerical calculations. For example, taking into account (6.154) and (6.124), we can obtain for eight-phase PDM [23]:

$$p_3 \approx p_{31} = p_{32} = \frac{1}{2}\left[Q\left(\sqrt{2}\,h\,\sin\frac{\pi}{16},\ \sqrt{2}\,h\,\cos\frac{\pi}{16}\right)\right.$$

$$+ Q\left(\sqrt{2}\,h\,\sin\frac{3\pi}{16},\ \sqrt{2}\,h\,\cos\frac{3\pi}{16}\right)\right]$$

$$- \frac{1}{4}e^{-h^2}\left[I_0\left(h^2\,\sin\frac{\pi}{8}\right) + I_0\left(h^2\,\sin\frac{3\pi}{8}\right)\right] \qquad (6.158)$$

where, let us remind, Q function is determined as (6.121). Formula (6.158) can be presented through the Laplace function (6.8) by the following approximate expression [23]:

$$p_3 \approx 0.68\mathsf{F}\,(0.39h) \qquad (6.159)$$

In Figure 6.10 the curves of noise immunity of the optimum incoherent reception of $m = 2^N$-ary (N-valued) PDM-1 signals are shown by dashed lines. Naturally, incoherent reception concedes to coherent reception in terms of noise immunity in the channel with constant parameters. It is necessary to emphasize that incoherent reception energy losses per an information bit increase more rapidly at increasing modulation multiplicity N than at coherent reception. Therefore at $N \geq 2$ the energy losses of incoherent reception compared to coherent reception become considerable and increase with increasing N. This loss can be compensated for by optimum incoherent processing within more than two chips, as is done, for example, in the case of PDM-2 (see Sections 4.5 and 6.4).

In light of the fact that approximate formulas for the error probability in multiposition phase systems are rather bulky and do not always give satisfactory estimations, attempts have been undertaken to develop different methods for analyzing noise immunity. These attempts use numerical calculations. Most of these methods are based on representation of the probability density of a phase or a phase difference of an additive mixture of a harmonic signal and the Gaussian white noise. In coherent demodulators of PM signals we calculate the values, proportional to trigonometric functions of the received signal phase relative to the reference oscillation, and in the optimum incoherent demodulators of PDM signals we calculate the values proportional to trigonometric functions of the adjacent chip phase difference. Therefore, having the formulas for the above-mentioned probability densities, we can calculate the error probability for different modulation multiplicity [24,25].

To calculate the error probability at coherent reception of the PM signals we can use the well-known expression for phase (φ) probability density of the sum of a harmonic signal and a noise in the form of Fourier series [13]:

$$W_1(\varphi) = \frac{1}{2\pi} + \sum_{n=1}^{\infty} a_n \cos n\varphi \qquad (6.160)$$

where a_n are the coefficients, depending on SNR h^2, namely,

$$a_n = \frac{\Gamma(1 + n/2)}{\pi n!} e^{-h^2} h^n \Phi\left(1 + \frac{n}{2}, n + 1, h^2\right) \qquad (6.161)$$

Here Γ is the gamma function; Φ is the hypergeometric function [1].

Using (6.161), we realize that at coherent reception, the error probability in the ith binary subchannel of the N-valued ($m = 2^N$-ary) PM system is equal to (at $N \geq i \geq 2$) [25]:

$$p_{Ni} = \frac{1}{2} + 2\sum_{n=1}^{\infty} (-1)^n \frac{a_{(2n-1)}2^{i-2}}{2n-1} \prod_{m=2}^{N-i+2} \cos\frac{2n-1}{2^m}\pi \qquad (6.162)$$

The phase difference probability density is equal to

$$W_2(\Delta\varphi) = \frac{1}{2\pi} + \pi\sum_{n=1}^{\infty} a_n^2 \cos n\Delta\varphi \qquad (6.163)$$

where a_n are also determined by (6.161). On the basis of (6.163) we can calculate the error probability in the ith binary channel of the N-valued system with PDM at the optimum incoherent reception [25]:

$$p'_{N_i} = \frac{1}{2} + 2\pi\sum_{n=1}^{\infty} (-1)^n \frac{a_{(2n-1)}^2 2^{i-2}}{2n-1} \prod_{m=2}^{N-i+2} \cos\frac{2n-1}{2^m}\pi \qquad (6.164)$$

As is obvious from the comparison of (6.162) with (6.164), the formulas for noise immunity of coherent and incoherent reception in this case are completely identical in structure and differ only by the degree of phase probability density decomposition coefficients. It gives us the opportunity to make up a standard program for numerical calculations of phase system noise immunity in the channels with constant and variable parameters.

Figure 6.14 shows the average bit error probability in $m = 2^N$-ary ($N = 1, 2, 3, 4, 5$) PM and PDM systems for coherent and optimum incoherent receptions, respectively, for rather large error probabilities. These results are in addition to the results shown in Figure 6.10.

Now we consider the noise immunity of autocorrelated reception of multiposition signals with PDM-1. Let one of the m-phase differences $\Delta\varphi_i$ be transmitted within the nth chip, and the received phase difference be equal to $\Delta_n\varphi_\xi$. Then according to the common algorithm of demodulation (2.62) the signal will be received correctly if the following inequalities take place:

$$\cos \Delta\varphi_i \cos \Delta_n\varphi_\xi + \sin \Delta\varphi_i \sin \Delta_n\varphi_\xi > \cos \Delta\varphi_j \cos \Delta_n\varphi_\xi + \sin \Delta\varphi_j \sin \Delta_n\varphi_\xi \tag{6.165}$$

where $j = 1, 2, \ldots, m; j \neq i$.

For autocorrelated reception, the trigonometric functions of the received phase difference included in (6.125) are

$$\cos \Delta_n\varphi_\xi = A \int_0^T x_n(t)\, x_{n-1}(t)\ dt \tag{6.166}$$

$$\sin \Delta_n\varphi_\xi = A \int_0^T x_n(t)\, x_{n-1}^*(t)\ dt$$

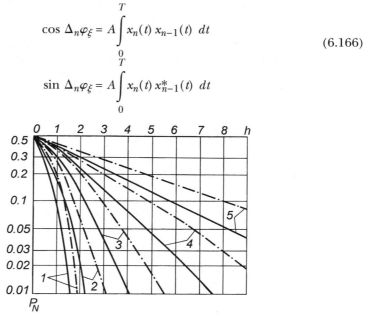

Figure 6.14 Comparative noise immunity of $m = 2^N$-ary ($N = 1, 2, 3, 4, 5$) systems with PM and PDM at coherent (solid lines) and optimum incoherent (dashed-dotted lines) processing.

Having substituted into (6.166) the expression for the received signals within the nth and $(n-1)$th chips

$$x_n(t) = a \sin(\omega t + \varphi_n) + \xi_n(t)$$

$$x_{n-1}(t) = a \sin(\omega t + \varphi_{n-1}) + \xi_{n-1}(t)$$

we obtain

$$\cos \Delta_n \varphi_\xi = A \left[\frac{a^2 T}{2} \cos(\varphi_n - \varphi_{n-1}) + \theta_1 + \theta_2 + \theta_3 \right]$$

$$\sin \Delta_n \varphi_\xi = A \left[\frac{a^2 T}{2} \sin(\varphi_n - \varphi_{n-1}) + \theta_4 + \theta_5 + \theta_6 \right]$$

(6.167)

where

$$\theta_1 = \int_0^T a \sin(\omega t + \varphi_n) \xi_{n-1}(t) \, dt$$

$$\theta_2 = \int_0^T a \sin(\omega t + \varphi_{n-1}) \xi_n(t) \, dt$$

$$\theta_3 = \int_0^T \xi_n(t) \xi_{n-1}(t) \, dt$$

(6.168)

$$\theta_4 = \int_0^T a \cos(\omega t + \varphi_n) \xi_{n-1}(t) \, dt$$

$$\theta_5 = \int_0^T a \cos(\omega t + \varphi_{n-1}) \xi_n(t) \, dt$$

$$\theta_6 = \int_0^T \xi_n(t) \xi_{n-1}^*(t) \, dt$$

The random values θ_1, θ_2, θ_4, and θ_5 are equivalent to the values (6.65a), and θ_3 and θ_6 to (6.65b). Therefore θ_1, θ_2, θ_4, and θ_5 are the independent normal random values with zero means and dispersions (6.66). At a great parameter FT, the values

θ_3 and θ_6 can also be considered to be normal random values with zero means and dispersions (6.69). Substituting (6.167) into (6.165), we obtain

$$\frac{a^2 T}{2} + \cos \Delta \varphi_i (\theta_1 + \theta_2 + \theta_3) + \sin \Delta \varphi_i (\theta_4 + \theta_5 + \theta_6)$$

$$> \frac{a^2 T}{2} \cos(\Delta \varphi_j - \Delta \varphi_i) + \cos \Delta \varphi_j (\theta_1 + \theta_2 + \theta_3) + \sin \Delta \varphi_j (\theta_4 + \theta_5 + \theta_6)$$

(6.169)

$j = 1, 2, \ldots, m; j \neq i$. We can present inequalities (6.169) in the form

$$a^2 T \sin^2 0.5 (\Delta \varphi_j - \Delta \varphi_i) > \theta$$

(6.170)

where

$$\theta = (\cos \Delta \varphi_j - \cos \Delta \varphi_i)(\theta_1 + \theta_2 + \theta_3)$$
$$+ (\sin \Delta \varphi_j - \sin \Delta \varphi_i)(\theta_4 + \theta_5 + \theta_6)$$

(6.171)

We consider the random value θ as being distributed by the Gaussian law. Then the probability of fulfilling the jth inequality of (6.170) is

$$p_j = 1 - \mathsf{F} \left[\frac{a^2 T \sin^2 0.5 (\Delta \varphi_j - \Delta \varphi_i)}{\sqrt{D(\theta)}} \right]$$

(6.172)

where $D(\theta)$ is the dispersion of the random value (6.171), which is easy to calculate, using (6.66) and (6.69):

$$D(\theta) = (2a^2 T N_0 + 2FT N_0^2) \sin^2 0.5 (\Delta \varphi_j - \Delta \varphi_i)$$

(6.173)

Having substituted (6.173) into (6.172) and entering $h^2 = a^2 T / 2N_0$, we obtain

$$p_i = 1 - \mathsf{F} \left(\frac{h \sin 0.5 |\Delta \varphi_j - \Delta \varphi_i|}{\sqrt{1 + FT / 2h^2}} \right)$$

(6.174)

We note that inequalities (6.170) and equality (6.174) are similar in structure to (6.138) and (6.142), respectively. The difference is that the absolute phases φ_i and φ_j have been replaced by the phase differences $\Delta\varphi_i$ and $\Delta\varphi_j$, and in the argument of the Laplace function there has appeared the specific for autocorrelated processing component, which depends on the parameter FT. Therefore, the probability of fulfilling inequalities (6.170) can be found in the same way, as the probability of fulfilling (6.138). Repeating this calculation with reference to (6.170) and replacing the phases φ_i and φ_j by the phase difference $\Delta\varphi_i$ and $\Delta\varphi_j$, we obtain that the error probability at autocorrelated reception of signals with $m = 2^N$-ary PDM is within the limits

$$\mathsf{F}(\gamma) < p_N < 2\mathsf{F}(\gamma)[1 - 0.5\mathsf{F}(\gamma)] \tag{6.175}$$

where

$$\gamma = \frac{h\,\sin(\pi/2^N)}{\sqrt{1 + FT/2h^2}}$$

For practical calculations we can use the simple estimation

$$p_N \approx 2\mathsf{F}\left(\frac{h\,\sin(\pi/2^N)}{\sqrt{1 + FT/2h^2}}\right) \tag{6.176}$$

Remember that (6.175) and (6.176) are correct only at large FT, when the assumptions of the Gaussian distribution and independence of the random values (6.168) are correct. These formulas are quite proximate for the practice at $FT \geq 10$. On fulfilling the specified condition the bit error probability can be calculated by the approximate formula

$$p_{N(\text{BER})} \approx \frac{2}{N}\mathsf{F}\left(\frac{h\,\sin(\pi/2^N)}{\sqrt{1 + FT/2h^2}}\right) \tag{6.177}$$

In concluding this section, we note that the theory of noise immunity of multiposition PM signal reception is rapidly developing at the present time. Analytical and numerical methods have been developed for calculating the error probability at coherent reception of different combinations of two-dimensional signals with QAM.

6.7 CARRIER-FREQUENCY-INVARIANT DEMODULATORS OF SECOND-ORDER PDM SIGNALS

In the noise immunity analysis of autocorrelated demodulators of signals with PDM-2, we expect to encounter difficulties connected with several nonlinear conversions in the appropriate processing algorithms. Therefore, as a rule, theoretical studies of this problem are restricted to obtaining error probability estimates.

A rather good lower estimate for the error probability of an absolute invariant autocorrelated demodulator of signals with PDM-2 is the error probability formula for an incoherent demodulator of these signals, shown in Figure 4.18(a). Remember that this demodulator is intended for operating in a channel with known carrier frequency, and its algorithm is based on calculating trigonometric functions of the second phase difference by the projections of the received signal onto orthogonal reference signals. The noise immunity of this suboptimum demodulator can be calculated according to the formula obtained by Sudakov for the probability of more than $\pi/2$ deviation of the second phase difference of a harmonic signal in the AWGN channel:

$$
p = \frac{1}{2} - \frac{\sqrt{\pi}\, h^3}{4} e^{-\frac{3}{2}h^2} \sum_{i=0}^{\infty} \left[I_i\left(\frac{h^2}{2}\right) + I_{i+1}\left(\frac{h^2}{2}\right) \right]^2
$$
$$
\times \left[I_{2i+0.5}\left(\frac{h^2}{2}\right) + I_{2i+1.5}\left(\frac{h^2}{2}\right) \right]\left[\frac{(-1)^i}{2i+1} \right] \tag{6.178}
$$

When the system parameter $2FT$ decreases, algorithm (5.36) for the autocorrelated reception of PDM-2 signals degenerates into the incoherent reception algorithm (4.48) of incoherent reception. Therefore, formula (6.178) can be a lower bound for noise immunity of the absolutely invariant demodulator for PDM-2 signals at any $2FT$. In addition, for the same reason, this estimate will be sufficiently precise for the practical applications at small $2FT$. Expression (6.178) is depicted as curve 8 in Figure 6.15. Obviously, the error probability of (6.178) is a little bit greater than the error probability for optimum incoherent reception of two orthogonal signals, equal to $0.5 \exp(-h^2/2)$.

Let us find an approximate expression for the noise immunity of the PDM-2 autocorrelated demodulator invariant to the signal frequency (see Fig. 5.16) at arbitrary system parameter $2FT$ and compare it with the lower bound (6.178). In this case the original algorithm is relation (5.36), according to which the following value is calculated in the demodulator:

$$
I = X_n X_{n-1} + Y_n Y_{n-1} \tag{6.179}
$$

where

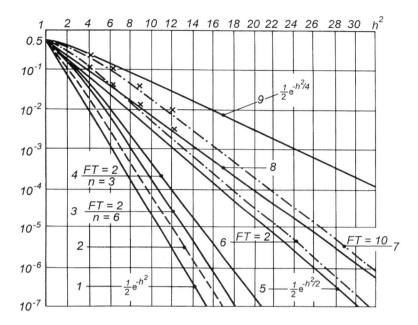

Figure 6.15 Comparative noise immunity of two-phase PDM-2 autocorrelated demodulators, invariant to a carrier frequency: (1) PDM-1, optimum incoherent; (2) PDM-1, autocorrelated; (3,4) PDM-2, relatively invariant, n-phase-autocorrelated; (5) FM, optimum incoherent; (6,7) PDM-2, absolutely invariant, autocorrelated (approximate formula); (8) PDM-2, absolutely invariant, autocorrelated (lower bound); (9) FM, absolutely invariant ($FT = 2$).

$$X_n = \int_0^T x_n(t)\, x_{n-1}(t)\ dt$$

$$X_{n-1} = \int_0^T x_{n-1}(t)\, x_{n-2}(t)\ dt$$

$$Y_n = \int_0^T x_n(t)\, x_{n-1}^*(t)\ dt$$ (6.180)

$$Y_{n-1} = \int_0^T x_{n-1}(t)\, x_{n-2}^*(t)\ dt$$

Random values similar to those of (6.180) have already been considered in Section 6.2 where we studied the noise immunity of an autocorrelated demodulator for PDM-1 signals [see (6.42) through (6.44), (6.64), and (6.65)]. If the signals within three adjacent chips are identical, that is, the second phase difference $\Delta^2 \varphi = 0$ has been transmitted, and the system parameter $2FT$ is rather great, the random values (6.180) can be approximately considered as independent normal random values with means, equal to signal energy, and with dispersions according to (6.70) equal to $D = EN_0 + N_0^2 FT/2$. For further calculations we represent (6.179), as before, as a squares sum:

$$I = 0.25\left[(X_n + X_{n-1})^2 + (Y_n + Y_{n-1})^2 - (X_n - X_{n-1})^2 - (Y_n - Y_{n-1})^2 \right] \qquad (6.181)$$

The random value, similar to that in square brackets, was considered earlier when we studied the noise immunity of an optimum incoherent demodulator [see (6.30)]. Repeating the same conversions as in the derivation of (6.40), we find that the required error probability is equal to

$$p_1^{(2)} \approx 0.5 \exp(-B^2/4D) \qquad (6.182)$$

where

$$B = \sqrt{m^2(X_n + X_{n-1}) + m^2(Y_n + Y_{n-1})} = 2E$$
$$D = 2EN_0 + N_0^2 FT \qquad (6.183)$$

Having substituted (6.183) into (6.182), we obtain:

$$p_1^{(2)} \approx 0.5 \exp\left(- \frac{h^2}{2 + FT/h^2} \right) \qquad (6.184)$$

The plots of (6.184) at $FT = 2$ and $FT = 10$ are shown in Figure 6.15 as curves 6 and 7. At $FT = 2$ formula (6.184) gives underestimated values of error probabilities, because curve 6 corresponding to it is mainly below the lower bound of (6.178) (curve 8). However the difference between them is not significant. Simulation of autocorrelated reception of binary PDM-2 signals has illustrated [26] that (6.184) gives a good approximation of the error probability at parameters $FT = 2$ to $FT = 10$. Figure 6.15 designates with crosses the points obtained by statistical modeling for the corresponding demodulator at $FT = 2$ and 10. As is obvious, (6.184) gives a good estimates at least in the range of error probabilities from 10^{-1} to 10^{-3}.

What should we pay for the property of absolute invariance to a carrier frequency, achievable at autocorrelated reception of signals with PDM-2? To answer this question, we have to compare the noise immunity of this reception method with the noise immunity of autocorrelated reception of signals with PDM-1, that is, formulas (6.184) and (6.59), at the same parameter $2FT$ and absence of frequency variations.

For the most typical value, $FT = 2$, this comparison can be easily made by comparing expression (6.178) with (6.60), which are presented in curves 2 and 8 in Figure 6.15. As is obvious from comparison of these curves, PDM-2 loses to PDM-1 about 3 dB by energy at the known signal frequency. This loss, however, rapidly decreases and turns into a gain at carrier frequency (or delay duration) deviation from the nominal value. For example, at the carrier frequency deviation by a quarter of a bit rate ($\Delta f = 1/4T$) the autocorrelated demodulator of PDM-1 becomes inoperable, and at a frequency deviation of about $\Delta f = 1/6T$ its noise immunity decreases by 3 dB. At the same time the autocorrelated demodulator of signals with PDM-2 is absolute insensitive to similar frequency deviations (or equivalent changes of the delay duration). Thus, at carrier frequency deviations of more than $1/6T$, where T is the signal chip duration, the autocorrelated PDM-2 demodulator exceeds the autocorrelated PDM-1 demodulator by the noise immunity.

For a detailed analysis of comparative noise immunity of autocorrelated demodulators of PDM-2 and PDM-1 signals we should know the error probability of the PDM-1 demodulator as a function of the carrier frequency variation. This function appears to be rather bulky, for example, as illustrated by the expression obtained in [27] for the case of binary PDM-1. At the same time the approximate relations, sufficient for practice, can be obtained if we use the known formulas for error probability without frequency variation, having substituted instead of the usual SNR h^2 the equivalent one: $h^2_{\text{equiv}} = h^2 \cos^2 \Delta \omega T$ where $\Delta \omega$ is frequency variation and T is the elementary signal duration. For example, at autocorrelated reception of binary PDM-1 signals, the energy loss of this estimate does not exceed 1 dB at $FT \leq 10$, $h^2 \geq 4$, and $\Delta \omega T \leq \pi/4$.

The PDM-2 gain is also rather significant in comparison with FM. At FM the insensitivity to a carrier frequency is achieved by means of the so-called "narrowband reception by an envelope" [4,5], which is a modification of the autocorrelated reception. The error probability at this reception method is equal to

$$p_{\text{FM}} \approx 0.5 \exp(-h^2/4) \tag{6.185}$$

The dependence corresponding to (6.185) is shown as curve 9 in Figure 6.15. Comparing it with curve 8 we can conclude that the energy gain of PDM-2 in comparison with FM at the uncertain carrier frequency is about 2.3 dB. This is an important advantage of PDM-2.

The greater effect can be achieved by using relatively invariant autocorrelated demodulators of PDM-2 signals. The error probability of these demodulators depends on a carrier frequency, but in average it appears to be less than the error probability of absolutely invariant demodulators.

Let us find the average error probability $p_1^{(2)}$ of a multiphase autocorrelated demodulator of PDM-2 signals, as considered in Section 5.4 (see Fig. 5.23). In this demodulator, the choice of a signal at the output of one of n autocorrelators is automatic. After the choice is made regarding the autocorrelator with greatest output signal and its demodulation, the additional errors can appear only as a result of multiplying the initial errors at the autocorrelator output. Therefore the required probability is equal to

$$p_1^{(2)} = 2p_{aut}(1 - p_{aut}) \qquad (6.186)$$

where

$$p_{aut} = \int_0^{2\pi} p_1^{(1)}(\varphi)\, W(\varphi)\, d\varphi \qquad (6.187)$$

and $p_1^{(1)}(\varphi)$ is the error probability of autocorrelation reception of binary PDM-1 signals as a function of spurious deviation of the phase difference φ between the direct and the delayed signals at the autocorrelator input; $W(\varphi)$ is probability density of the phase difference φ deviation.

According to (6.186) and (6.187) we can calculate error probabilities at any number of autocorrelators and any system parameter $2FT$. The probability $p_1^{(1)}(\varphi)$ depends on $2FT$, and function $W(\varphi)$ in (6.187) depends on a number of autocorrelators.

As follows from Figure 5.24, the signal level at the autocorrelator output changes under the cosine law, when the frequency is changing, therefore the equivalent SNR for the autocorrelator is related to the SNR at the demodulator input as follows:

$$h_{aut}^2 = h^2 \cos^2 \varphi \qquad (6.188)$$

Thus, to find the function $p_1^{(1)}(\varphi)$ we should substitute the value (6.188) instead of h in one of the formulas (6.60) through (6.63) depending on the system parameter $2FT$. Hence, in the case $FT = 2$ we have

$$p_1^{(1)}(\varphi) = 0.5(1 + 0.25h^2 \cos^2 \varphi) \exp(-h^2 \cos^2 \varphi) \qquad (6.189)$$

Let us further assume that a spurious deviation of the phase difference, equal to $\varphi = \Delta\omega T$, where $\Delta\omega$ is the deviation of a carrier frequency from nominal, is uniformly distributed within the interval $(0, 2\pi)$:

$$W(\varphi) = 1/2\pi \tag{6.190}$$

When using n autocorrelators, the interval of changing the spurious phase deviation $(0, 2\pi)$ consists of $2n$ identical intervals of the length π/n (see Fig. 5.24), inside of which the SNR changes under law (6.188). Taking into account this fact, and having substituted (6.189) and (6.190) into (6.187), we obtain the following expression for the average error probability of the n-channel demodulator of the binary PDM-2 signals at $FT = 2$:

$$p_{\text{aut}} = \frac{n}{2\pi} \int\limits_{-\pi/2n}^{\pi/2n} \exp(-h^2 \cos^2 \varphi) \left(1 + \frac{h^2 \cos^2 \varphi}{4} \right) d\varphi \tag{6.191}$$

Having substituted (6.191) into (6.186), we can obtain the required general expression for the error probability. Curves 3 and 4 of Figure 6.15 show the results of computer calculations according to (6.191) and (6.186) for cases in which three and six autocorrelators ($n = 3, 6$) are used [28]. The comparison of these curves with curves 1 and 2 indicates that when the number of processing channels is more than 3, the considered relatively invariant demodulator of PDM-2 signals differs little in its noise immunity from the optimum incoherent and autocorrelated demodulators of PDM-1 signals.

Thus, the absolute invariant to a carrier frequency demodulator of PDM-2 signals has almost the same noise immunity as the optimum incoherent demodulator of FM signals (at an exactly known carrier frequency), and the relatively invariant demodulator of PDM-2 signals approximates its noise immunity to the optimum incoherent demodulator of PDM-1 signals.

The indicated unusual combination of high noise immunity and of insensitivity to carrier frequency deviations predetermines the significant practical interest in the autocorrelated processing of signals with PDM-2.

6.8 QUASICOHERENT RECEPTION OF PHASE AND PHASE-DIFFERENCE MODULATION SIGNALS

So far in this chapter, noise immunity of coherent signal processing has been considered assuming that reference oscillations in the coherent demodulator are absolutely precise. Actually, they are formed of the received signal or of a special synchronization signal ("training signal") by appropriate nonlinear filtering (see Sections 3.2, 3.3, 3.6) at interference influence and, therefore, are not precise samples.

In this section the noise immunity performances of real quasicoherent demodulators are studied. The purpose of this study is to show how the dynamics of error probability change depending on the training signal duration (averaging interval) and to find the limit to which the error probability tends as a result of unlimited increases in this interval.

Here as an object for study we have chosen the algorithms considered in Chapter 3 for quasicoherent reception of multiposition PM signals, when coherent processing is realized on the basis of quadrature projections of the received signal onto reference oscillations with an arbitrary and unknown initial phase. Let us consider an arbitrary m-ary system of equiprobable and equipower phase-modulated signals within the time interval $(0, T)$:

$$S_j(t) = a \sin(\omega t + \varphi_j) \qquad j \in [1, m] \qquad (6.192)$$

The algorithm of the ideal coherent reception can be presented in the following form:

$$i = \text{ind max } V_{jn} \atop j \qquad (6.193)$$

$$V_{jn} = \frac{2}{T} \int\limits_{nT}^{(n+1)T} x(t) S_j(t) \, dt$$

where $x(t)$ is the sum of signal (6.192) and a noise. We present the value V_{jn} by the projections X_0 and Y_0 of the received signal $x(t)$ onto the orthogonal reference oscillations with a random phase φ_0 and with a frequency ω, coinciding with the frequency of useful signals (6.192):

$$X_{0n} = \frac{2}{T} \int\limits_{nT}^{(n+1)T} x(t) \sin(\omega t + \varphi_0) \, dt$$

$$Y_{0n} = \frac{2}{T} \int\limits_{nT}^{(n+1)T} x(t) \cos(\omega t + \varphi_0) \, dt \qquad (6.194)$$

After elementary conversions

$$V_{jn} = X_{0n}[X_1 \cos(\varphi_j - \varphi_1) - Y_1 \sin(\varphi_j - \varphi_1)]$$
$$+ Y_{0n}[Y_1 \cos(\varphi_j - \varphi_1) + X_1 \sin(\varphi_j - \varphi_1)] \qquad (6.195)$$

where

$$X_1 = \frac{2}{T} \int\limits_0^T S_1(t) \, \sin(\omega t + \varphi_0) \, dt$$

$$Y_1 = \frac{2}{T} \int\limits_0^T S_1(t) \, \cos(\omega t + \varphi_0) \, dt$$

(6.196)

The first signal projections (6.196) are replaced by their estimates $\tilde{X}_1$ and $\tilde{Y}_1$ which depend on system algorithms as a whole. If the information transmission is preceded by the training signal (synchronization signal), consisting of the chips, for example, of the first signal variant, that takes place in the systems with pulse transmission, then we have

$$\tilde{X}_1 = \frac{1}{N} \sum_{k=1}^N X_{0k}$$

$$\tilde{Y}_1 = \frac{1}{N} \sum_{k=1}^N Y_{0k}$$

(6.197)

If there is no training sample in the form of a special synchronization signal that takes place in the systems with the continuous transmission mode, the estimation is made up of N information chips, preceding the given chip and taking into account the demodulator decisions, that is,

$$\tilde{X}_1 = \frac{1}{N} \sum_{k=n-N}^{n-1} (X_{0k} \cos \Delta\varphi_k + Y_{0k} \sin \Delta\varphi_k)$$

$$\tilde{Y}_1 = \frac{1}{N} \sum_{k=n-N}^{n-1} (Y_{0k} \cos \Delta\varphi_k - X_{0k} \sin \Delta\varphi_k)$$

(6.198)

where $\Delta\varphi_k$ is the difference between the signal phase, in favor of which a decision has been made in the demodulator within the kth chip, and the phase of the first signal variant. Note that the estimates (6.197) are maximum probable, unbiased and effective, and the estimates (6.198) have the same qualities at rather small error probabilities.

Thus, relations (6.197) and (6.198) in a combination with (6.193) and (6.195) are the common algorithms for quasicoherent reception of arbitrary PM signals in the presence or absence of special training signals, respectively. The special algorithms for quasicoherent reception of PM signals for the most important, in a practical sense, two-, four- and eight-phase systems have been considered in Section 3.2.

Let us find the noise immunity estimates of the considered algorithms at different length N of the averaging interval and show that at $N \to \infty$ the corresponding error probability is equal to the error probability of the ideal coherent reception, if the initial phase is kept constant [29]. As an object for the analysis we take the 2^M-phase system with the phases

$$\varphi_i = (i - 1)\,\pi/2^{M-1}$$

where $i = 1, \ldots, 2^M$, in which the information transmission is preceded by a synchronization signal (training sample) of N-chip length.

When the Gray code is used, the error probabilities of the first two binary subchannels of such a system are identical and equal to

$$p_1 = p_2 = 1/2^{M-1} \sum_{i=1}^{2^{M-1}} p_{1,i} \tag{6.199}$$

where $p_{1,i}$ is the conditional error probability in the first binary subchannel of the 2^M-phase system at transmission of the signal with the information phase $\varphi_{1,i}$.

Using the designations entered above, it is easy to demonstrate that this probability is equal to the probability of the following inequality:

$$\cos(\pi/2^M)(Y_{0n}\tilde{X}_1 - X_{0n}\tilde{Y}_1) < -\sin(\pi/2^M)(X_{0n}\tilde{X}_1 + Y_{0n}\tilde{Y}_1) \tag{6.200}$$

at the condition that a signal with phase φ_i has been transmitted. The random values X_{0n} and Y_{0n} are the Gaussian random values with the means

$$\begin{aligned} m(X_{0n}) &= a\,\cos[\varphi_0 + [(i-1)\,\pi/2^{M-1}]] \\ m(Y_{0n}) &= a\,\sin[\varphi_0 + [(i-1)\,\pi/2^{M-1}]] \end{aligned} \tag{6.201}$$

and $\tilde{X}_1$ and $\tilde{Y}_1$ are the Gaussian random values with the means

$$\begin{aligned} m(\tilde{X}_1) &= a\,\cos\,\varphi_0 \\ m(\tilde{Y}_1) &= a\,\sin\,\varphi_0 \end{aligned} \tag{6.202}$$

The quadrature forms in (6.200) can be reduced to the canonical form

$$\Theta_1^2 + \Theta_2^2 < \Theta_3^2 + \Theta_4^2 \tag{6.203}$$

by means of the corresponding linear conversions [29].

As a result the left and right parts of inequality (6.203) have noncentral χ^2 distribution with two degrees of freedom and with noncentrality parameters

$$\lambda_1 = m^2(\Theta_1) + m^2(\Theta_2) = h^2\left[N + 1 + 2\sqrt{N}\sin((2i - 1)\pi/2^M)\right]$$

$$\lambda_2 = m^2(\Theta_3) + m^2(\Theta_4) = h^2\left[N + 1 - 2\sqrt{N}\sin((2i - 1)\pi/2^M)\right]$$

where h^2 is the ratio of signal element energy to the spectral density of noise power. Thus, the probability that the inequality (6.203) takes place is equal to

$$p_{1,i} = Q(\sqrt{\lambda_2/2}, \sqrt{\lambda_1/2}) - 0.5\exp(-(\lambda_1 + \lambda_2)/4)I_0(0.5\sqrt{\lambda_1\lambda_2}) \qquad (6.204)$$

where $Q(x,y)$ is the Marcum function, and $I_0(x)$ is the modified Bessel function of the zero order [22].

Having substituted (6.204) into (6.199), we obtain the required error probability in the first two subchannels of the considered 2^M-ary PM system:

$$p_1 = p_2 = 1/2^{M-1}\sum_{i=1}^{2^{M-1}}[Q(z_{1i}, z_{2i}) - 0.5\exp(-(z_{1i}^2 + z_{2i}^2)/2)I_0(z_{1i}z_{2i})]$$

$$(6.205)$$

where

$$z_{1i} = h\sqrt{(N + 1)/2 - \sqrt{N}\sin((2i - 1)\pi/2^M)}$$

$$z_{2i} = h\sqrt{(N + 1)/2 + \sqrt{N}\sin((2i - 1)\pi/2^M)}$$

From (6.205) for the cases of two- and four-phase PM we obtain (at $M = 1$ and $M = 2$, respectively):

$$p_1(M = 1) = Q\left(\frac{(\sqrt{N} - 1)h}{\sqrt{2}}; \frac{(\sqrt{N} + 1)h}{\sqrt{2}}\right)$$

$$- \frac{1}{2}\exp\left[-\frac{(N + 1)h^2}{2}\right]I_0\left[\frac{(N - 1)h^2}{2}\right] \qquad (6.206)$$

$$p_1(M = 2) = Q\left(\frac{h}{\sqrt{2}}\sqrt{N + 1 - \sqrt{2N}}; \frac{h}{\sqrt{2}}\sqrt{N + 1 + \sqrt{2N}}\right)$$

$$- \frac{1}{2}\exp\left[-\frac{(N + 1)h^2}{2}\right]I_0\left[\frac{\left(\sqrt{N^2 + 1}\right)h^2}{2}\right] \qquad (6.207)$$

Expressions (6.205), (6.206), and (6.207) give the error probability as a function of the averaging interval N.

At $N = 1$ the noise immunity of quasicoherent algorithms of PM signal reception coincides with the noise immunity of the optimum incoherent reception of PDM signals. It is obvious that at $M \geq 2$ this statement follows from (6.205). In the case $M = 1$, we should take into account that $Q(0,x) = \exp(-x^2/2)$. We can also prove that at $N \rightarrow \infty$ the value (6.205) coincides with the error probability in the first and second binary subchannels of the ideal coherent receiver of 2^M-ary PM signals [29].

For the cases of two- and four-phase PM the calculation results for formulas (6.206), (6.207) are depicted in Tables 6.8 and 6.9, respectively. In the first column of both tables, error probabilities are shown; in the last ($N = \infty$), the SNR h^2 is shown, which is necessary to provide for the appropriate error probability at the coherent ideal reception. The other columns give the values for h^2 necessary to

Table 6.8
Energy Parameters of Quasicoherent Reception for Two-Phase PM Signals

p_1	$N = 1$ h^2	Δh^2 (dB)	$N = 2$ h^2	Δh^2 (dB)	$N = 3$ h^2	Δh^2 (dB)	$N = 5$ h^2	Δh^2 (dB)	$N = 10$ h^2	Δh^2 (dB)	$N = \infty$ h^2
10^{-1}	1.61	2.92	1.20	1.63	1.05	1.08	0.945	0.61	0.876	0.28	0.821
10^{-2}	3.91	1.60	3.09	0.58	2.92	0.33	2.82	0.18	2.76	0.08	2.71
10^{-3}	6.21	1.14	5.14	0.32	4.98	0.19	4.89	0.10	4.83	0.05	4.77
10^{-4}	8.52	0.90	7.28	0.22	7.12	0.13	7.03	0.07	6.97	0.03	6.92
10^{-5}	10.80	0.75	9.45	0.17	9.30	0.10	9.21	0.05	9.15	0.025	9.09
10^{-6}	13.10	0.65	11.70	0.13	11.50	0.08	11.40	0.04	11.35	0.02	11.30
10^{-7}	15.40	0.57	13.90	0.11	13.70	0.07	13.60	0.035	13.57	0.016	13.52

Table 6.9
Energy Parameters of Quasicoherent Reception for Four-Phase PM Signals

p_1	$N = 1$ h^2	Δh^2 (dB)	$N = 2$ h^2	Δh^2 (dB)	$N = 3$ h^2	Δh^2 (dB)	$N = 5$ h^2	Δh^2 (dB)	$N = 10$ h^2	Δh^2 (dB)	$N = 20$ h^2	Δh^2 (dB)	$N = \infty$ h^2
10^{-1}	3.15	2.83	2.37	1.59	2.12	1.10	1.92	0.68	1.78	0.34	1.71	0.17	1.64
10^{-2}	9.57	2.47	7.29	1.30	6.61	0.87	6.10	0.52	5.75	0.26	5.58	0.13	5.41
10^{-3}	16.62	2.41	12.71	1.24	11.55	0.83	10.69	0.49	10.10	0.24	9.82	0.12	9.55
10^{-4}	23.94	2.38	18.31	1.22	16.66	0.81	15.45	0.48	14.61	0.24	14.21	0.12	13.83
10^{-5}	31.37	2.37	24.02	1.21	21.86	0.80	20.28	0.47	19.19	0.23	18.68	0.12	18.19
10^{-6}	38.90	2.36	29.79	1.20	27.12	0.80	25.17	0.47	23.83	0.23	23.20	0.11	22.60
10^{-7}	46.47	2.35	35.59	1.20	32.42	0.79	20.10	0.47	28.50	0.23	27.75	0.11	27.03

provide the appropriate error probability of the quasicoherent demodulator with different intervals of averaging N, and also the energy losses Δh^2, decibels of this demodulator in comparison with the ideal coherent one. At $N \geq 10$ the energy loss does not exceed 0.3 dB. The three-chip interval of averaging for binary signals and the five-chip interval for 4-ary signals can be considered sufficient practically.

The diagrams, illustrating how rapidly the noise immunity of the quasicoherent demodulator approximates the noise immunity of the ideal coherent demodulator, are depicted in Figure 4.20 in Section 4.5.

Thus, the algorithms considered for quasicoherent processing of PM signals provide the same noise immunity as the optimum incoherent reception of PDM signals in the beginning of a communications session, and then they approximate ideal coherent processing during several elementary signals.

References

[1] Abramowitz, M., and I. A. Stegun, *Handbook on Mathematical Functions with Formulas, Graphs and Mathematical Tables*, Washington, DC: National Bureau of Standards, Applied Mathematics Series, 1964.

[2] Bronshtein, I. N., and K. A. Semendyaev, *Mathematics Handbook*, Moscow: Science Publ., 1986.

[3] Gradshtein, I. S., and I. M. Rizhik, *Tables of Integrals, Sums, Series and Products*, Moscow: Science Publ., 1971.

[4] Korzhik, V. I., L. M. Fink, and K. N. Schelkunov, *Noise Immunity of Discrete Message Transmission Systems: Handbook*, Moscow: Radio & Svyaz, 1981.

[5] Fink, L. M., et al., *Signal Transmission Theory*, Moscow: Radio & Svyaz, 1986.

[6] Proakis, J. G., *Digital Communications*, New York: McGraw-Hill Book Company, 1989.

[7] Kotelnikov, V. A., *Potential Noise Immunity Theory*, Moscow: Gosenergoizdat, 1956.

[8] Bobrov, N. P., "Noise Immunity of Single-Valued System for PDM Signal Transmission," *Electrosvyaz*, No. 3, 1959, pp. 27–31.

[9] Fink, L. M., *Discrete Message Transmission Theory*, 2nd ed., Moscow: Sov. Radio, 1970.

[10] Clark, G. C., and J. B. Cain, *Error-Correction Coding for Digital Communications*, New York: Plenum Press, 1987.

[11] Lawton, J. G., "Theoretical Error Rates of Differentially Coherent and Kineplex Data Transmission Systems," *Proc. IRE*, No. 2, 1959.

[12] Andronov, I. S., and L. M. Fink, *Discrete Messages Transmission over Parallel Channels*, Moscow: Sov. Radio, 1971.

[13] Levin, B. R., *Theoretical Foundations of Statistical Radio Engineering*, Moscow: Sov. Radio, Vol. 1–3, 1976.

[14] Gut, R. E., and M. Ya. Lesman, "Noncentral χ^2 Distribution," in *Statistical Methods in Communications Theory*, Scientific Works of Educational Communications Institutes, Leningrad: LEIS, 1987, pp. 133–136.

[15] Gut, R. E., "On Noise Immunity of Autocorrelated Reception of Signals with Single-Valued PDM," *Radiotechnika*, No. 10, 1972, pp. 90–92.

[16] Gut, R. E., et al., "Noise Immunity of Autocorrelated Reception of Signals with Two-Valued PDM," Scientific Technical Conference on Coding and Information Transmission Theory, Odessa, 1988, pp. 23–26.

[17] Okunev, Yu. B., and E. Z. Finkelshtein, "Noise Immunity of Autocorrelated Reception of Signals with PDM," *STC LEIC Proc.*, No. 2, 1965, pp. 25–38.

[18] Solonina, A. I., "Fast Algorithm of Approximation by Recurrent Splines," in *Theory of Information Transmission in Communications Channels*, Scientific Works of Educational Communications Institutes, Leningrad: LEIC, 1983, pp. 45–150.

[19] Gallager, R. G., *Low-Density Parity-Check Codes*, Cambridge, MA: The MIT Press, 1963.

[20] Okunev, Yu. B., and L. M. Fink, "Noise Immunity of Various Receiving Methods for Binary System with Second Order Phase Difference Modulation," *Telecom. Radioeng.*, Vol. 39, No. 9, 1984.

[21] Okunev, Yu. B., N. M. Sidorov, and L. M. Fink, "Noise Immunity of Incoherent Reception of Signals with Single Phase Difference Modulation," *Telecom. Radioeng.*, Vol. 40, No. 12, 1985.

[22] Price, R., "Error Probabilities for Adaptive Multichannel Reception of Binary Signals," *Trans. Info. Theory*, Vol. IT-8, No. 5, No. 6, 1962.

[23] Khvorostenko, N. P., *Statistical Theory of Discrete Signal Demodulation*, Moscow: Svyaz, 1968.

[24] Kislyuk, L. D., and L. Ya. Lipkin, "Noise Immunity of an Incoherent Digital Demodulator of DPSK Signals," *Electrosvyaz*, No. 2, 1975, pp. 61–64.

[25] Girshov, V. S., "Noise Immunity of Coherent Reception of Multiposition Signals with PDM," *Radiotechnika*, No. 1, 1988, pp. 47–49.

[26] Okunev, Yu. B., and N. M. Sidorov, "Noise Immunity of a Carrier-Frequency-Invariant Demodulator of DPSK-2 Signals," *Telecom. Radioeng.*, Vol. 41, No. 8, 1986.

[27] Gut, R. E., Yu. B. Okunev, and N. M. Sidorov, "Effect of Carrier Detuning on the Noise Immunity of Autocorrelation Reception of DPSK Signals," *Telecom. Radioeng.*, Vol. 41, No. 9, 1986, pp. 96–98.

[28] Shchelkunov, K. N., and E. S. Barbanel, "Application of the Phase Difference Modulation in Optical Communication Systems," *Telecom. Radioeng.*, Vol. 35, No. 4, 1979.

[29] Messel, A. F., and Yu. B. Okunev, "Effective Quasicoherent Algorithms of Phase Modulated Signal Reception," *Radioelectronika*, No. 5, 1991.

Conclusion

The developers of contemporary digital communications systems are faced with complex system engineering problems. Among these problems is the need to develop a modem that can provide the required bit rate and error bit rate during information transmission over communications channels. The author hopes that the great number of algorithms and schemes for PM and PDM modems considered in this book, as well as the results of analyses of their performance, will help engineers make the best decision that will result in new inventions.

At the same time, as is often the case, the expansion of opportunities does not always simplify the choice. As the ancients said, the more we know, the more we don't know, and only complete ignorance involves an indisputable uniqueness.

Each of the methods of phase modem implementation considered here has certain advantages and deficiencies, which were discussed in detail. So how does one make a final decision? Certainly, it is impossible to make recommendations without knowing the specific situation. However, one common recommendation is certainly fair in all situations: The basis for choosing the best option has to be the systems approach.

Despite the complexity of the systems approach methodology, the key to understanding its essence is an extremely simple and almost obvious postulate: The complex systems should not be considered as optimum merely because they consist of separate optimum subsystems. This postulate is a warning for the developer against hastily choosing the optimum subsystems to build a complex system. Algorithms, the performance, and the parameters of every unit, which is a component of a complex system, must satisfy the technical project requirements (restrictions), providing maximum global criterion (goal function) for the system as a whole.

The preceding statement is especially true of the modem, which is a component of a complex communications system. Therefore, the choice of modem algorithm and scheme should be determined by the architecture, performance, conditions, and purposes of system operation. During modem development, the most important point is to take into account the characteristics of units with which the modem is directly connected; that is, on the one hand, the channel equipment (transmitter, receiver, communications line, transmission system, and so on) and, on the other hand, the encoder and decoder.

Then the systems approach specifies the expediency to combine a modem and a codec in one unit, as a *codem*. This conclusion of systems theory is completely coordinated with the tendencies of digital system engineering: forming signal-code structures on the basis of PM and PDM signals and signal processing as a whole.

Hilbert Transform

The Hilbert transform is extensively used in signal theory and for optimal signal processing, including synthesis of coherent and incoherent algorithms of signals reception[1–5]. This transform is an integral linear operation that is usually realized by a linear filter with a unity-equal amplitude-frequency characteristic and $\pi/2$-equal phase-frequency characteristic.

Mathematical definition of the Hilbert transform is formulated as follows. If we have a time-infinite signal $S(t)$, $-\infty < t < \infty$, and integral

$$\int_{-\infty}^{\infty} |S(t)|^k \, dt$$

converges at least for some $k \geq 1$, then the Hilbert transform of signal $S(t)$ is

$$S^*(t) = \frac{1}{\pi} \int_{-\infty}^{\infty} \frac{S(\tau)}{t - \tau} \, d\tau \qquad (A.1.1)$$

$S^*(t)$ is said to be the Hilbert conjugate signal with $S(t)$. Because the integrand in (A.1.1) has a singular point, we should define more exactly what this integral means in terms of the Cauchy principal value, namely,

$$S^*(t) = \lim_{\epsilon \to 0} \frac{1}{\pi} \int_{\epsilon}^{\infty} \frac{S(t - \tau) - S(t + \tau)}{\tau} \qquad (A.1.2)$$

If signal $S(t)$ is defined within a limited time interval, it is used in [2] the following definition of the Hilbert transform for time interval $0 \leq t \leq T$:

$$S^*(t) = \lim_{\epsilon \to 0} \frac{1}{\pi} \int_{\epsilon}^{\infty} \frac{S(t - \tau) - S(t + \tau)}{tg(\pi\tau/T)} \, d\tau \qquad (A.1.3)$$

The Hilbert transform can be used for signals with any spectrum. For a narrowband signal, when $\Delta f \ll f$, where Δf is the signal frequency bandwidth and f is the signal carrier, transformations (A.1.2) and (A.1.3) lead to very similar results [2].

We should mention the following important properties of the Hilbert transform [1,2,4,5]:

- The double Hilbert transform of a signal gives the same signal with a negative sign.
- A signal and its Hilbert transform are orthogonal.
- If a signal is an even function, then its Hilbert transform is an odd function and vice versa.
- If $S(t) = S_1(t) \, S_2(t)$, frequency spectrums of $S_1(t)$ and $S_2(t)$ do not overlap, and the spectrum of $S_1(t)$ is lower than the spectrum of $S_2(t)$, then $S^*(t) = S_1(t) \, S_2^*(t)$.

The Hilbert transform of a signal, consisting of harmonic components, is reduced to $\pi/2$ phase shifting these components. If signal $S(t)$, defined within the time interval $(0, T)$, is represented by a Fourier series

$$S(t) = a/2 + \sum_{k=1}^{\infty} (a \cos k\omega t + b \sin k\omega t) \qquad (A.1.4)$$

where $\omega = 2\pi/T$, then the Hilbert conjugate signal $S^*(t)$ will be equal to

$$S^*(t) = a/2 + \sum_{k=1}^{\infty} [a \cos(k\omega t + \pi/2) + b \sin(k\omega t + \pi/2)]$$
$$= a/2 + \sum_{k=1}^{\infty} (a \sin k\omega t - b \cos k\omega t) \qquad (A.1.5)$$

Thus, in practice a Hilbert converter is implemented as a linear filter that shifts phases of all signal components by 90 degrees. This filter is called the *wideband phase shifter* [6]. In the simplest case, when the input signal is a harmonic oscillation $S(t) = a \cos \omega t$, the Hilbert conjugate signal is also a harmonic signal $S^*(t) = -a \sin \omega t$, and the Hilbert converter is a fixed signal-frequency phase shifter.

Various phase shifters, executing Hilbert converter function, are important parts of two-dimensional signal modems, including PM, PDM, FM, and QAM modems.

In signal theory the Hilbert transform is used for constructive determination of the so-called "analytic signal," a complex signal in which the imaginary part is equal to the Hilbert transform of the real part:

$$S_j(t) = S(t) + jS^*(t) \tag{A.1.6}$$

The analytic signal allows us to express the amplitude, frequency, phase, and other parameters of arbitrary signals. For example, the envelope of an arbitrary signal $S(t)$ is equal to

$$A(t) = \sqrt{S^2(t) + S^{*2}(t)} \tag{A.1.7}$$

and its instant phase $\Phi(t)$ is determined as follows:

$$\sin \Phi(t) = S^*(t)/A(t) \qquad \cos \Phi(t) = S(t)/A(t) \tag{A.1.8}$$

For an arbitrary frequency ω the instant initial phase is $\varphi(t) = \Phi(t) - \omega t$ and the complex signal and its components can be represented in the following way:

$$S_j(t) = A(t) \exp[j(\omega t + \varphi(t))]$$
$$S(t) = A(t) \cos[\omega t + \varphi(t)] \tag{A.1.9}$$
$$S^*(t) = A(t) \sin[\omega t + \varphi(t)]$$

In addition, any signal can be represented by its quadrature components $X(t)$ and $Y(t)$:

$$S(t) = X(t) \cos \omega t - Y(t) \sin \omega t \tag{A.1.10}$$

which is equal to

$$X(t) = S(t) \cos \omega t + S^*(t) \sin \omega t$$
$$Y(t) = S^*(t) \cos \omega t - S(t) \sin \omega t \tag{A.1.11}$$

Formulas (A.1.7) through (A.1.11), based on the Hilbert transform, are often used in algorithms of digital signal processing.

References

[1] Franks, L. E., *Signal Theory*, Englewood Cliffs, NJ: Prentice-Hall, 1969.

[2] Korzhik, V. I., L. M. Fink, and K. N. Schelkunov, *Noise Immunity of Discrete Message Transmission Systems: Handbook*, Moscow: Radio & Svyaz, 1981.

[3] Proakis, J. G., *Digital Communications*. New York: McGraw-Hill Book Company, 1983.

[4] Hahn, S. L., *Hilbert Transforms in Signal Processing*, Norwood, MA: Artech House, 1996.

[5] Poularikas, A. D., *The Transforms and Applications Handbook*, Boca Raton, FL: CRC Press, 1996.

[6] Avramenko, V. L., Yu. P. Galyamichev, and A. A. Lanne, *Electrical Delay Lines and Phase-Shifters*, Moscow: Svyaz Publ., 1973.

Abbreviations and Symbols

ADC	analog-to-digital converter
AFC	automatic frequency control
AM	amplitude modulation
APM	amplitude-phase modulation
AWGN	additive white Gaussian noise
BER	bit error rate
CDMA	code division multiple access
CO	controlled oscillator
DSP	digital signal processing
F	filter
FDMA	frequency division multiple access
FM	frequency modulation
INV	inverter
LPF	low-pass filter
O	oscillator
OA	operational amplifier
PD	phase detector
PDCU	phase difference computation unit
PDM	phase-difference modulation = differential phase shift keying (DPSK)
PDM-1	first-order phase-difference modulation

PDM-2	second-order phase-difference modulation
PDM-k	phase-difference modulation of the kth order
PM	phase modulation = phase shift keying (PSK)
QAM	quadrature amplitude modulation = APM
ROS	reference oscillation selector
SMF	switched matched filters
SNR	signal-to-noise ratio
TDMA	time division multiple access
a	signal amplitude (envelope)
C_k^i	binomial coefficient
$D(y)$	dispersion of a random value y
E, E_s	signal energy
F	frequency passband; frequency bandwidth
$\mathsf{F}(y)$	Laplace function: $\mathsf{F}(y) = \dfrac{1}{\sqrt{2\pi}} \displaystyle\int_y^\infty \exp\left(-\dfrac{x^2}{2}\right) dx$
f	carrier frequency
h^2	signal-to-noise ratio: $h^2 = E/N_0 = P_s T/N_0$
I_n	modified Bessel function of the first kind and nth order
invar	invariant
J_i	binary information symbol transmitted over the ith binary sub-channel of a multiposition system
$m(y)$	mean of a random value y
$m = 2^N$	number of levels (phases or phase-differences) in the PM or PDM signal; number of points (positions) in signal constellation
$N = \log_2 m$	modulation multiplicity of PM or PDM signals (N-valued PM or PDM signals)
$n(t)$	white Gaussian noise
N_0	spectral density of white noise power
P_s	signal power
p	probability; error probability
p_0	error probability of the binary perfect PM system
$p_1^{(k)}$	error probability of the two-position PDM system of the kth order
$p_N^{(k)}$	error probability of the N-valued PDM-k system
$Q(\alpha, \beta)$	Marcum Q-function: $Q(\alpha, \beta) = \displaystyle\int_\beta^\infty t \exp\left(-\dfrac{\alpha^2 + t^2}{2}\right) I_0(\alpha t) dt.$
$\mathrm{sgn}(a)$	sign function: $\mathrm{sgn}(a) = \begin{cases} +1 & \text{for } a > 0 \\ -1 & \text{for } a < 0 \end{cases}$

$S(t)$	transmitted signal
$S_i(t)$	ith variant of the transmitted signal
S_j^*	Hilbert transformation of the signal S_j
T	elementary signal duration (baud or signal chip duration)
$x(t)$	received signal
$x^*(t)$	Hilbert transformation of the received signal
$x_n(t)$	nth element (chip) of the received signal
X_i, Y_i	projections of the ith element of the received signal onto orthogonal reference oscillations with an arbitrary initial phase:

$$X_i = \int_{(i-1)T}^{iT} x(t) \, \sin(\omega t + \varphi) \, dt$$

$$Y_i = \int_{(i-1)T}^{iT} x(t) \, \cos(\omega t + \varphi) \, dt$$

(x_n, x_{n-1})	scalar product (convolution) of adjacent signal elements
γ	information parameter of a signal
φ	signal phase
$\omega = 2\pi f$	signal carrier frequency
$\Delta^k \gamma$	kth-order differences of parameter γ
Δf	filter frequency passband
$\Delta_n^k f$	k-order finite difference of function $f(t)$ at the nth time instant
$\Delta \varphi$	phase difference
$\Delta^2 \varphi$	second-order phase difference
$\Delta \varphi_0$	initial phase difference
$\Delta^k \varphi_0$	initial k-order phase difference
$\Delta \varphi_i$	ith variant of the transmitted phase difference
$\Delta_n \varphi$	phase difference at the nth chip time instant
$\Delta^k \varphi$	phase difference of the kth order
$\Delta \varphi_\xi$	phase difference in presence of interference ξ
$\xi(t)$	interference

Bibliography

Abramowitz, M., and I. A. Stegun, *Handbook on Mathematical Functions With Formulas, Graphs and Mathematical Tables,* National Bureau Standards, Applied Mathematics Series, 1964.

Andronov, I. S., and L. M. Fink, *Discrete Message Transmission Over Parallel Channels,* Moscow: Sov. Radio, 1971.

Banket, V. L., and V. M. Dorofeev, *Digital Methods in Satellite Communications,* Moscow: Radio & Svyaz, 1988.

Benedetto, S., E. Biglieri, and V. Castellani, *Digital Transmission Theory,* Englewood Cliffs, NJ: Prentice-Hall, 1987.

Bingham, J. A. C., *The Theory and Practice of Modem Design,* New York: Wiley & Sons, 1988.

Blokh, E. K., and V. V. Zyablov, *Linear Cascade Codes,* Moscow: Science Publ., 1982.

Clark, G. C., and J. B. Cain, *Error-Correction Coding for Digital Communications,* New York: Plenum Press, 1987.

Fink, L. M., *Discrete Message Transmission Theory,* 2nd ed., Moscow: Sov. Radio, 1970.

Fink, L. M., et al., *Signal Transmission Theory,* Moscow: Radio & Svyaz, 1986.

Franks, L. E., *Signal Theory,* Englewood Cliffs, NJ: Prentice-Hall, 1969.

Gallager, R. G., *Information Theory and Reliable Communication,* New York: John Wiley & Sons, 1968.

Ginzburg, V. V., "Multidimensional Signals for Continuous Channels," *Problems of Information Transmission,* Vol. 20, No. 1, 1984, pp. 28–46.

Ginzburg, V. V., and A. A. Kayatskas, *Demodulator Synchronization Theory,* Moscow: Svyaz, 1974.

Gitlin, R. D., J. F. Hayes, and S. B. Weinstein, *Data Communications Principles,* New York: Plenum Press, 1992.

Gold, B., and L. R. Rabiner, *Theory and Application of Digital Signal Processing,* Englewood Cliffs, NJ: Prentice-Hall, 1975.

Goldenberg, L. M., B. D. Matyushkin, and M. N. Polyak, *Digital Signal Processing,* Moscow: Radio & Svyaz, 1990.

Golomb, S., et al., *Digital Communications with Space Applications,* Englewood Cliffs, NJ: Prentice-Hall, 1964.

Gradshtein, I. S., and I. M. Rizhik, *Tables of Integrals, Sums, Series and Products,* Moscow: Science Publ., 1971.

Khvorostenko, N. P., *Statistical Theory of Discrete Signal Demodulation,* Moscow: Svyaz, 1968.

Klovsky, D. D., *Discrete Message Transmission over Radio Channels,* Moscow: Radio & Svyaz, 1982.

Klovsky, D. D., B. Ya. Kontorovich, and S. M. Shirokov, *Models of Continuous Communications Channels on the Base of Stochastic Differential Equations,* Moscow: Radio & Svyaz, 1984.

Kolesnik, V. D., and G. Sh. Poltyrev, *Information Theory,* Moscow: Science Publ., 1982.

Korzhik, V. I., and L. M. Fink, *Noise Immunity Coding of Discrete Messages in the Channels with Random Structures,* Moscow: Svyaz, 1975.

Korzhik, V. I., L. M. Fink, and K. N. Schelkunov, *Noise Immunity of Discrete Message Transmission Systems,* Moscow: Radio & Svyaz, 1981.

Kotelnikov, V. A., *Potential Noise Immunity Theory,* Moscow: Gosenergoizdat, 1956.

Lee, E. A., and D. G. Messershmitt, *Digital Communication,* Boston: Kluwer Academic Publishers, 1988.

Levin, B. R., *Theoretical Foundations of Statistical Radio Engineering, Vols. 1–3,* Moscow: Sov. Radio, 1976.

Levin, L. S., and M. A. Plotkin, *Digital Systems of Information Transmission,* Moscow: Radio & Svyaz, 1982.

Lin, S., and J. Costello, *Error Control Coding,* Englewood Cliffs, NJ: Prentice-Hall, 1983.

Lindsey, W. C., *Synchronization Systems in Communications,* Englewood Cliffs, NJ: Prentice-Hall, 1983.

Lucky, R. W., J. Salz, and E. J. Weldon, *Principles of Data Communications,* New York: McGraw-Hill, 1968.

Makarov, S. B., and I. A. Tsikin, *Discrete Message Transmission by Radio Channels with Limited Bandpass,* Moscow: Radio & Svyaz, 1988.

Massey, J. L., *Threshold Decoding,* Cambridge, MA: MIT Press, 1963.

McWilliams, F. J., and N. J. A. Sloane, *The Theory of Error-Correcting Codes,* Amsterdam: North-Holland Publ. Co., 1977.

Neyfakx, A. E., *Convolution Codes for Digital Information Transmission,* Moscow: Science Publ., 1979.

Nikolaev, B. I., *Serial Digital Information Transmission over Channels with Memory,* Moscow: Radio & Svyaz., 1988.

Okunev, Yu. B., *Digital Transmission of Information by Phase Modulated Signals,* Moscow: Radio & Svyaz, 1991.

Okunev, Yu. B., *Telecommunications Systems With Invariant Performances,* Moscow: Svyaz, 1973.

Okunev, Yu. B., *Theory of Phase-Difference Modulation,* Moscow: Svyaz, 1979.

Okunev, Yu. B., and L. A. Yakovlev, *Telecommunications Systems With Spread Spectrum Signals,* Moscow: Svyaz, 1968.

Okunev, Yu. B., and V. G. Plotnikov, *Principles of the System Approach to Telecommunication Design,* Moscow: Svyaz, 1978.

Oppenheim, A. V., and R. W. Schafer, *Digital Signal Processing,* Englewood Cliffs, NJ: Prentice-Hall, 1979.

Papoulis, A., *Probability, Random Variables, and Stochastic Processes,* New York: McGraw Hill Book Company, 1965.

Peterson, W. W., and E. J. Weldon, *Error-Correcting Codes,* Cambridge, MA: MIT Press, 1972.

Petrovich, N. T., *Discrete Information Transmission in Channels with Phase Keying,* Moscow: Sov. Radio, 1965.

Pirogov, A. A., *Voicecoder Telephony,* Moscow: Svyaz, 1974.

Pritchard, W. L., H. G. Suyderhold, and R. A. Nelson, *Satellite Communication System Engineering,* Englewood Cliffs, NJ: Prentice-Hall, 1993.

Proakis, J. G., *Digital Communications,* New York: McGraw-Hill Book Company, 1983.

Shannon, C. E., "A Mathematical Theory of Communication," *Bell System Tech. Journal,* Vol. 27, 1948.

Simon, M. K., S. M. Hinedi, and W. C. Lindsey, *Digital Communication Techniques,* Englewood Cliffs, NJ: Prentice-Hall, 1995.

Smith, D. R., *Digital Transmission Systems,* New York: Van Nostrand Reinhold Company, 1985.

Spilker, J. J., *Digital Communication by Satellite,* Englewood Cliffs, NJ: Prentice-Hall, 1977.

Stiffler, J. J., *Theory of Synchronous Communications,* Englewood Cliffs, NJ: Prentice-Hall, 1971.

Tikhonov, V. I., and N. K. Kulman, *Nonlinear Filtering and Quasicoherent Reception of Signals,* Moscow: Sov. Radio, 1975.

Tsikin, I. A., *Discrete-Analog Signal Processing,* Moscow: Radio & Svyaz, 1982.

Turkin, A. I., *Recurrent Reception of Composite Signals,* Moscow: Radio & Svyaz, 1988.

Van Trees, H. L., *Detection, Estimation, and Modulation Theory, Vol. II: Nonlinear Modulation Theory,* New York: John Wiley & Sons, 1975.

Viterbi, A. J., *Principles of Coherent Communication,* New York: McGraw-Hill, 1966.

Viterbi, A. J., and J. K. Omura, *Principles of Communication Engineering,* New York: John Wiley, 1979.

Wozencraft, J. M., and I. M. Jacobs, *Principles of Communication Engineering,* New York: John Wiley, 1965.

Zayezdny, A. M., D. Tabak, and D. Wulich, *Engineering Application of Stochastic Processes,* New York: John Wiley, 1989.

Zayezdny, A. M., Yu. B. Okunev, and L. M. Rakhovich, *Phase-Difference Modulation,* Moscow: Svyaz, 1967.

Ziemer, R. E., and R. L. Peterson, *Introduction to Digital Communication,* New York: Macmillan Publishing Company, 1992.

Zuko, A. G., et al., *Noise Immunity and Efficiency of Information Transmission Systems,* Moscow: Radio & Svyaz, 1985.

About the Author

Dr. Yuri Okunev was born in St. Petersburg, Russia. He obtained his M.S. and Ph.D. in electrical engineering at the St. Petersburg State University of Telecommunications.

For more than 20 years Dr. Okunev was head of the Digital Communications Research Laboratory at the above university. During this period his research interests were focused on theories of modulation, optimal signal processing, and spread spectrum signals, as well as on general system engineering. At the laboratory he participated in the development and implementation of advanced radio communications and data transmission systems for military and commercial applications. He also taught at the university and supervised doctoral students who studied digital communications and systems engineering. In 1990 he became one of the founders of Radiotelecom, Inc., which was one of the first private production companies in the field of telecommunications in Russia.

In 1995 Dr. Okunev joined Bell Laboratories of Lucent Technologies in New Jersey, where he worked on problems of development and applications of wireless technology, particularly CDMA technology, for satellite systems and systems based on the use of high-altitude aeronautical platforms. Currently, he is with General DataComm, Inc. in Connecticut, where he develops high bit rate modems for data transmission systems.

Dr. Okunev is the author of several monographs and textbooks in the field of digital communications and radio systems engineering. He is a member of IEEE and of the New York Academy of Sciences.

Index

The Artech House Telecommunications Library

Vinton G. Cerf, Series Editor

Principles of Secure Communication Systems, Second Edition, Don J. Torrieri

Principles of Signaling for Cell Relay and Frame Relay, Daniel Minoli and
    George Dobrowski

Principles of Signals and Systems: Deterministic Signals, B. Picinbono

Private Telecommunication Networks, Bruce Elbert

Radio-Relay Systems, Anton A. Huurdeman

RF and Microwave Circuit Design for Wireless Communications,
    Lawrence E. Larson

The Satellite Communication Applications Handbook, Bruce R. Elbert

Secure Data Networking, Michael Purser

Service Management in Computing and Telecommunications, Richard Hallows

Smart Cards, José Manuel Otón and José Luis Zoreda

Smart Highways, Smart Cars, Richard Whelan

Successful Business Strategies Using Telecommunications Services,
    Martin F. Bartholomew

Super-High-Definition Images: Beyond HDTV, Naohisa Ohta, et al.

Television Technology: Fundamentals and Future Prospects, A. Michael Noll

Telecommunications Technology Handbook, Daniel Minoli

Telecommuting, Osman Eldib and Daniel Minoli

Telemetry Systems Design, Frank Carden

Teletraffic Technologies in ATM Networks, Hiroshi Saito

*Toll-Free Services: A Complete Guide to Design, Implementation, and
    Management*, Robert A. Gable

Transmission Networking: SONET and the SDH, Mike Sexton and Andy Reid

Troposcatter Radio Links, G. Roda

Understanding Emerging Network Services, Pricing, and Regulation,
    Leo A. Wrobel and Eddie M. Pope

Understanding GPS: Principles and Applications, Elliot D. Kaplan, editor

Understanding Networking Technology: Concepts, Terms and Trends, Mark Norris

UNIX Internetworking, Second Edition, Uday O. Pabrai

Videoconferencing and Videotelephony: Technology and Standards,
    Richard Schaphorst

Wireless Access and the Local Telephone Network, George Calhoun

Wireless Communications in Developing Countries: Cellular and Satellite Systems,
    Rachael E. Schwartz

Wireless Communications for Intelligent Transportation Systems,
  Scott D. Elliot and Daniel J. Dailey

Wireless Data Networking, Nathan J. Muller

Wireless LAN Systems, A. Santamaría and F. J. López-Hernández

Wireless: The Revolution in Personal Telecommunications, Ira Brodsky

Writing Disaster Recovery Plans for Telecommunications Networks and LANs,
  Leo A. Wrobel

X Window System User's Guide, Uday O. Pabrai

For further information on these and other Artech House titles, contact:

Artech House
685 Canton Street
Norwood, MA 02062
617-769-9750
Fax: 617-769-6334
Telex: 951-659
email: artech@artech-house.com

Artech House
Portland House, Stag Place
London SW1E 5XA England
+44 (0) 171-973-8077
Fax: +44 (0) 171-630-0166
Telex: 951-659
email: artech-uk@artech-house.com

WWW: http://www.artech-house.com